Radiation Chemistry
of Monomers,
Polymers, and Plastics

Radiation Chemistry of Monomers, Polymers, and Plastics

Joseph E. Wilson

Professor of Chemistry
Bishop College
Dallas, Texas

MARCEL DEKKER, INC. New York 1974

CHEMISTRY

MARCEL DEKKER, INC.
305 East 45 Street, New York, New York 10017

LIBRARY OF CONGRESS CATALOG CARD NUMBER 83-82706

ISBN 0-8247-6095-6

Current printing (last digit):
10 9 8 7 6 5 4 3 2 1

PRINTED IN THE UNITED STATES OF AMERICA

this book is dedicated to

Milton K. Curry, Jr.

for unfailing support and encouragement

CONTENTS

PREFACE

This book was written to provide an up-to-date reference work
on the radiation chemistry of polymers, suitable for use by academic
and industrial scientists. The radiation chemistry of polymers provides a research field that is full of fresh and stimulating dis-
coveries, from the finding that two film surfaces in close contact
can be tightly linked together by irradiation, to the observation
that emulsion polymers produced by irradiation are more stable because a charge is developed on the particles. The rapid expansion
of knowledge in the field has been greatly aided by concurrent
developments in photochemistry, spectroscopy, radiochemistry, and
reaction kinetics. Progress has also been promoted by improved
techniques in ESR spectroscopy, mass spectrometry, and isotope di-
lution, and the relative ease with which large sources of radiation
can be obtained.

The first several chapters of the book cover essential information on radiation properties, measurement, and detection, and on
the primary chemical results of the interaction of radiation with
matter. These are followed by a chapter on homogeneous polymerization with emphasis on recent new insights, and one on solid state
polymerization that includes an extensive alphabetical table of
monomers that have been irradiated and polymerized in solid form.
The book includes several such extensive tables, each being located
in the chapter pertinent to its subject matter.

Chapters 7 and 8 discuss radiation cross-linking in general and
in particular relation to selected polymeric examples, with special
attention to the recent developments in the upgrading of polymer
properties and the curing of polymeric coatings. Chapters 9 and 10

ix

cover radiation-induced grafting on films and fibers, respectively,
and highlight recent information on how properties can be modified
and controlled by grafting.

While in many fields the "state of the art" has outstripped
fundamental knowledge of the principles involved, the reverse
appears to be true in the application of radiation to polymer technology. This may be partially due to the fact that there is a very
large number of polymer chemists, and a considerable group of
radiation chemists, but only a very limited number of scientists
trained in both radiation and polymer chemistry. The writer would
be very gratified if the present book could aid in a modest way to
advance polymer technology by acquainting radiation chemists with
polymer chemistry, and providing basic radiation chemical facts to
polymer chemists. In further pursuit of this concept, the book
includes a Glossary that defines fundamental radiation terms for
polymer chemists, and polymer technical terms for radiation
chemists.

The present book, due to space limitations, is confined almost
entirely to fundamentals and principles, and does not include any
discussion of industrial applications except in very brief form or
by way of illustrating a fundamental principle.

For aid in preparing the book, the writer is indebted to
colleagues and associates too numerous to mention. For contributing
to my background of training in the subject matter, mention should
be made of Dr. W. Albert Noyes, Jr., who introduced me to the
mysteries of photochemistry in my graduate student days at the
University of Rochester. Also, Dr. Farrington Daniels, under whom
I worked, was responsible for my initiation to radiation and
nuclear chemistry at the old Metallurgical Laboratory on the campus
of the University of Chicago.

The manuscript was typed in major part by Mrs. Sue Adams, whose
assistance I acknowledge with thanks. I am also indebted to my wife,
Dorothy, for her patience with a preoccupied husband during the
period of writing.

The initiation of the book was due in no small measure to
Dr. Milton K. Curry, Jr., President of Bishop College, for providing an atmosphere in which scholarly efforts are encouraged.

                                              Joseph E. Wilson

Chapter 1

TYPES AND SOURCES OF RADIATION

I. DEFINITION OF RADIATION CHEMISTRY

Radiation chemistry is defined as the study of chemical effects caused by the passage of ionizing radiation through matter. Ionizing radiation comes from substances undergoing nuclear transformations, from outer space in the form of cosmic rays, and from particle accelerators. It includes, α-, β-, and γ-rays from radioactive nuclei, charged particles such as protons and deuterons, and x-rays of wavelength less than approximately 250 Å. Table 1-1 lists some of the more common types of ionizing radiation.

Chemical reactions caused by ultraviolet and visible radiation make up the field of photochemistry. The principle difference between radiation chemistry and photochemistry is that the energy of the ionizing radiation employed in radiation chemistry is usually much greater than that of the light rays employed in photochemistry.

In photochemistry, each absorbed photon excites one reacting molecule to a definite, usually known, excited electronic state. These excited molecules are distributed rather homogeneously throughout the reaction system, except for a slight decrease in concentration in the direction of the light beam owing to absorption of the light. In contrast, each particle or ray of ionizing radiation produces a large number of ionized and excited molecules

1

TABLE 1-1

Characteristics of Nuclear Radiation

Name	Symbol	Charge	Rest mass[a]	Spin[b]	Magnetic moment[c]
Neutron	n	0	1.0086654	1/2	-1.913
Proton	p	+1	1.0072766	1/2	+2.793
Positron	e^+ or β^+	+1	0.005486	1/2	+1836
Electron	e^- or β^-	-1	0.005486	1/2	-1836
mu-Meson	μ	+1,-1	0.113	1/2	—
pi-Meson	$\pi^\pm$	+1,-1	0.150	0	—
pi-Meson	π^o	0	0.145	0	—
Photon	γ	0	0	1	0
Neutrino	ν	1	$<2 \times 10^{-7}$	1/2	<0.3

[a] In atomic mass units (C^{12} = 12.00000).

[b] In units of $h/2\pi$.

[c] In units of the nuclear magneton ($eh/4\pi\, mc$), where m is the proton mass.

along its track. The ionizing radiation is not selective and may interact with any molecule in its path and raise it to any of its possible ionized and/or excited states. The inhomogeneity of the latter type of reaction is especially marked in the liquid or solid phase, where there is more of a tendency for the excited species to remain close to the tracks where they were originally formed.

Since the beginnings of radiation chemistry, a wide variety of reactions has been investigated. In recent years, research has grown in radiation-induced polymerization and in the radiation-activated grafting of monomers on polymeric films and fibers.

II. HISTORY OF RADIATION CHEMISTRY

With a background of interest in the phosphorescence of
uranium salts, Henri Becquerel reported in February of 1896 that
crystals of uranyl salts emit a radiation which blackens a photo-
graphic plate after penetrating black paper, glass, and other
substances [1]. The radiation was also emitted by uranous salts,
solutions of uranium salts, and other forms of uranium, in each
case with an intensity proportional to the uranium content.

The results in this new field of science were summarized in
1898 by Pierre and Marie Curie, who stated that the rays from
uranium were an atomic phenomenon characteristic of the element
but not dependent on its chemical or physical state. The Curies
carried out extensive research in this field and coined the name
"radioactivity" for the phenomenon. They found that some natural
uranium ores were more radioactive than pure uranium. The chemical
decomposition and processing of these ores led them to the discovery and isolation of polonium and radium. Madam Curie determined the atomic weight of radium to be 226.5 (compared to a recent
value of 226.05).

Becquerel demonstrated that uranium left in the dark, and not
supplied with energy in any way, continued to give off radiation
for years with little or no decrease in intensity. The Curies had
noted a heating effect of radium, amounting to the production of
about 100 cal/h/g of radium. The source of this power was then
unknown, and an article in the St. Louis Post Dispatch of
October 4, 1905 formulated some interesting speculations about use
of this mysterious new power as an instrument of war.

J. J. Thompson and associates observed that radiation could
discharge an electroscope through ionization of the air molecules.
A technique was originated for using the amount of air ionization
as a measure of radiation intensity. Rutherford studied the properties of the rays, employing an instrument similar to an

electrometer. Studies of ray absorption in metal foils indicated
that two components or types of rays were often present. One
component was blocked by a few thousandths of a centimeter of
aluminum and was called α-radiation. The other component required
a thickness of the order of 100 times as much to stop it and was
given the name of β-radiation.

By 1904, it was recognized that α-rays were actually particles,
and W. H. Bragg showed that such particles had a definite range or
distance of passage through matter. Several radioactive substances
were found to give out α-rays with different characteristic ranges.
Alpha- and beta-rays were shown to be streams of particles through
magnetic and electrostatic deflection experiments. The evidence
indicated that β-rays were electrons moving at speeds approaching
the velocity of light. From α-particle deflections the ratio of
mass to charge was computed to be about half that of the hydrogen
ion, with the charge positive. It was later concluded that the α-
particle was a helium ion.

In the meantime a still more penetrating radiation from radioactive materials had been discovered. This latter radiation was
not deflected by a magnetic field: This radiation was named γ-rays
and was recognized as being electromagnetic and similar to x-rays
in nature. The general understanding of radioactivity was advanced
in 1903, when Rutherford and Soddy stated that radioactive elements
were changing spontaneously from one element to another, and that
the radioactive process was a subatomic transformation inside the
atom.

Equally important was the growing accumulation of information
regarding the promotion of chemical reactions by radiation. Ramsay
and Soddy in 1903 showed that the gas evolved in the irradiation
of water was a mixture of hydrogen and oxygen. Their data enabled
Bragg in 1907 to compare the chemical and ionizing results of α-
particles for the first time. A few years later, Madam Curie
postulated that the primary effect of radiation on matter is the
production of ions, which then leads to chemical reactions.

Lind [2] carried out experiments relating the amount of ionization to the amount of chemical change. Lind defined the *ion-pair*
or *ionic yield* (M/N) as the ratio of the number of molecules reacting to the number of ion pairs produced (N). For the irradiation
of oxygen with α-rays, Lind found a yield of 0.5 molecules of ozone
per ion pair. Lind was also involved in developing the "ion-
cluster" theory, which postulated that an ion could act as a
nucleus to which a cluster of molecules was attracted and held by
polarization forces [3]. Since the total cluster could then
presumably react, ion-pair yields greater than one were easily
accounted for.

Objections to the ion-cluster theory were raised by Eyring *et
al.* [4]. Eyring pointed out that radiation supplies enough energy
to produce excited molecules as well as ions. Both excited molecules and ions lead to free radicals. Free radical mechanisms were
used successfully in explaining several reactions initiated by
radiation.

An important characteristic of radiation-induced reactions was
noted by Fricke [5]. When the reactant is a solute in a solution,
energy is absorbed by both solute and solvent. If the solution is
dilute, the direct action of radiation on the solute may be rather
unimportant. However, indirect action may take place, in which the
solvent absorbs energy from the radiation and then reaction takes
place between the "activated" solvent and the solute. For example,
recent work shows that irradiation of water solutions produces
hydrogen and hydroxyl radicals, and that the reactions of these
radicals lead to the products of the total process.

The ion-pair yield M/N is less useful in liquids because N,
the number of ions formed, is not known. The value of N can be
computed by assuming a value of W, the average energy loss in producing an ion pair in the liquid. However, the value of W is
usually uncertain. These uncertainties led to an expression of
yield in terms of the *G value*, defined as the number of molecules
changed for each 100 eV of energy absorbed. For example, G(X)

would indicate the number of molecules of a product X formed on irradiation per 100 eV of energy absorbed, while G(-Y) would equal the number of molecules of Y destroyed per 100 eV absorbed. The G value is now the usual expression employed to indicate radiation-chemical yields.

The relation between the G value and the ion-pair yield is given by the following equation:

$$G = \frac{M}{N} \times \frac{100}{W} \qquad\qquad (1\text{-}1)$$

The value for W in air is about 32.5 eV, and if that value is used in the above equation,

$$G \cong \frac{3\,M}{N} \qquad\qquad (1\text{-}2)$$

III.   RADIATION SOURCES

There are two main types of radiation sources: (a) radio-isotopes, and (b) devices such as x-ray tubes and electron accelerators.

A.   Sources Employing Radioisotopes

Recent years have seen a great increase in the production and usage of ionizing radiation sources. Most of the applications of such sources are in the research field, but some of them, such as the sterilization of sutures and treatment of polyethylene, are in the commercial stage.

There are three main techniques for producing isotopic
sources. One method is to place in a nuclear reactor an element
that can react with a neutron to form a radioisotope. These reactions can be of the (n,γ) type,[*] where the product of the reaction
is a different isotope of the original element irradiated. On
the other hand, the reaction can be of the (n,p) or (n,α) type,
in which case a new element is formed instead of the starting
element. The rate of forming the desired product will vary with
the cross section of the original nuclide, the half-life of the
product nuclide, the cross section of the product nuclide, and the
neutron flux. A neutron flux of 10^{12} to 10^{14} $n/cm^2/sec$ is needed
to produce radioisotopes at an acceptable rate.

A second source of radioisotopes is spent fuel elements from
nuclear reactors. After removal from the reactor, these elements
are stored for a short period of time to allow decay of short-lived
fission products. The remaining fission products include about 34
radioisotopes, each having its own radiation and decay properties.
The mixture of radiation energies presents a problem in some
applications, but constitutes a potential and increasingly useful
radiation source.

A third technique used in obtaining radiation sources is the
selective extraction and concentration of radioisotopes from the
waste material remaining after the processing of the fuel elements.

The isotope most frequently used as a radiation source is
cobalt-60. This has come about because of the advantageous properties of cobalt-60: (1) availability, (2) high-energy γ-rays, and
(3) a 5.27-year half-life. The reaction used in forming cobalt-60
is a neutron capture by cobalt-59. Two capture reactions actually
take place:

---

[*] An (n,γ) reaction is one in which the reactant interacts with a
neutron (n) to form the product plus a gamma (γ)-ray.

$$\ce{^{59}_{27}Co + ^{1}_{0}n -> ^{60}_{27}Co} \quad (10 \text{ min}) \qquad\qquad (1\text{-}3)$$

$$\ce{^{59}_{27}Co + ^{1}_{0}n -> ^{60}_{27}Co} \quad (5.3 \text{ years}) \qquad\qquad (1\text{-}4)$$

About 99% of the 10-min isomer is transformed by isomeric
transition to the 5.3-year isomer. The rate of production of the
5.3-year isomer at any instant equals the rate of neutron capture
by cobalt-59 to form this isomer directly, plus the rate of decay
of the 10-min isomer to the 5.3-year isomer, minus the rate of
decay of the 5.3-year isomer already present and the rate at which
this isomer interacts with neutrons to become cobalt-61.

If one assumes that the material present at time zero is 100
at. % cobalt-59, the above computation indicates 9.52 at. %
cobalt-60 present after 1 year in a neutron flux of 10^{14} n/cm/sec.

The figure for neutron flux used here is the "depressed flux."
The cobalt has a high neutron absorption capacity and depresses the
flux in the section of the reactor where it is placed. The nominal
neutron flux cannot, therefore, be used in calculations. Rather, a
flux depression factor must be computed for the specific size of
cobalt piece being irradiated.

The flux depression factor at a particular point depends on
the concentration of cobalt at that point in the reactor. In the
case of a large piece of cobalt the concentration is dependent on
the geometry of the piece. When small pieces are used, the arrangement of the pieces determines the cobalt concentration. Large
pieces of cobalt are usually in the shape of rods, tubes, or rec-
tangular strips. Smaller pieces may be in the form of pellets,
disks, or small segments of a rod. One set of experiments with
cobalt cylinders, for example, indicated a flux depression to 44%
of its nominal value [6].

High neutron fluxes are required for the production of cobalt-60. The shape and size of a cobalt-60 source are determined by the wishes of the user and the characteristics of the reactor where it is produced. In the design of large scale sources it is important to take into account the self-absorption of the source. Depending on the shape of the source and the encapsulating metal, this can range from a few percent up to 40%.

Cesium-137 is a radioisotope of considerable interest that can be obtained from used fuel rods. When used fuel elements are removed from nuclear reactors, they are allowed to stand for some time so that the short-lived radioactivity can decay. They are then dissolved and processed chemically to reclaim fissionable material, leaving a solution of radioactive fission fragments. The fragments are separated into groups of chemically similar isotopes and finally into individual isotopes, such as cesium-137 and strontium-90.

Cesium-137 is a source of γ-rays and has a half-life of about 30 years. Cesium-137 always contains some cesium-134, which is formed by (n,γ) reaction with cesium-133 resulting from fission. The amount of cesium-134 will depend on the irradiation conditions and the length of time the spent fuel was allowed to age. Cesium-137 sold commercially generally contains about 5% cesium-134.

Cesium-137 may be produced in the form of cesium chloride in specific activities up to 25 Ci/g [7]. The CsCl powder is pressed at 20,000 psi to a density of 3.9 g/cm^3 in the form of pellets. The pellets are enclosed in two concentric stainless-steel capsules, each of which is hermetically sealed by welding. The decay scheme shows that the γ-radiation arises from a daughter isotope of the cesium-137:

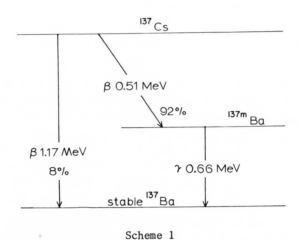

Scheme 1

In England, cesium-137 is prepared in the form of cesium
sulfate compressed to a density of 3 g/ml and having an activity
of 18 Ci/g. Activities above 2.5 Ci are double sealed, first in a
10% indio platinum cell and then in a monel metal outer capsule.
All seals are of hard solder.

Recent price reductions at Oak Ridge National Laboratory have
made cesium-137 available at a reasonable price. Disadvantages of
cesium-137 include solubility of cesium chloride in water, slight
instability of ^{137}Cs to radiation, and corrosive action of CsCl on
its encapsulating container. Most investigators seeking an isotopic radiation source select either cobalt-60 or cesium-137.

In applications of cobalt-60 as a radiation source, an activity of 1 to 5 Ci/g is usually employed. Sources of 100 to 10,000
Ci are often utilized in university laboratories, and sources of
about 1000 Ci are used for cancer therapy. Large sources of
100,000 to 1,000,000 Ci have been designed for industrial irradiation of substances on a commercial scale.

Gamma sources are enclosed in thick shields of dense material
to protect nearby personnel. In one source design the cobalt-60
is arranged so as to irradiate a cavity, with the cavity being

enclosed inside a considerable mass of shielding material. Some
type of movable shielding is provided so that the sample can be
placed into the cavity without exposing the cobalt-60 to the
operator. Such sources commonly contain from 1000 to 48,000 Ci of
cobalt and can be built compactly using lead shielding so as to provide complete safety to persons nearby. The sample may be introduced into the source inside a moving drawer, with spiral tubes
through the drawer to allow wires and tubes leading into the irradiated cavity without escape of radiation. The drawer is essentially a double-ended plug, with the sample cavity in the center,
which moves up and down the center axis of the total unit.

A source design used frequently is that of Ghormley and
Hochnadel [8]. The lead shield is constructed in two sections and
has a bottom section consisting of a large block of lead with two
cubical cavities into which samples may be placed. The top section
is a heavy lead cylinder containing the cobalt and is mounted on a
dolly that can be rolled into position directly over either cavity.
The cobalt-60 is enclosed in peripheral holes in a hollow brass
cylinder, which is fastened to the end of a vertical rod. The rod
comes out of an axial hole in the upper shield to a parallel tube
running in guides and fitted with a handle and ratchet for vertical
positioning of the source. When the source is lowered (into either
cavity) the parallel tube locks the dolly against horizontal motion
by engaging one of the two holes in a horizontal strip of angle
iron attached to the lower shield.

The two cavities are similar, the lower 6 inches of each being
closed with lead plugs and having an 8 x 8 x 10-in. cavity. The
source can be rolled over a cavity and lowered, leaving the other
cavity open for examination or removal of a prior sample. The
entire unit weighs about 6 tons and provides about 10 in. of lead
shielding between the cobalt and the operator. This allows the
operator to work safely in the room where the source is located.
Using 300 Ci of cobalt-60, the central portion of the irradiation
cavity had a rather uniform intensity of 300,000 rad/h.

A design that eliminates the need for expensive shielding is
the Burton cobalt-60 irradiator [9]. This device consists essentially of a bundle of 7-ft-long vertical brass tubes which are
buried in the ground. There is an inner circle of tubes of about
5/8 in. each in diameter and an outer circle of tubes about 1 in.
each in diameter. The whole bundle has a roughly circular cross
section of about 5 in. in diameter, with 12 inner tubes and 12
outer tubes arranged concentrically. Rod-shaped capsules of
cobalt-60 are lowered to predetermined depths in selected inner
tubes from a specially designed shipping container. Samples can be
irradiated at different intensities at three vertical levels in the
other tubes and in the central core by appropriate cobalt-60 load-
ing patterns. The maximum intensity at the lowest vertical level
was reported to be 66,000 rad/h, corresponding to a surface radiation level of less than 0.2 mrad/h. This type of facility could
be constructed for an estimated cost of something less than $1000,
plus the cobalt purchase and shipping costs.

More elaborate irradiation facilities are the "pool" and
"cave" types. These have been developed to remove the restrictions
of limited irradiation volume and (in most cases) fixed intensity.
In the pool-type facility, the radioisotope source is stored and
used with auxiliary experimental equipment under a depth of water
sufficient to reduce the intensity at the water surface to toler-
able levels. By placing a number of sources in the pool, the
intensity can be varied by using different combinations of sources
and by changing the distance between source and experiment. Because of the large size of the facility, very large experiments can
be irradiated and many different experiments can be carried out
simultaneously. One technique for positioning the experiment is to
insert a long pipe, sealed at the bottom, into the irradiation
zone. The pipe, since is is open, permits access to wires and
tubes carrying circulating liquid. A 4-in.-diam. tube can accommodate sizeable Dewar flasks and furnaces so that controlled

temperatures can be maintained. Some of the canisters can be
rotated at 1 rpm by a motor-driven gear arrangement.

In the cave-type design, the cobalt-60 is stored in a shielded
container and, when necessary, moved out of the container and close
to the sample in a small shielded room or cave. The cobalt-60 may
be stored in the floor or walls of the cave. In this case it is
necessary to provide sufficient shielding for the room and also for
the entrance, which is usually built in the form of a maze. The
usual shielding for cave-type sources is concrete, but the distance
between the active source and the outside of the shielding also
attenuates the radiation because of the inverse-square law regarding intensity. In the cave source it is possible to vary the distance between the source and the sample and thus irradiate samples
at different radiation intensities. Most very large cobalt-60
sources (over 10,000 Ci) are of the cave type.

Great ingenuity has been shown in the design of special
sources for industrial use. Weinstock [10] has described large-
plane β-ray sources (^{90}Sr, ^{147}Pm, and ^{114}Ce) for potential use in
graft copolymerization, prepared by applying a vitreous enamel
containing a high concentration of the desired isotope to a stainless steel sheet. Methods of dose calculation for this source
design have been worked out [10].

Heger [11] described a cobalt-60 source suitable for irradiation of a moving sheet of polymeric film or fabric (Fig. 1-1). The
cut-away shows the grid construction of individual encapsulated
units of cobalt-60. The radiation dose received by the substrate
could be controlled in two ways: (1) by computerized rearrangement
and replacement of the individual cobalt-60 capsules and (2) by
regulating the speed of movement of the sheet through the source
area. The dotted outer line in Fig. 1-1 indicates lead shielding
surrounding the source and auxiliary apparatus. The amount of lead
shielding required could be reduced by locating the source below
ground.

FILM IN   FILM OUT

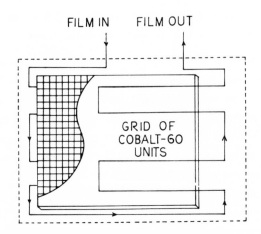

FIG. 1-1. Simplified drawing of cobalt-60 source described by
Heger [11], suitable for irradiating moving sheet of polymeric film
or fabric. Cut-away shows grid construction of individual cobalt-
60 encapsulations.

B.   Types of Radiation

The principal types of ionizing radiation that have been used
to initiate chemical reactions include α-, β-, and γ-rays and
neutrons.

1. Alpha Rays

Alpha-rays or -particles consist of helium atoms that have
lost both electrons and have a double positive charge, He^{2+}. These
particles are emitted by certain radioactive nuclei and constituted
one of the first types of radiation available for study. The
energy with which these particles are emitted depends on the radioactive species and ranges up to about 10 MeV. One-million electron
volts (1 MeV) is the kinetic energy imparted to an electron or
other singly charged particle when it is accelerated through a
potential difference of one-million volts. Under similar

circumstances, a doubly charged helium nucleus would receive an increase in energy of 2 MeV.

When α-particles pass through absorbers, they lose energy by excitation and ionization of the absorber atoms. Formerly, absorption studies of the radiation from radioisotopes were important in determining the energy of the radiation. More recently, energy determinations are carried out by the measurement of deflections in electric and magnetic fields, by the use of detectors whose output is sensitive to energy, or by the use of crystal diffraction. However, the absorption of radiation in matter is still useful for the reduction in energy of high energy particles and for the detection of a certain type of radiation in the presence of others with somewhat different absorption characteristics.

In passage through matter, α-particles lose energy mainly by interaction with electrons. This may result in the dissociation of molecules or in the excitation and ionization of atoms and molecules. Ionization is the effect most easily measured and most often used for the detection of α-particles.

Since α-particles travel only a short distance in matter before being reduced to thermal energies, it is possible to have a known number of α-particles of known energy expend their total energy inside an ionization chamber and, hence, allow measurement of the total ionization per α-particle. Such research indicates that about 35 eV of energy are dissipated for each ion pair produced in air. Corresponding figures for other gases amount to 36 eV in hydrogen, 32.5 eV in oxygen, and 26.4 eV in argon.

Example 1-1. Calculate the charge of positive sign produced in hydrogen by a 5-MeV α-particle. Assume its entire energy is imparted to the hydrogen, and that the energy loss per ion pair formed is 36 eV.

Solution. Pairs produced equal

$$E/W = \frac{5 \times 10^6}{36} = 1.39 \times 10^5 \text{ ion pairs}$$

The positive (or negative) charge produced is

$$E_e/W = (1.39 \times 10^5)(1.60 \times 10^{-19}) = 2.23 \times 10^{-14} \text{ C}$$

Much of the energy lost by α-particles is represented by the kinetic energy given to the electrons removed from atoms or molecules that collide with the α-particles. It can be demonstrated from conservation of momentum considerations that the maximum velocity which an α-particle of velocity v can impart to an electron is about 2v. Hence, the maximum energy which an electron can receive from the impact of a 4-MeV α-particle is approximately 2 keV. The average energy obtained by electrons from α-particles in their penetration of matter is of the order of magnitude of 100 to 200 eV. Many of these secondary electrons, or δ-rays, possess sufficient energy to ionize other atoms. Actually, about 60 to 80% of the ionization produced by α-particles is caused by secondary ionization. Delta-ray tracks are sometimes seen in cloud chamber pictures of α-ray tracks.

When the velocity of the α-particle has been reduced to the point where it is comparable to the velocity of the valence electron in an atom of the stopping material, a new process becomes significant: The α-particle starts making elastic collisions with the atoms instead of exciting the atomic electrons. Such ion-atom collisions result in what is known as "nuclear stopping" instead of the "electronic stopping" that takes place at higher α-particle velocities. Also, when the velocity of the α-particles becomes comparable to that of an electron in a He^+ ion, the α-particles will start picking up electrons from the atoms in the stopping material, and the average charge of the particles will

change from 2+ to 1+. Generally speaking, heavy ions passing
through matter will be stripped of all orbital electrons whose orbital velocity is less than that of the heavy ion.

Since an α-particle is not significantly deflected in a single
collision with an electron, α-particle paths are approximately
straight lines. Also, because roughly 10^5 collisions are needed to
bring an α-particle of a few million volts initial energy to rest,
the ranges of all α-particles of the same initial energy are
very close to the same value. Ranges of α-particles are usually
measured by absorption methods, either employing solid absorbers
or more accurately with gaseous absorbers.

There are small variations in the range of α-particles in a
given medium, caused by statistical fluctuations in the number of
collisions and in the energy loss per collision. This phenomenon
is illustrated by the solid curve in Fig. 1-2, where the number
of α-particles is presented graphically as a function of distance
traveled. If all the particles had the same range, the "number"
curve would drop vertically to N=0 at the distance equal to the
range, R_o. Actually, some straggling takes place, and the mean
range (R_o) is determined from the point of inflection of the curve.

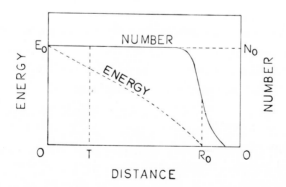

FIG. 1-2. Energy and number of α-particles as a function of
distance traveled.

As the α-particle penetrates more deeply into matter, its
energy decreases. This is demonstrated by the dotted curve in
Fig. 1-2, where the mean energy of the α-particles is plotted
versus distance traveled. When α-particles penetrate a thickness
less than the range, they are slowed down somewhat. For example,
α-particles in Fig. 1-2, which had penetrated a distance, T, would
have an average energy of the amount indicated at the intersection
of the two dotted curves.

The energy lost by an α-particle in its passage through matter
results in the formation of a large number of excited molecules
and ions in the particle track. An α-particle from polonium-210,
having an energy of about 5.3 MeV, produces about 150,000 ion pairs
and an even larger number of excited molecules when it is slowed
down by passage through air. The extent of the resulting chemical
reaction is a function not only of the number of active species
formed but also of their concentration in the particle track, which
depends, in turn, on the rate of energy loss as the particle slows
down. This rate of energy loss is generally referred to as the
linear energy transfer (LET), which is defined as "the linear rate
of loss of energy (locally absorbed) by an ionizing particle
traversing a material medium" [12]. Values of LET are usually
given in kiloelectron volts per micron (keV/μ). The LET is a
function of the energy of the α-particle and increases as the
particle slows down. An approximate LET value can be estimated by
dividing the initial energy of the particle by its average range.
Values of the "average LET" computed in this way indicate differences between the various types of radiation.

Extensive investigations have been made of the range-energy
relationships of α-particles in various materials. An empirical
formula [13] which permits the computation of the range R_A of α-
particles in a material of atomic weight A is

$$R_A \, (\text{mg/cm}^2) = 0.56R \, (\text{cm}) A^{1/3} \tag{1-5}$$

where R is the range, expressed in centimeters, of the α-particle
in air at 15°C and 760 mm Hg.

Example 1-2. An α-particle with a range of 1.41 cm in air would
have what range in stainless steel, expressed mg/cm^2?

Solution. Using 56 as the effective atomic weight of steel, the
range in steel according to Eq. (1-5) is

$$R_A (mg/cm^2) = 0.56R(cm)A^{1/3} = 0.56(1.41)(56)^{1/3} = 3.0 \text{ mg/cm}^2$$

Alpha-particle sources have been constructed in several ways.
A small type of source of about 1 Ci has been constructed by
electrodeposition of polonium-210 onto a 2-in. diam gold foil disk
mounted in a stainless-steel holder. A thin mica disk is cemented
in place as a protective device over the polonium deposit, and a
rod-shaped steel handle is screwed into the back of the holder.
The mica window will absorb part of the energy of the emitted α-
particles, but not enough to lower the energy substantially. If
desired, the energy of the α-particles can be further decreased
(and the LET increased) by positioning additional sheets of mica
between the source and the substrate to be irradiated. The additional mica will cause some straggling, and the energies of the
emitted particles will be distributed over a broader band.

2. Beta-Rays

Beta-rays are fast electrons given off by radioactive substances. Positrons may also be emitted in radioactive decay
reactions. These latter particles are sometimes called β-plus-
particles. The other main source of positrons is the pair-production process. In pair production, a high energy γ-ray

disappears and is replaced by a positron and an electron.

When a positron comes to rest, it combines with an electron
and the pair is annihilated. The mass energy of the two particles
is converted into two photons. The photons, called annihilation
radiation, have 0.51 MeV of energy each and travel off in opposite
directions. However, the fate of β-particles (electrons) is
usually not annihilation, because the number of positrons available
for taking part in the process is relatively small.

In contrast to α-particles, the β-particles from a particular
isotope are not all emitted with the same energy but with energies
ranging from zero up to a maximum value (E_β) characteristic of the
particular element. The range in energy results from the fact that
in β decay the energy released is carried off partly by β-particles
and partly by antineutrinos, the total (E_β) being divided between
the two particles. Antineutrinos have no charge and produce no
significant effect on the matter through which they pass. Their
existence was postulated to allow conservation of energy, momentum,
and spin in the β-decay process.

The energy spectrum of β-particles from phosporous-32 shows a
maximum energy (E_β) of 1.71 MeV and an average energy (E_β) of
0.70 MeV. The average β-particle energy is computed from the
expression

$$E_\beta = \frac{\int_0^{E_\beta} EN(E)dE}{\int_0^{E_\beta} N(E)dE} \qquad (1\text{-}5a)$$

where N(E) is the number of particles with energy between E and
E + dE. The average energy is usually about one-third of the
maximum energy.

The interaction of β-particles with matter is similar to that
of α-particles in some ways. The processes of energy loss are
qualitatively identical for both. The average energy loss per ion
pair formed is nearly the same for β-particles as for α-particles

(35 eV for electrons in air). The primary ionization by β-rays produces only 20-30% of the total ionization; the remainder is caused by secondary ionization.

However, there are several differences between the interactions of two types of particles with matter. For a given energy, the velocity of an electron is much larger than that of an α-particle, and hence the specific ionization is less for electrons.

Since an electron may lose much of its energy in a single collision, straggling is much more pronounced than for α-particles. As an originally homogeneous electron beam passes through matter, the straggling is increased by the scattering of electrons in different directions, which produces widely different path lengths for electrons traversing the same thickness of absorber. Nuclear scattering produces most of the large-angle deflections, although energy loss takes place almost completely through interaction with electrons.

Another mechanism of energy loss must be considered for electrons of high energy: the emission of radiation (bremsstrahlung) when an electron is accelerated in the electric field of a nucleus. The ratio of energy loss by bremsstrahlung to energy loss by ionization equals roughly EZ/800, where E is the electron energy in megaelectron volts and Z is the atomic number of the element traversed by the radiation. Hence, in air and other low density materials the percent of energy loss by bremsstrahlung is very small, whereas in an element such as lead the energy loss is significant for electrons of 1 MeV energy.

As a result of the scattering of β-particles and the broad continuous spectrum of β-energies from a single radioactive decay, there is an approximately exponential drop-off in β-activity with increasing absorber thickness. Curves of activity versus thickness of absorber traversed are generally plotted on semilogarithmic paper. The approximately exponential decrease is true of both numbers and specific ionizations of β-particles. The shape of the

absorption curve depends to some extent on the shape of the β-ray
spectrum and also on the geometrical relationship of the radioactive sample, absorber, and detector. If the absorber and sample
are as close as possible to the detector, the absorption curve
closely approximates a straight line. Otherwise, some curvature
toward the axes is usually present.

Absorption measurements are usually employed to find the upper
energy limit of the β-ray spectrum. Such determinations are useful,
but are not as accurate as those made with electron spectrographs.
The absorber is generally employed to find the range in the absorber
of the most energetic β-particles. If the absorption curve were
exactly exponential, it would eventually turn down toward minus
infinity on the usual semilog plot. In practice the absorption
curve generally does not turn toward minus infinity, because of the
presence of even more penetrating radiation than the β-rays. Such
radiation in the form of bremsstrahlung will always result from the
deceleration of the β-particles in the sample and in the absorber,
even when neither nuclear γ-radiation nor characteristic x-rays
are present.

When elements of low atomic number are used as absorbers, the
difference in slope between β- and γ-ray absorption curves is
especially noticeable. In such cases the absorption curve shows a
sharp break where the β-ray component turns into the photon "tail,"
as seen in Fig. 1-3. Because of the clarity of this phenomenon,
β-ray absorption curves are always run using absorbers of low
atomic number. Plastic or aluminum absorbers are generally employed,
and beryllium absorbers are especially helpful in distinguishing β-
rays from soft x-rays. It should be noted that β-ray ranges in
milligrams per square centimeter are essentially independent of the
absorber composition, but absorption of γ-rays increases rapidly
with atomic number. Comparative penetration figures for β-rays,
γ-rays, and other types of radiation are presented in Table 1-2.

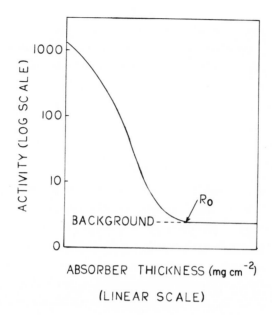

FIG. 1-3.   General absorption curve for monoenergetic
electrons.

TABLE 1-2

Approximate Ranges for 1 MeV Radiations

Radiation type	Range variable	Penetration	
		In water (cm)	In air (cm)
β or electron	Range	0.5	400
γ- or x-ray	Half-thickness	10	7000
Neutron	Half-thickness	1.5	Large
Proton	Range	0.002	2.3
α	Range	0.0005	0.2
Fission fragment (100 MeV)	Range	0.01	2.5

In measuring β- and γ-rays, most gas-filled ionization
chambers are approximately 100 times as efficient for β-rays as for
1 MeV γ-rays. A β-ray absorption curve for a β-spectrum accompanied by gamma rays will usually have a γ-ray tail with intensity of
the order of 1% of the initial β-activity. A tail composed of
bremsstrahlung only is generally about 0.05% of the initial β-
activity. In the absorption of positron (β$^+$) particles, there is
always present a background of annihilation radiation, amounting to
roughly 1% of the original β$^+$-intensity.

Various techniques have been employed to obtain the maximum
range for a β-particle spectrum from an experimental absorption
curve. The visual inspection method would indicate a range of R_o
in Fig. 1-3, selected as the point at which the β-activity is no
longer detectable above the background. The visual method usually
gives too small a value for the range, but its accuracy improves as
the background decreases. An improved estimate of the range can
often be obtained by subtracting the penetrating background radiation from the total absorption curve. A more precise method for
determining β-ray curves for "standard" substances measured with
the same experimental arrangement.

Average LET values for β-particles have also been found
useful; computed by dividing the maximum energy for the particle
by the path length (see Table 1-3). Average LET values for the
whole β-ray energy spectrum will be larger than those shown,
because of the increased contribution for lower energy β's. The
table shows clearly that the average LET decreases as the maximum
energy increases.

Much of the absorption curve for a specific β-particle source
can be expressed by an equation of the following type:

$$\text{Relative intensity} = e^{-\mu_m d} \tag{1-6}$$

where μ_m is the mass absorption coefficient in square centimeters

TABLE 1-3

Average LET Values for β-Particles

Isotope	Maximum energy of β-particles (MeV)	Path length in air (cm)	Average LET in water (keV/μm)
Hydrogen-3	0.018	0.65	2.6
Sulfur-35	0.167	31.	0.52
Strontium-90	0.544	185.	0.27
Phosphorous-32	1.71	770	0.21
Yttrium-90	2.25	1020.	0.20

per gram and d is the absorber thickness in grams per square centimeter. An empirical equation that can be used to estimate the mass absorption coefficient is the following

$$\mu_m = \frac{22}{E_m^{1.33}} \tag{1-7}$$

where E_m is the maximum energy of the β's in MeV from a particular source. This equation does not apply for $E_m < 0.5$ MeV and $E_m > 6$ MeV.

Example 1-3. Using Eqs. (1-6) and (1-7), estimate the fraction of β-particles from ^{32}P absorbed by a Geiger-Mueller tube window having a thickness of 25 mg/cm^2. The E_m value for β's from ^{32}P is 1.71 MeV.

Solution.

$$\mu_m = \frac{22}{(1.71)^{1.33}} = 10.7 \ cm^2/g$$

$$\text{Fraction absorbed} = 1 - e^{-\mu_m d}$$

$$= 1 - e^{-(10.7)(25 \times 10^{-3})}$$

$$= 1.00 - 0.76 = 0.24$$

A strontium-90 β-ray source similar in construction to the α-ray source discussed above is available in a range of activities from 5 mCi to 1 Ci. In this source the active disk contains strontium-90 bonded in silver and covered with silver, with no need for the overlay mica window. The strontium-90 is in equilibrium with its daughter element, yttrium-90, and β-particles from each of these isotopes are emitted. Neither isotope gives off γ-rays, but bremsstrahlung may be produced by action of the β-rays on the materials in the source. Such secondary radiation would not contribute significantly to the energy absorbed by a substance being irradiated, but would have great penetrating power and might result in a health hazard in the area of the source.

3. Gamma-Rays

Gamma- and x-rays are types of electromagnetic radiation that differ only in the manner that they are produced. Gamma-rays originate in nuclear reactions, whereas x-rays are produced by the excitation or removal of orbital electrons or by the deceleration of electrons. A photon of either type of radiation has energy equal to

$$E = \frac{hc}{\lambda} = h\nu \tag{1-8}$$

where h is Planck's constant, c is the velocity of light, and λ and

ν are the wavelength and frequency of the radiation, respectively.
Wavelengths for this radiation range from 3×10^{-9} to 3×10^{-4} cm,
corresponding to energies from about 40 keV to 4 MeV.

The emission of γ-rays by a nucleus is a means through which
the energy of excitation of the nucleus may be removed. The existence of excited states of the nucleus may accompany the decay of
radioisotopes or result from induced nuclear transmutations. The
x-rays arising from a certain variety of nuclear change are composed of photons with either a single energy or a group of discrete
energies. For example, cobalt-60 gives an equal number of γ-
photons of 1.332 and 1.173 MeV energies.

The excitation or removal of orbital electrons that takes
place prior to the formation of x-rays may result from several
phenomena, including the inelastic scattering of other electrons,
the internal-conversion process, or electron capture. The energy
which is released when the excited electrons revert to the ground
state takes the form of x-ray photons, which may be regarded as a
form of fluorescent radiation. The energy of the photons which
are emitted in this way is a function of the emitting element,
ranging from a few electron volts for lighter elements to about
0.1 MeV for the transuranium elements.

The x-ray formation which accompanies the deceleration of
electrons is the bremsstrahlung discussed earlier. Such x-rays
are produced whenever high energy electrons strike targets, especially targets containing elements of high atomic number. The
energy spectrum of the bremsstrahlung photons ranges from zero to
the maximum energy possessed by the bombarding electrons. Hence,
such radiations can contain photons of energies up to the BeV
region, when produced in high energy particle accelerators.

While α- and β-particles lose energy gradually through a
series of small energy losses, each γ-photon loses the greater
part of its energy through a single interaction. Thus while monoenergetic α- and β-particles are slowed down by thin absorbers,

some of the incident γ-photons are completely absorbed while the
rest are transmitted with their initial quantity of energy unchanged. For the passage of x- or γ-rays through an absorber of
thickness d, the following relation holds true

$$\frac{I}{I_o} = e^{-\mu d} \tag{1-9}$$

where I/I_o is the fraction of photons remaining in the beam after
passage through the absorber, and μ is the total absorption coefficient of the material for γ-rays of the particular energy
chosen.

Gamma-rays do not have a definite range in matter. Rather,
the rays have a half-thickness value for a particular absorbing
material, which is the thickness of absorber required to reduce
the intensity of the γ-radiation to one-half its initial value.
The half-thickness $X_{\frac{1}{2}}$ is given by the equation,

$$X_{\frac{1}{2}} = \frac{\ell n 2}{\mu} = \frac{0.693}{\mu} \tag{1-10}$$

which can be used to calculate the absorption coefficient from the
half-thickness.

Example 1-4. Calculate the half-thickness value for the 0.66 MeV
γ-radiation from cesium-137 in lead, knowing that the absorption
coefficient has a value of 1.22 cm^{-1} [14].

Solution.

$$X_{\frac{1}{2}} = \frac{0.693}{1.22} = 0.57 \text{ cm}$$

Table 1-4 presents average values of the LET of the secondary
electron produced by the absorption of certain γ-rays in water.

Gamma-rays do not form a track of charged particles, and hence the LET value is not directly applicable to the γ-radiation itself. The values shown were computed by employing an average value for the secondary electrons.

TABLE 1-4

Average LET Values for γ-Radiation

Isotope	Photon energy (MeV)	Average LET in water (keV/μm)
Cesium-137	0.66	0.39
Cobalt-60	1.25 (avg.)	0.27

Apart from absorption by matter, the intensity of α-, β-, and γ-rays from point sources decreases in proportion to the square of the distance from the source, in accordance with the inverse-square law. To compute the total decrease in intensity, absorption by matter and the inverse-square law are taken into account independently.

The properties and uses of x- and γ-rays depend on their initial energies. "Soft" x-rays of less than 100 keV are strongly absorbed by thin layers of light elements, whereas very "hard" x- and γ-rays are able to pass through rather thick layers of heavy metals.

x-Ray photons in the energy range from 10 to 100 keV are given off by the usual type of x-ray tube employed in crystallographic studies. Higher intensity rays are emitted by therapeutic units intended for skin and low depth treatment. Using such tubes, operating with electron currents up to 50 mA, it is possible to irradiate thin layers of materials at very high dose rates up to 10^6 rad/min.

x-Rays of 100 keV have a much greater penetration ability, but are produced by only a small number of commercially available units. Such radiation can be used to irradiate aqueous solution layers of 1 cm thickness with an intensity change of less than 40% in passing through the solution.

"Hard" x-rays containing photons of 200 to 300 keV are often used in industrial radiography and in medical treatments. These units have also been used frequently in radiation chemistry, although the design of the tubes never produces beams of very high intensity. Such radiations have very great penetrating power. For example, an x-ray beam of this type may be reduced by only 15% by passing through a 1 cm thickness of aqueous solution.

4. Neutrons

The behavior of neutrons in matter is not at all similar to that of α-, β-, or γ-rays. Because the neutron carries no charge, it exerts no Coulombic forces on either the orbital electrons or the nuclei. Hence, for neutrons to affect matter, they must either strike the nucleus or come close enough to interact with nuclear forces. By contrast, nuclear interaction plays only a small part in the absorption of charged particles or γ-rays.

A nuclear reaction, such as that of a nucleus $^{A}X_{Z}$ with a neutron can be indicated by the equation,

$$^{A}_{Z}X + ^{1}_{0}n \rightarrow ^{A+1}_{Z}Y^{*} \tag{1-11}$$

where the asterisk indicates that the resulting nucleus is in an excited state. The excitation energy, including the original kinetic energy of the neutron, is distributed among the components of the nucleus. The product nucleus remains in an excited state for a very brief period, perhaps 10^{-12} to 10^{-20} sec. The excess energy may be lost by giving off one or more particles or rays.

For example, if a proton is given off, the reaction is referred to
as an (n,p) reaction. Other possible reactions that could take
place would include (n,γ), (n,α), (n,2n), or fission.

If the neutron is reemitted, the phenomenon is referred to
as a scattering process. The scattering is termed inelastic or
elastic, respectively, if the nucleus involved is left in an excited or unexcited state. In both cases the neutron has lost some
of its original energy, the loss being greater for the case of
inelastic scattering.

For any excited nucleus formed by a reaction such as that in
Eq. (1-11), several different additional nuclear transformations
may be energetically possible. The relative probabilities for
further reaction are functions of the amount of excitation energy
and the energy levels that may exist in the initial and final
nuclei. Hence, the probability for each conceivable absorption
process depends in a complex way on the energy of the incident
neutron and the composition of the absorber. These probabilities
may be quite different for different isotopes of the same element,
since each nuclide has its own particular nuclear characteristics.

The importance of the neutron energy in determining the course
of such nuclear reactions has been one reason for classifying
neutrons by energy into groupings known as the thermal neutrons,
intermediate neutrons, fast neutrons, and relativistic neutrons
[15]. Thermal (slow) neutrons are in thermal equilibrium with the
surroundings and have energies of about 0.025 eV at room temperature; intermediate neutrons, with energies slightly above those of
thermal neutrons and ranging up to 10 keV; fast neutrons, with
energies from 10 keV to 10 MeV; and relativistic neutrons with
energies greater than 10 MeV.

Elastic scattering is the most highly probable method of
interaction between fast neutrons and matter and is also significant for intermediate neutrons. The nucleus struck by the neutron
is not left in an excited state. The smaller the mass of the

nucleus, the greater the fraction of the neutron's kinetic energy
taken by the nucleus.

The initial kinetic energy of the neutron is shared with the
nucleus according to the laws of conservation of energy and momentum. The fraction of energy transferred to a nucleus of atomic
weight A varies from zero up to a maximum which has been shown to
be

$$\left(\frac{\Delta E}{E_o}\right)_{max} = \frac{4A}{(A+1)^2} \tag{1-12}$$

For elastic scattering of neutrons with energies below about 10 MeV,
all energy transfers between zero and the maximum, $4AE_o/(A+1)^2$,
are equally probable.

The fractional energy transfer is greatest for hydrogen, A=1,
and drops off as the atomic weight becomes larger. In biological
materials and other hydrogenous substances, one of the most important interactions of fast neutrons is elastic scattering with
hydrogen. The recoiling hydrogen nuclei then cause ionization and
excitation of the irradiated material.

Inelastic scattering is not energetically possible for thermal
or intermediate neutrons. It occurs if the neutron is absorbed by
the nucleus and then a neutron of lower energy reemitted. Inelastic scattering is not possible for neutrons of energy below
that of the lowest excited state of the nucleus (usually a few
hundred keV) but becomes increasingly important as neutron energy
increases. As neutron energy increases above 10 MeV, inelastic
scattering becomes about as probable as elastic scattering. The
possible inelastic scattering processes include (n,n), (n,nγ), and
(n,2n). In the (n,nγ) process the excitation energy is removed by
a γ-ray. In the (n,n) reaction, the nucleus remains in a meta-
stable state. When the incident neutron has an energy of 10 MeV
or more, the (n,2n) reaction can take place.

For thermal energy neutrons, the most probable interaction process is simple capture of the neutron by the nucleus to yield an isotope of the target element (n,γ). The γ-rays which are given off in the (n,γ) reaction generally have energies of several megaelectron volts. The probability for this reaction is generally low at high neutron energies, but may rise in a series of resonance peaks at energies in the intermediate range, increasing again in the thermal range in inverse proportion to neutron velocity.

Other possible types of neutron interaction with nuclei include ejection of charged particles, fission, and various high energy processes. Examples of charged-particle ejection include (n,p), (n,d), (n,α), (n,t), etc. In view of the fact that the charged particle must overcome the Coulombic barrier before escaping from the nucleus, this type of reaction is more probable for light nuclides and fast neutrons. Exceptions to this rule include (n,α) reactions, which are exothermic enough to allow escape from the Coulombic barrier even with thermal neutrons. Examples of such exceptions are the ^{6}Li (n,α) and ^{10}B (n,α) reactions. In another type of reaction, fission, the compound nucleus splits into two fission fragments and one or more neutrons. Fission takes place with thermal neutrons in ^{235}U, ^{239}Pu, and ^{233}U and with fast neutrons in many heavy nuclides. The last type of interaction, high energy processes, refers to the phenomenon of capture in a nucleus of neutrons with energies of 100 MeV or more, causing the emission of a shower of many types of particles.

The recoil protons resulting from the elastic scattering of fast neutrons in paraffinic materials are often employed for detecting such neutrons. Approximately seven protons leave a thick paraffin layer for each 10^4 incident neutrons of 1 MeV energy, and for other neutron energies the ratio of protons to neutrons is approximately proportional to neutron energy [16]. The energy of the fastest recoil protons equals that of the incident neutrons.

Ionization caused by protons or α-particles formed in (n,p) or (n,α) reactions has also been employed for detecting neutrons.

Ionization chambers may be lined with boron or filled with BF_3 gas, and the ionizing α-particles from the ^{10}B (n,α) ^{7}Li may be detected. Also, fission fragments may be detected in an ionization chamber lined with fissionable material and exposed to incident neutrons. In addition, neutron-capture reactions leading to the formation of radioactive products have often been used to detect neutrons through the induced activity.

A quantitative interpretation of the interaction of neutrons with matter can be made by employing the idea of nuclear cross sections. The rate at which a nuclear reaction takes place in a certain material is a function of the number and energy of the incident neutrons and the type and quantity of nuclei in the material.

The dependence on neutron energy target type is included in the value of the nuclear cross section, σ. The cross section has the units of area and can be determined by a simple transmission experiment. If N is the number of target particles per cubic centimeter, then NdX is the number of target particles in an element of volume of 1 cm^2 in cross section and dX cm in thickness. The fraction of the area covered by target particles, σNdX, is equal to the fractional reduction, $-dI/I$, of a beam of incident particles.

$$\frac{-dI}{I} = \sigma NdX \qquad (1\text{-}13)$$

Integration gives

$$I = I_o e^{-N\sigma X} \qquad (1\text{-}14)$$

where I_o is the number of incident particles and I is the number of particles which go through the sample. The number of collisions per unit time can be computed from

$$I_o - I = I_o(1 - e^{-N\sigma X})\qquad(1-15)$$

The cross-sectional area for a nucleus of 10^{-12} cm would be roughly 10^{-24} cm^2. A cross-sectional area of 10^{-24} cm^2 has been given the unit name of one "barn."

Example 1-5. The capture cross section of ^{197}Au for thermal neutrons is 99 barns. Assume that a thermal neutron flux of 10^9 neutrons cm^{-2} sec^{-1} strikes a foil of ^{197}Au 0.01 mm thick and 1 cm^2 in area. Compute the number of ^{198}Au nuclei formed per second, knowing that the density of ^{197}Au is 19.3 g cm^{-3}.

Solution.

$$N = \frac{19.3}{197}(6.02 \times 10^{-23}) = 5.89 \times 10^{22}$$

$$I_o - I = I_o(1 - e^{-N\sigma X})$$

$$= 10^9[1 - e^{-(5.89 \times 10^{22})(99 \times 10^{-24})(10^{-3})}]$$

$$= 6.0 \times 10^6 \text{ nuclei/sec}$$

A variety of nuclear (capture) reactions can be produced by thermal neutrons, that is, neutrons whose energy distribution is approximately that of gas molecules at ordinary temperatures. However, nuclear reactions other than capture can only take place when the neutron has an energy in excess of the threshold energy for the reaction, and they are usually not important at neutron energies below several megaelectron volts. The energy of the emitted particle is a function of that of the absorbed neutron, and also of the energy evolved or absorbed by the particular

nuclear reaction. In biological systems and in organic compounds
containing only C, H, O, and N, the most significant reactions are
^{1}H(n,γ)^{2}H, which produces a 2.2 MeV γ-ray; and ^{14}N(n,p)^{14}C, which
gives off a 0.66 MeV proton. Capture by oxygen and carbon in such
materials is less important because the capture cross sections of
oxygen and carbon are small [17].

Fast neutrons generally lose significant quantities of energy
through inelastic collisions, particularly with heavy nuclei. This
interaction loses effectiveness for intermediate neutrons. Further
slowing down usually results through a series of elastic collisions
with nuclei.

Therefore, the slowing down of neutrons by the moderator of a
nuclear pile is caused mainly by elastic scattering and is brought
about most efficiently by materials of low atomic weight. Hence,
the average number of collisions to lower the energy of a neutron
from 2 MeV (the average energy of neutrons from fission) to the
thermal level is about 18 in hydrogen, over 100 in graphite, and
approximately 2000 in lead [18]. The cross section for elastic
scattering becomes larger as the energy of the neutron drops.

A thermal neutron in a hydrogen-containing material will most
probably be eventually captured by a proton to form a deuteron.
However, the cross section for this capture is rather small compared to the cross section for scattering, so that a thermal
neutron experiences about 150 collisions on the average before
capture. Water and paraffin are especially effective in slowing
down neutrons, since the capture cross section for oxygen and
carbon are much smaller than that for hydrogen. Heavy water is
more efficient in slowing neutrons than ordinary water, because
the capture cross section for deuterium is smaller than that for
hydrogen. Many more collisions are needed in carbon to reduce
neutrons to thermal energies than in hydrogen, but after reaching
thermal energy the neutrons can exist for a longer time in carbon
(graphite). However, the lifetime of a thermal neutron before
capture is only a fraction of a second in either carbon or hydrogen.

Neutrons would not exist as such for very long, even if they were in a medium where they would not be captured. Free neutrons are unstable and tend to decay to give protons and electrons. Robson observed neutrons in free flight and in a vacuum and found that they disintegrate with a half-life of 13 min, releasing energy of 0.78 MeV.

C. Laws of Radioactive Disintegration

For any radioactive element, the number of atoms disintegrating within a short period of time is proportional to the number present,

$$\frac{-dN}{dt} = \lambda N \tag{1-16}$$

where λ is the rate constant in units of reciprocal time and N is the number of atoms. Integration yields

$$N = N_o e^{-\lambda t} \tag{1-17}$$

where N_o is the number of atoms originally present, and N is the number present after time t.

In experimental work, the quantities N and dN/dt are hardly ever measured directly, but instead the activity A in counts per minute is determined. The activity A is proportional to the rate of decay, the proportionality constant depending on the measuring instrument and on the geometrical arrangement of the sample and the detector. Assuming that counting geometry and related factors are held constant throughout a sequence of measurements,

$$A = A_o e^{-\lambda t} \tag{1-18}$$

where A_o is the activity at time zero. Since Eq. (1-17) represents a first-order reaction, it is correct to say that

$$t_{\frac{1}{2}} = \frac{0.693}{\lambda} \qquad\qquad (1\text{-}19)$$

where $t_{\frac{1}{2}}$ is the half-life or time for half of the radioisotope to disintegrate.

In studying radioactivity, individual disintegration events are observed and counted. There is some statistical fluctuation in the rate, and the standard deviation of the observed number of counts is approximately equal to the square root of the number of counts. For 100 counts the standard deviation is about ±10. For 10,000 counts the standard deviation is about ±100, and the percentage uncertainty is ±1%. Hence the larger the number of counts, the smaller the percentage uncertainty.

The amount of any radioisotope required to provide 3.7×10^{10} disintegrations per second is called a *curie (Ci)*. The number of atoms making up a curie will be different for different radio-isotopes.

Example 1-6. How many grams of carbon-14 equals 1 Ci, if the half-life of carbon-14 is 5720 years?

Solution. First compute the rate constant for carbon-14 decay:

$$\lambda = \frac{0.693}{t_{\frac{1}{2}}} = \frac{0.693}{(5.760)(365)(24)(60)(60)} = 0.383 \times 10^{-11} \ \sec^{-1}$$

Then equate rate of decay to 3.7×10^{10} disintegrations per second

$$\frac{-dN}{dt} = \lambda N = 3.7 \times 10^{10} = \frac{(0.383 \times 10^{-11})(6.02 \times 10^{23})(g\ ^{14}C)}{(14.0)}$$

$$g\ ^{14}C = 0.224 \ \text{in 1 Ci}$$

D. x-Ray Machines and Particle Accelerators

High energy particle accelerators that were developed for
nuclear research have frequently found use in the study of
radiation-chemical problems. For radiation-chemical purposes,
accelerators are suitable which can produce a very intense beam of
radiation at moderate energies. Accelerators can often provide an
intensity several orders of magnitude greater than that available
from radioisotope sources. The x-ray machine, Van de Graaff
accelerator, and linear accelerator have been used most often in
radiation chemistry.

1. x-Ray Generators

As discussed above, x-rays are electromagnetic radiation in
the wavelength range from about 3×10^{-9} to 3×10^{-11} cm. They
result when high speed electrons are quickly slowed down, as when
they pass through the electric field of an atomic nucleus. Loss
of electron energy by this means results in the production of
radiation known as bremsstrahlung. The other method of electron
energy loss is by collision, and the ratio of loss by bremsstrahlung to loss by collision can be expressed approximately by

$$\frac{(de/dX0)_{rad}}{(de/dX)_{coll}} \cong \frac{EZ}{1600 \; m_o C^Z} \tag{1-20}$$

where E and m_0 are the energy and rest mass of the electron,
respectively, and Z is the atomic number of the target element.
Therefore, loss by radiation and collision is approximately equal
for elements of high atomic number such as lead and tungsten when
E equals about 10 MeV.

The photons of bremsstrahlung radiation possess energies
ranging from about zero to the maximum energy of the incident electrons. The energy of each bremsstrahlung photon is derived from
the electron producing it, and hence depends on how much the

electron was slowed down. If the electron was brought to rest, the
resulting photon of bremsstrahlung radiation will possess the full
amount of the electron's initial energy. The usual plot obtained
for intensity versus x-ray energy is presented in Fig. 1-4.

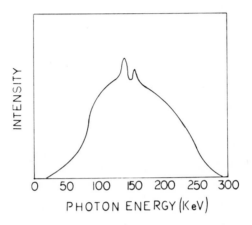

FIG. 1-4. Generalized spectrum of x-rays caused by electrons
striking metallic target; intensity not shown to scale.

The peaks in Fig. 1-4 are not bremsstrahlung radiation, but
are due to radiation that is *characteristic* of the target element.
Such characteristic radiation is produced when electrons in the
outer shells of the target atom fall into a vacancy in an inner
shell. The energy difference between the initial shell and the
final shell, which represents energy lost by the electron, may be
emitted as an x-ray. Another possibility is that the energy may be
employed in a sort of internal photoelectric process, in which an
additional extranuclear electron may be given out with kinetic
energy equal to the characteristic x-ray energy minus its own
binding energy. Electrons emitted in this way are referred to as
Auger electrons. In a heavy atom, many readjustments may take
place and several Auger electrons may be emitted.

Each characteristic x-ray photon corresponds to a particular energy change accompanying an electron move from one shell to another. For example, in using a tungsten target, transitions between outer shells and the K shell produce x-ray photons having energies between 58 and 70 keV. Transitions between outer shells and the L shell yield a series of lines near 9 keV. Target elements of lower atomic weight produce corresponding x-ray lines, but of lower energies. Frequently characteristic x-ray radiation makes up only a minor portion of the whole x-ray spectrum, with most of the energy attributable to bremsstrahlung.

An x-ray tube generally involves a tungsten wire cathode and a large anode located inside an evacuated glass envelope. The cathode, which is heated, produces a stream of electrons that are accelerated toward the anode by the applied potential. The maximum energy of the x-ray photons produced is determined by the size of the applied potential. It is advantageous to make the anode of an element of large Z, so that a large fraction of the electron energy is changed into x-rays. The anode should have a high melting point, since energy lost by collision is converted into heat in the target. The target element is often tungsten, which is embedded in a heavy copper support to help carry off the thermal energy produced.

Many of the x-ray photons produced are of relatively low energy. It is desirable to filter out these low energy photons, so as to make the resultant beam more uniform in energy and penetrating ability. Such filtration can be accomplished by locating metal filters between the x-ray tube and the sample. The filters and the window of the x-ray tube absorb the less energetic photons and also reduce the intensity of the high energy rays, so that care must be taken not to use excessive filtration.

A complete description of an x-ray source should include the applied voltage, composition of target and window materials, thickness and type of filters, and the *half-value layer* of the resultant

beam in terms of some standard absorber such as aluminum. The
half-value layer (HVL) is defined as the thickness of absorber
required to reduce the intensity of the beam to half its initial
value. For the beam from most x-ray tubes, it has been observed
that the HVL increases as the filter thickness increases. However,
for monoenergetic x-rays, the HVL remains constant as the beam is
attenuated.

Generally speaking, x-ray machines are operated in a potential
range from 40 to 300 keV. Even higher energy x-ray photons can be
produced by stopping electrons from an accelerator with a tungsten
target. In each case, the maximum energy of the photons in the
x-ray spectrum will equal the maximum energy of the incident
electrons.

2. Van de Graaff Accelerator

In the Van de Graaff accelerator, a high potential is accumulated on a conducting sphere by continually transferring static
charges from a moving belt to the sphere. The belt is made of an
insulating material such as rubber or paper andis driven by a
motor and pulley system. The belt passes through a narrow gap,
where positive (or negative) charges are spread onto it through
a sharp-toothed comb, which is connected to a source of 10,000 to
30,000 V dc. The charges are carried along on the moving belt to
the interior of an insulated conducting metal sphere, where they
are transferred from the belt to the sphere by another sharp-
toothed comb connected to the sphere. The sphere continues to
build up charge until the loss of charge from the surface by
corona discharge balances the rate of acquiring charge from the
moving belt.

While Van de Graaff's first machine involved two spheres, most
of the recent installations use a single electrode (sphere) with
acceleration of the ions between that electrode and ground potential, because of the advantages of operating auxiliary equipment
at ground potential.

The voltage of an electrostatic generator is limited by the breakdown of the gas surrounding the charged electrode. The breakdown potential can be increased by operating at high pressure, hence, most electrostatic generators are completely enclosed in steel tanks in which air pressures of 10 atm or more are maintained. Operation can be further improved by the use of high pressure nitrogen instead of air. Pressure-type generators are available, which are capable of accelerating protons or other positive ions to energies of 2 to 6 MeV.

Leading from the highly charged electrode is an evacuated "acceleration tube," which may proceed downward vertically or outward horizontally, depending on the particular design employed. Depending on the polarity of the high voltage, electrons or positive ions are impelled along the tube by an electric field and accelerated to high energy in a well-collimated beam. The beam current is provided by additional charge, which is carried to the high voltage electrode by the belt. Stable operation at any particular current is maintained by balancing the particle beam intensity with the charging current. All parameters for producing and focusing the beam are controlled from a remotely located operating console.

When a beam of electrons is desired, electrons are given off from a heated tungsten or tantalum hairpin filament. Positive ions are produced in a concentrated plasma in a suitable gas discharge. Ionization is usually produced by radio-frequency power. Ions are removed from the plasma by an extraction probe and then focused prior to acceleration by the uniform field of the tube. A wide variety of ions have been accelerated by this technique, including helium, lithium, carbon, and bromine.

Depending on research objectives, the particle beam may be stopped by a selected target, confined in the vacuum region of experimental apparatus, or brought into the atmosphere through a thin metal foil. In any case, the evacuated region of the acceleration tube must be hermetically sealed. The tube may be

permanently sealed, or there may be a demountable tube system
connected to a vacuum pump.

Constant potentials as high as 10 million volts can be used
for particle acceleration, but most machines operate below 6 MV.
Voltage controls enable the attainment of highly constant voltage
with only a few hundred volts of ripple.

Positive-ion accelerators are generally controlled very precisely by the employment of a magnetic system of deflecting and
analyzing the accelerated ion beam. The beam constituents are
separated according to their momenta, and the emerging fraction is
monitored with regard to its deflection angle. The entering and
emerging beams are collimated by means of a system of slits. The
magnetic field is held constant, resulting in an optical system
with a resolution of $E/\Delta E = 1000$ or better. Any deviation of the
beam center line from its correct deflection angle sends a signal
to the control tube of the corona system opposite the Van de Graaff
terminal. The resulting modulation of the corona current holds the
high voltage at its selected value. The accelerated electron or
ion beam from a Van de Graaff machine is intrinsically homogeneous
in energy to within a few electron volts.

The maximum generated current available for particle generation is a function of the width and speed of the charging belt.
Although electron beams as high as 6 mA dc have been produced by
such generators, the usual output currents are commonly below 1 mA
in size. The Van de Graaff should be classed as a radiation source
of moderate power, below 5 kW of continuous beam output.

One desirable aspect of the Van de Graaff's operation is that
its output can be either dc or pulsed. For pulsed operation, the
intrinsic Van de Graaff limitations in continuous-beam intensity
are no longer applicable. The amount of charge available in a
pulse is determined by the capacitance of the high voltage
terminal, rather than by the available power in the voltage generating system. Electron-beam pulses of several hundred milliamps
peak can be accelerated, with pulse lengths of several microseconds.

The homogeneous character of the accelerated electron or ion beam is well adapted for postacceleration control by deflection or focusing with magnetic or electric fields. By use of a modulated magnetic field, the beam can be scanned linearly across a target or window to provide uniform radiation distribution over a selected area, or scanned spirally or circularly over a large area to allow improved heat dissipation. Solenoidal magnetic lenses or alternating-gradient strong-focusing lenses can be used to focus the beam to small dimensions or to pipe the beam to distant locations for special research applications.

The monoenergetic electrons from a Van de Graaff generator generally make up a well-collimated beam having a Gaussian intensity distribution across the diameter [19]. Such electrons can be brought from the vacuum region out into the air with little energy loss through a thin window of aluminum or titanium. The scattering of electrons by air increases as distance from the thin window increases. To reduce power loss in the window material, the electrons can be scanned across widths up to 48 in. The distribution is again influenced by air scattering at considerable distances from the window. Concentrated beams are generally restricted to about 100 $\mu A/cm^2$ unless cooling methods are used to avoid heating and deterioration of the window foil.

x-Rays can be produced with a Van de Graaff by directing the electron beam onto targets of large atomic weight such as tungsten or gold [20]. The total production of x-rays, over the 4π angle, is roughly linear with atomic number and exponential with electron energy. The energy of the x-ray photons forms a continuous energy spectrum, ranging from zero to a value equal to the energy of the incident electrons. The intensity versus energy plot passes through a peak at about 0.33 to 0.50 of the maximum energy, depending on the target thickness, type of filtration, and electron energy.

Electrons and positive ions from a Van de Graaff generator can be used to produce neutrons [21]. Although use of ions has

advantages, electrons are often used so as to avoid the need for
changing the polarity of the Van de Graaff. By bombarding beryllium with x-rays resulting from electrons over 1.67 MeV in energy,
neutrons of various energies are produced. Moderation of these
neutrons in paraffin or water surrounding the beryllium results in
the formation of thermal neutrons, which can be employed in various
activation experiments.

3. Linear Accelerator

In the Van de Graaff machine for accelerating ions, the entire
high potential corresponding to the final energy of the ions must
be provided, and certain limitations are introduced owing to insulation problems. A device free of some of these problems employs
successive accelerations of the particles through small potential
differences. It is called a linear accelerator, because the particles are accelerated in a straight line. In such an accelerator,
the particle arrives at suitable gaps in the accelerating structure
at the correct time in the period of the radio-frequency excitation, or else moves along keeping in step with the traveling wave
of the electromagnetic field.

The traveling wave variety of linear accelerator for electrons
has been shown to be a source of high power beams at reasonable
energy, with highly efficient conversion from radio-frequency power
to electron-beam power [22]. Such a device is a very economical
source when power at energies above 2 or 3 MeV is required.

The principle of operation can be summarized in simple terms.
When radio-frequency power of suitable wavelength is propagated
inside a hollow smooth-walled metal tube, the phase velocity of
such a wave along the tube is always found to be greater than the
velocity of light. However, an electron cannot travel faster than
the velocity of light, and hence could not move along the tube at
the phase velocity of the radio-frequency wave. By placing metallic irises into the smooth-walled guide tube, the phase velocity of
the electromagnetic wave can be reduced to the desired value.

Such a waveguide is referred to as a "loaded" or "corrugated" waveguide tube. The corrugation allows a component of the electric field to remain in a direction suitable for the acceleration of electrons along the axis of the tube. The distance between the metallic irises is usually 0.2 to 0.25 of the wavelength of propagation in the waveguide.

More important dimensions are the radii of the central iris hold, a, and of the waveguide tube, b. These may be expressed as fractions of the free space wavelength, λ, of the radio-frequency power, that is, as a/λ and b/λ. A small value for a/λ gives high accelerating fields for given power, but with increased attenuation. Hence, in machines where high energy but rather little current is needed, a relatively small a/λ would be employed. For maximum power output and increased efficiency, larger values of a/λ are required. Values of a/λ employed in practice generally range from 0.1 to 0.2. For each value of a/λ, a value of b/λ can be selected to give a suitable velocity.

Hence, it is arranged to inject a beam of relatively slow electrons at one end of the waveguide tube, and then by a progressive variation of a and b along the guide tube the wave velocity is increased toward the velocity of light. The field produced for a given power input can be controlled so that electrons injected along the axis will be collected about some discrete phase of the wave and move along with the wave and abstract from it.

For electrons with energies over 1 MeV, the velocity is very slightly less than the velocity of light. If the electron acquires higher energy, the energy received from the wave is employed in increasing electron mass instead of velocity. Hence, in the later stages of the acceleration, the wave velocity is changed only slightly or may be held constant.

In a corrugated waveguide of the type described above, there is not only an electric field suitable for electron acceleration along the axis, but there is also a radial electric field and a

circumferential magnetic field.   The resultant of these fields is a
dispersal of the electron beam outward from the axis of the waveguide tube.   This dispersion may be overcome by the axial magnetic
field of special solenoids surrounding the tube.

Power input to the accelerator is always on a pulsed basis.
Theoretical considerations indicate that the free space wavelength
of the radio-frequency power should be approximately 10 cm, so that
the operation of the equipment becomes similar to that of pulsed
radar.   The magnetron oscillator or the high power klystron
amplifier can be used as power sources.   Operation is commonly in
pulses of a few microseconds duration, repeated hundreds of times
per second.   Since the pulse power used is several megawatts, the
average power available is of the order of kilowatts.

Electron linear accelerators are simpler to operate than
proton linear accelerators, because electrons having energies of
even a few megaelectron volts move along with close to the velocity
of light and can travel down a waveguide with the accelerating
wave.   In the linear ion accelerator, positive ions are accelerated
by passing through a sequence of tubes of increasing length, the
particles obtaining an increment of energy each time they cross a
gap between two tubes.   Such accelerators have been used to
accelerate the nuclei of light elements from helium to argon to
energies of approximately 10A MeV, where A equals the atomic weight
of the nucleus involved.

Electron linear accelerators of 1 GeV have been constructed at
Stanford University and Orsay, France.   The Stanford accelerator is
220 ft long, and radio-frequency power is imparted to its waveguide
by 22 klystrons, each delivering 17,000 kW at a frequency of
approximately 3000 MHz.

For acceleration of electrons to the multi-GeV range, the
linear accelerator is probably the device of choice, because it
avoids large energy losses by radiation characteristic of circular
electron accelerators.   On the other hand, acceleration of protons

to energies over 100 MeV is economically favored by circular
accelerators such as the cyclotron.

Radiation characteristics for various sources are summarized
in Table 1-4. The cyclotron, betatron, Cockroft-Walton accelerator,
and resonant transformer are less commonly used in radiation
chemistry but are included in the table for the sake of completeness.

E. Choice of Radiation Source

In the selection of a radiation source for a particular usage,
one of the most important considerations is the type of radiation
desired and whether it is obtainable from a given source (see
Table 1-4). In comparing radioisotopes with machine sources, it
should be noted that a radioisotope decays continuously and cannot
be stopped. The intensity of radiation available from an isotope
decreases with time, the rate of decrease depending on the half-
life of the particular species.

With regard to machine sources, any accelerator supplying
high energy electrons can be made an x-ray source by stopping the
electron beam with a heavy metal target such as tungsten. The
x-rays produced in this fashion will have a continuous spectrum of
energies from zero up to the energy of the incident electrons.
Values for x-ray energy given in the table refer to their peak
energy.

"Pulsed" or "continuous" beam indicates that radiation may be
received from a source as a beam of constant (continuous) intensity,
or may be broken up into a series of pulses interspersed by periods
in which no radiation is received. Pulsed beams may frequently
give instantaneous radiation intensities far greater than are
obtained from a continuous beam. Energies for positive ions in

TABLE 1-5

Radiation Characteristics from Various Sources

Source	Radiation (or particles)	Energy (MeV)	Comments
x-Ray machine	x-Rays	0.05 – 0.03	Pulsed beam unless use constant potential power supply; continuous energy spectrum
Linear electron accelerator	Electrons	3 – 600	Pulsed beam; monoenergetic radiation
	x-Rays	3 – 600	Pulsed beam; continuous energy spectrum
Van de Graaff accelerator	Electrons and positive ions	1 – 5	Continuous beam; monoenergetic radiation
	x-Rays	1 – 5	Continuous beam; continuous energy spectrum
Linear ion accelerator	Positive ions	4 – 400	Pulsed beam; monoenergetic radiation
Cyclotron	Positive ions	10 – 20	Continuous beam; monoenergetic radiation

Betatron	Electrons	10 - 300	Pulsed beam; monoenergetic radiation
	x-Rays	10 - 300	Pulsed beam; continuous energy spectrum
Resonant transformer	x-Rays	0.1 - 3.5	Pulsed beam; continuous energy spectrum
Cockroft-Walton accelerator	Positive ions	0.1 - 1.5	Continuous beam; monoenergetic radiation
Radioactive sources	α-ray	0. - 5.5	Monoenergetic
	β-ray	0.02 - 2.25	Continuous energy spectrum
	λ-ray	0.5 - 2	Monoenergetic, or a few discrete energies

Table 1-4 refer to singly charged ions. Ions having two or more
positive charges would receive corresponding multiples of the
energies shown in the table. For example, a helium nucleus with
two positive charges would be accelerated to twice the energy of
a proton by the same accelerating potential.

In selecting a source for an industrial application, both the
efficiency of power conversion and the power available for radiation are important. The production of x-rays is generally an
efficient process. At low voltages, the efficiency of conversion
from electron-beam power to x-ray beam power would probably be no
greater than a few percent. If electrons of greater energy are
employed, the efficiency of conversion is increased. For 8 MeV
electrons, the conversion efficiency is about 25%, but now the
resulting radiation is extremely penetrating, which may or may not
be desirable depending on the experiment. For the radiation just
mentioned, the x-ray intensity would be reduced by only 50% after
traversing 16 cm of unit density material, and full utilization of
the x-ray power would therefore be difficult. It may be concluded
that inefficiency in the use of machine-made x-rays is unavoidable
in either the conversion or in the absorption. For this reason,
machine power can generally be utilized most efficiently by using
the particle beam directly.

Selection of the proper type of an accelerated particle for a
particular usage depends on a variety of factors. The maximum
range of an electron measured in centimeters into unit density
material generally equals about half the value of its energy
expressed in megaelectron volts. For example, a 4 MeV electron
will penetrate approximately 2 cm into water. Hence, for certain
practical applications, electrons of a few megaelectron volts
energy give satisfactory penetration. The thickness of the substrate treated can readily be doubled by separate treatments from
opposite sides. With electrons up to 10 MeV, material up to 4 in.
in thickness can be treated, but above 10 MeV, nuclear transitions
in the substrate may become significant.

For practical usages, the penetrating range of protons, α-particles, and heavier ions is generally too small. However, when the material to be irradiated is a gas, the greater energy loss of the heavier particles may be an advantage. For gaseous irradiation, the acceleration of the heavier particle by a Van de Graaff generator or resonant transformer may be the best approach. When only small penetration of liquid or solid is needed, the Van de Graaff or resonant transformer may supply adequate electron beams. Such is the case when no more than 3 MeV electrons are required, which can be furnished by either of these two devices. At such electron energies, they both supply relatively cheap power, and their efficiency of conversion of power from the main frequency input to high energy electron-beam power is quite good. The economically important considerations of cost per installed kilowatt and running power cost are both favorable. However, for electron energies above 4 MeV, both the Van de Graaff and resonant transformer run into severe insulation problems and are unlikely to satisfy the demand for cheap power at energies above 4 MeV. Perhaps the most suitable and economical source for electrons (and positive ions) above 4 MeV is the linear accelerator. For the latter, the conversion efficiency from radio-frequency power to electron-beam power can be quite high.

REFERENCES

1. Becquerel, H., *Compt. Rend.*, 122, 422, 501, 559, 689, 762, 1086 (1896).

2. Lind, S. C., *Monatsh. Chem.*, 33, 295 (1912).

3. Lind, S. C., *The Chemical Effects of Alpha Particles and Electrons,* Chemical Catalog Company, New York, 1928.

4. Eyring, H., Hirschfelder, J. O., and Taylor, H. S., *J. Chem. Phys.*, 4, 479, 570 (1936).

5. Fricke, H., *Cold Spring Harbor Symp.*, 2, 241 (1934).

6. Levine, J. S., and Hughes, D. J., *Nucleonics,* 11, No. 7 (1953).

7. Rupp, A. F., *Proc. Int. Conf. Peaceful Uses of At. Energy, Geneva (1955), P/135*, Vol. 14, p. 128, United Nations, New York, 1956.

8. Ghormley, J. M., and Hochnadel, C. J., *Rev. Sci. Inst.'*, 22, 473 (1951).

9. Burton, M., *Nucleonics*, 13, 74 (1955).

10. Weinstock, J. J., *Chem. Eng. Progr., Symp. Ser.*, 60(53), 62 (1964).

11. Heger, A., *Deutsche Textiltechnik*, 17(5), 311 (1967).

12. *Report of the International Commission on Radiological Units and Measurements (ICRU), 1959*, Nat. Bur. Std. Handbook 78, 1961.

13. Price, W. J., *Nuclear Radiation Detection*, McGraw-Hill, New York, 1958, p. 9.

14. Grodstein, G. W., *X-Ray Attenuation Coefficients from 10 keV to 100 MeV*, U.S. Nat. Bur. Std., Circ. 583 (1957).

15. *Measurement of Absorbed Dose of Neutrons, and of Mixtures of Neutrons and Gamma Rays*, U.S. Nat. Bur. Std. Handbook 75, 1961.

16. Friedlander, G., Kennedy, J. W., and Miller, J. M., *Nuclear and Radiochemistry*, 2nd ed., John Wiley, New York, 1964, p. 117.

17. Spinks, J. W. T. and Woods, R. J., *An Introduction to Radiation Chemistry*, John Wiley, New York, 1964, p. 48.

18. Whyte, G. N., *Principles of Radiation Dosimetry,* John Wiley, New York, 1959.

19. Industrial Processing Applications of Electron-Beam Radiation, *The Engineer*, 15, 22 (1960).

20. Bly, J. H. and Burrell, E. A., *Symposium on Nondestructive Testing in the Missile Industry*, ASTM STP No. 278, American Society for Testing Materials, Philadelphia, Pa., 1960.

21. Burrell, E. A., *Materials Research and Standards*, 2, No. 1.
     9-16 (1962).

22. Saxon, A., *Proc. Phys. Soc. London B*, 67, 705 (1954).

ADDITIONAL PROBLEMS

1-1. Compute the range in air of a 5 MeV proton, knowing that
the range of a 9.9 MeV deuteron in air is about 68 cm.

Suggestion: Use the relations,

$$R_{zm}(E) = \frac{M}{M_o} R_{zm_o}(E'), \quad E' = EM_o/M$$

where R_{zm} and R_{zm_o} are the ranges of particles of mass M and M_o,
respectively, and E and E' are the energies of the same particles.

Answer: 34 cm in air.

1-2. An α-emitter, ^{210}Po is plated on the surface of a nickel
planchet. At what thickness of plated layer would the addition of
more ^{210}Po make no increase in the intensity of α-radiation coming
from the surface?

Suggestion: Estimate the thickness of ^{210}Po required to
absorb the α-radiation, assuming a maximum range of 3.8 cm for such
radiation in air. Use the expression,

$$R_A(mg/cm^2) = 0.56R(cm)A^{1/3}$$

where R_A is the range of the radiation in an element of atomic
weight A, and R is the range in centimeters of the same radiation
in air.

Note that a layer R_A, mg/cm^2 in thickness, will absorb all
radiation from the bottom of the layer, and hence making the layer
thicker would not increase the intensity coming from the top
surface.

Answer: R_A = 9.3 mg/cm^2

1-3. Show that 1 g of radium produces 3.7 x 10^{10} disintegrations per second, since it has a half-life of 1620 years and an atomic weight of 226.

Suggestion: First compute the decay constant from the half-life and then note that the number of disintegrations per second equals N.

1-4. An experimental animal is injected with a sample of radioactive sodium. How long will it take for the radioactivity of the sodium to drop to 10% of its original value?

Suggestion: Assuming a half-life of 14.8 h, first compute the rate constant from the relation,

$$\lambda = \frac{0.693}{t^{\frac{1}{2}}}$$

Answer: 2.05 days.

1-5. Calculate the atom ratio of ^{210}Pb to ^{210}Bi in secular equilibrium with each other, noting that the half-lives are 22 years and 5 days, respectively.

Suggestion: When secular equilibrium exists between mother and daughter isotopes, the following relation holds:

$$\frac{N_1}{N_2} = \frac{t_{H1}}{t_{H2}}$$

where N_1 and N_2 are the number of atoms of mother and daughter, respectively, and t_{H1} and t_{H2} are the corresponding half-lives.

Answer: 1606/1.

1-6. When an electron and a positron combine, they are annihilated and produce two photons,

$\beta^+ + \beta^- \rightarrow 2h\nu$

Compute the total energy of the two photons produced in mega-electron volts.

Suggestion: Since mass is converted to energy make use of the relation,

$E = mc^2$

where m is the total mass converted and c is the velocity of light.

Answer: 1.02 MeV.

1-7. Based on the relation,

$I = I_o e^{-\mu d}$

and assuming a half-thickness of 1 cm in lead for a 1 MeV γ-ray, compute the value of the absorption coefficient.

Answer: 0.693.

1-8. Radium decays to produce one atom of gaseous radon for each original atom of radium. The half-lives of radium and radon are 1620 years and 3.82 days, respectively. Compute the number of milliliters of radon at 25°C and 1 atm in equilibrium with 1 mg of radium.

Suggestion: Use the secular equation of Problem 1-5 to compute the gram-atoms of radon. Then compute volume of radon by means of the perfect gas law, PV = nRT

Answer: 6.99×10^{-7} ml.

1-9. A γ-ray photon must have an energy of 2.21 MeV in order to disintegrate a deuteron. What is this energy equivalent to in atomic mass units (amu)?

Suggestion: Make use of the commonly employed relation,

E(MeV) = 931 amu

Answer: 0.00240 amu.

1-10. The density of ordinary cadmium is 8.65 cm^2. Assume that it contains 12% ^{113}Cd, which has a capture cross section for thermal neutrons of 2.0 x 10^4b. If neutrons are captured only by ^{113}Cd, what fraction of a beam of thermal neutrons striking a 0.01-cm foil of cadmium will pass through?

Suggestion: Recall

$$\frac{I}{I_o} = e^{-N\sigma X}$$

where the symbols are defined as discussed above.

Answer: 0.329

Chapter 2

FUNDAMENTAL EFFECTS OF THE IRRADIATION OF MATTER

I. INTRODUCTION

The study of the irradiation of monomers and polymers requires some understanding of the effects of irradiation on matter in general. The nature of the effects depends on the type of radiation used, whether electrons, heavy charged particles, neutrons, or γ-rays. More detailed information on the effects of the various types of radiation can be obtained from other references [1-4].

II. ELECTRONS

Electrons given off by nuclei in the process of radioactive decay are known as β-particles. Particles of the same mass but having a positive charge, known as positrons, are also given off in certain radioactive decay process. It is convenient to discuss together the effects of electrons and positrons on matter, since they are similar in nature.

The most important processes by which electrons interact with matter are the emission of electromagnetic radiation (bremsstrahlung) and elastic and inelastic collisions. Electrons differ from

heavy particles in that they do not have straight paths and definite ranges. Instead, the paths of electrons are quite crooked,
and the ranges of monoenergetic electrons vary considerably. The
crooked paths are due to collisions with orbital electrons or
nuclei of atoms along the path.

At high energies, electrons lose energy mainly by radiation
emission, and at low energies mainly by inelastic collisions.

A. Energy Loss of Electrons through Radiation

When a charged particle such as an electron passes close to
the nucleus of an atom, according to classic theory it will be
decelerated and will radiate energy at a rate, -dE/dX, which is
proportional to $z^2Z^2m^2$, where Z and z are the charges on nucleus
and particle, respectively, and m equals the mass of the particle.

Bremsstrahlung emission for electrons is small for energies
below 100 keV, but becomes the main method of energy loss at
electron energies between 10 and 100 MeV. The bremsstrahlung
spectrum is identical to the continuous x-ray spectrum and extends
from zero to the energy of the incident electron.

Bethe and Ashkin [2] have shown that the rate of energy loss
by an electron due to radiation can be expressed as

$$-(\frac{dE}{dX})_r \sim Z^2N(E + m_0c^2) \tag{2-1}$$

where Z is the atomic number, N equals the number of electrons
per cubic centimeter, and m_0c^2 is the equivalent rest mass of an
electron (0.51 MeV).

B. Energy Loss of Electrons through Inelastic Collisions

Electrons also lose energy by coulombic interaction with the
electrons of the absorbing material. This process causes excitation and ionization of the absorbing material and is the main
process by which electrons are slowed down at electron energies
below those leading to bremsstrahlung.

For electron energies up to 0.5 MeV the rate of energy loss
due to collision can be expressed

$$-(\frac{dE}{dX})_c \sim \frac{1}{v^2} \tag{2-2}$$

where v is the velocity of the incident electron. The rate of
energy loss passes through a minimum at about 1 MeV. The rate of
energy loss is proportional to the specific ionization, I_m,

$$-(\frac{dE}{dX})_c = wI_m \tag{2-3}$$

where w equals the energy loss per ion pair formed.

The energy loss per unit path length, $-(\frac{dE}{dX})_c$, is called the
stopping power or *specific energy loss*. Another quantity, the *mass
stopping power*, is defined by

$$_mS = -(\frac{dE}{dX})_c \times \frac{1}{\rho} \text{ ergs cm}^2/g \tag{2-4}$$

where ρ is the density of the absorber.

The ratio of energy loss by radiation to loss by collision in
the megaelectron volt range of energies is given approximately by

$$\frac{(dE/dX)_r}{(dE/dX)_c} \quad \frac{EZ}{800} \tag{2-5}$$

C. Energy Loss of Electrons by Elastic Collisions

Scattering of charged particles by the electrostatic field
of a nucleus may take place. This does have some importance for
electrons because of their small mass. Elastic scattering is
greatest for electrons of low energy and absorbers of high atomic
number.

D. Range of Electrons in Absorbing Materials

Empirical equations have been developed to relate the range
and energy of electrons. Katz and Penfold [5] observed that in
aluminum the maximum range of β-rays and the extrapolated range of
monoenergetic electrons could be represented by

$$R \ (mg/cm^2) = 412E^n \tag{2-6}$$

where E is the maximum β-ray energy or the energy of the monoenergetic electrons, and

$$n = 1.265 - 0.0954 \ lnE \tag{2-7}$$

for energies from 0.01 to 215 MeV. This equation applies to
materials other than aluminum, especially to other light elements.

A relation given by Feather has been widely used:

$$R = 0.543E - 0.160 \tag{2-8}$$

where E is the maximum β-energy in MeV and R is the range in
aluminum in g/cm^2. Equation (2-8) holds fairly well up to at
least 15 MeV.

Scattering of electrons is much more marked than the scattering of heavy particles. A considerable fraction of the electrons striking an absorber may be reflected as a result of scattering processes. The reflected intensity increases with increasing thickness of reflector, up to a thickness equal to about one-third of the range in the particular material. The ratio of the measured activity of a β-source with reflector to that without reflector is called the back-scattering factor. Above about 0.6 MeV, the factor is independent of electron energy and ranges from about 1.3 for aluminum to 1.8 for lead. The precise value of the factor measured is dependent on the geometrical arrangement of source, reflector, and detector.

III. HEAVY CHARGED PARTICLES

Charged particles such as protons and α-particles also interact with matter by emission of bremsstrahlung, inelastic collisions, and eleastic collisions. For such particles, bremsstrahlung emission is only important for energies of 1000 MeV or more, and elastic collisions can usually be neglected. Hence, the usual interaction mode of interest is inelastic collision.

The energy loss rate for the heavy particle in passing through matter is a function of the speed and charge of the particle. In the group containing alphas, protons, and deuterons, the charges differ by a factor of 2 and the velocities will not be greatly different for similar energy ranges. Hence, the range of one particle can be estimated from that of another, and the proton is sometimes used as a prototype in such calculations.

A. Absorption of Heavy Particles by Matter

When protons or similar particles pass through matter, they lose energy through excitation and ionization of the atoms of the absorber, with ionization being the more significant of the two effects. The energy loss comes about through the interaction of the coulombic fields of the incident particle with those of the electrons of the absorber atoms. The deflections of the incident particle are relatively very small because of its much larger mass.

When a charged particle of mass much greater than that of an electron passes through matter, the linear rate of the particle's energy loss along its path is given by [6]

$$\frac{dE}{dX} = \frac{4\pi e^4 z^2 NB}{m_0^2} \tag{2-9}$$

where B is the atomic stopping power and is expressed by

$$B = Z(\ln \frac{2m_0 v^2}{I(1 - \beta^2)} - \beta^2) \tag{2-10}$$

where m_0 = rest mass of the particle

E, ze, v = energy, charge, and velocity of the particle

N = number of absorber atoms per cubic centimeter

Z = atomic number of absorber

β = v/c, where c equals velocity of light in a vacuum

I = mean excitation and ionization potential of absorber atoms, which must be determined experimentally for each element

Because of the $1/v^2$ factor, the rate of energy loss increases as the velocity decreases (in the nonrelativistic energy range). This can be explained by the longer time the particle spends in the vicinity of the bound electrons as it slows down. Similarly, when two particles of equal energy but different mass are compared, the

heavier will have the smaller velocity and therefore a larger
linear rate of energy loss. As an example, the ion density along
the path of an α-particle is several hundred times as great as
that along the path of an electron of the same energy. Equation
(2-9) does not apply for particles of very low energy (below 0.1
MeV). Equation (2-9) indicates that for a constant velocity
dE/dX is approximately proportional to NZ, the electron density
in the absorbing material.

Experimental investigations of α-particle absorption have
measured the number of ion pairs produced per unit of path length,
a quantity called the specific ionization. A plot of specific
ionization as a function of the distance of the particle from the
end of its path (residual range) is known as a "Bragg curve." A
Bragg curve for initially homogeneous α-particles is presented
in Fig. 2-1. In this curve the residual range in air is used as
a measure of particle energy.

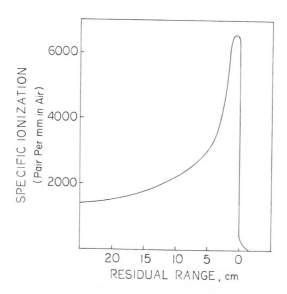

FIG. 2-1. Typical Bragg curve for initially monoenergetic
α-particles.

The curve indicates a maximum rate of energy loss at very low energies, and a decrease at higher energy corresponding approximately to an inverse dependence on the energy. This performance can easily be explained in a qualitative way. If the heavy charged particle decreases its charge by picking up electrons as it passes through matter, the coulombic interaction with electrons and the rate of energy loss will decrease. This phenomenon takes place on the low energy side of the maximum and eventually slows the particle to the point where energy is lost by elastic collisions with the atoms. Generally speaking, a heavy charged particle will pick up an electron having an orbital velocity no greater than the heavy particle's velocity. Near the end of its path, a heavy charged particle may pick up and lose electrons many times. For an α-particle, these variations in charge would be expected to occur in the last few millimeters of its range.

The decrease in rate of energy loss on the high energy side of the maximum is related to the decreased time of coulombic interaction between the charged particle and the electrons of the absorber. If the velocity of the charged particle is v, then the time that the particle spends within a certain distance from the atom is inversely proportional to v. Therefore, the impulse on the electrons in the atom is also inversely proportional to v.

The energy lost by the particle for each ion pair formed is called w and is expressed in units of electron volts per ion pair. Measured values of w depend on several variables, including the nature of the absorber and the particle type and energy. The values of w for all gases are in a range from about 25 to 50 eV per ion pair formed. The International Commission on Radiological Units has endorsed a value of 34.0 eV per ion pair for β-particles in air [7]. It should be noted that the energy loss by a particle in forming an ion pair in a gas is much greater than that required to ionize an atom of the gas. The excess energy is employed in the dissociation of gas molecules, in the excitation of atoms and molecules, and is imparted to the electrons removed in the form of kinetic energy.

The quantity w is rather dependent on the nature of the gas.
However, it is somewhat independent of the type and energy of the
particle. Values of w for solid or liquid absorbers are much less
than for gases.

B. Range of Heavy Particles in Matter

Monoenergetic heavy particles are observed to travel approximately the same distance in a particular absorber before coming to
rest. The straggling that takes place is a consequence of the
statistical nature of the energy loss process. The energy losses
that occur as the particle slows down consist of a large number of
individual energy transfers of various sizes. There are variations
in the number of transfers per unit path length, as well as in the
size of each energy transfer.

Ranges of charged particles have been measured by the use of
a collimated beam of particles from a thin radioactive source,
that is, a source thin enough that the loss of energy within the
source itself is negligible. Measurements of the number of
particles reaching a given distance in an absorber produce curves
similar to Fig. 2-2 for α-particles. The dotted line is obtained

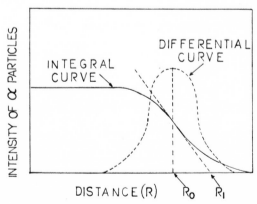

FIG. 2-2. Integral and differential range distribution curves
for α-particles.

by differentiating the solid curve with respect to distance and
represents a normal distribution of the ranges of individual particles about the mean range, R_0. The abscissa of the mean range
is found at the maximum of the differential curve. The extrapolated range, R_1, is obtained from the integral curve by a
straight-line extrapolation from the point on the curve determined
by R_0. The difference between the mean and extrapolated range is
known as the straggling, S, and is generally about 1 to 5% of the
mean range.

Reasonably accurate range-energy relations for a variety of
conditions and absorbers are available, based on a combination of
experimental results and theory. The mean range $R(E_0)$ of a
particle of kinetic energy E_0 can be expressed by

$$R(E_0) = \int_{E_0}^{0} \frac{dE}{dE/dX} \tag{2-11}$$

where dE/dX is given by Eqs. (2-9) and (2-10). Measurements of R
versus E permit the determination of values of I for use in
Eq. (2-10), so that range-energy curves can be computed from
Eq. (2-11).

Empirical equations have been used to express the range in air
of α-particles as a function of kinetic energy. An equation [8] of
better than 1% accuracy from 4 to 11 MeV and better than 4%
accuracy from 11 to 15 MeV is the following:

$$R = (0.005E + 0.285)E^{3/2} \tag{2-12}$$

where E is the α-particle energy in megaelectron volts and R is the
mean range in centimeters under normal conditions (15°C and 760 mm).

The range in an absorber, compared to that in air for the same
particle, is dependent on A, Z, and the particle energy E. The
following equation holds approximately for protons, deuterons, and
α-particles with energies from 0.1 to 1000 MeV in any elementary
absorber [9]:

$$\frac{R_z}{R_a} = 0.90 + 0.0275Z + (0.06 - 0.0086Z) \log \frac{E}{M} \qquad (2\text{-}13)$$

where R_z is the range in element Z in milligrams per square centimeter, R_a is the range of the same particle in air, M is the mass number of the particle, and E is the particle energy in megaelectron volts. The equation applies as written for absorbers with $z > 10$, but for lighter elements $(0.90 + 0.0275Z)$ should be replaced by 1.00, except for helium and hydrogen where 0.82 and 0.30 are employed, respectively. For elements heavier than air, the accuracy is increased by replacing R_z with $R_z + (0.01Z/2)$, where Z is the atomic number of the incident particle. Using these adjustments, Eq. (2-13) approximates range-energy curves for light element absorbers and for aluminum, copper, silver, and lead within a few percent in the range of particle energies from 1 to 100 MeV.

Often when ranges are available for only one type of heavy particle, the ranges of other particles can be easily computed. This is possible because for particles of the same velocity the linear rate of energy loss can be considered proportional to the square of the charge. The range is proportional to $E/(-dE/dX)$, and by considering particles having the same velocity the equations

$$\text{Range of particle } A = \frac{m_A Z_B^2}{m_B Z_A^2} \times \text{range of particle B} \qquad (2\text{-}14)$$

$$\text{Energy of particle A} = \frac{m_A}{m_B} \times \text{energy of particle B} \qquad (2\text{-}15)$$

can be derived, thus relating the ranges of particles A and B of mass and atomic number m_A and Z_A and m_B and Z_B, respectively, in the same absorber.

Example 2-1. How is the range of an α-particle related to the range of a proton of one-quarter the energy in a given absorber?

Solution.   For an α-particle (m = 4, Z= 2) and a proton (m = 1, Z = 1), Eqs. (2-15) and (2-15) yield

$$\text{Range of } \alpha\text{-particle} = \frac{(4)\ (1)^2}{(1)\ (2)^2} \times \text{range of proton}$$

$$\text{Energy of } \alpha\text{-particle} = \frac{4}{1} \times \text{energy of proton}$$

That is, in a certain absorber the range of an α-particle equals the range of a proton with one-quarter the energy.

Ranges for multiply charged particles in air computed from proton ranges are found to be smaller by a small constant value than the ranges measured experimentally.   This small correction constant, about 0.20 cm, results from the fact that capture and loss of electrons at low energies affects multiply charged particles more than protons.

The early work of Bragg showed that for a particle of a given energy the quantity $R\rho/\sqrt{A}$ is approximately constant for different absorbers, where ρ is absorber density and the other symbols have their usual meanings.   Hence, it is approximately correct to write

$$\frac{R_1}{R_0} = \frac{\rho_0 \sqrt{A}_1}{\rho_1 \sqrt{A}_0} \tag{2-16}$$

where the subscripts 0 and 1 refer to different absorbers.   This equation applies to protons and α-particles, but not to electrons.

Equation (2-16) can be employed to derive an approximate equation for R_1, the range in any substance in terms of R_{air}. the range in air at 15°C and 760 mm pressure.   By using values for air of

$$\rho_0 = 1.226 \times 10^{-3} \text{ g/cm}^3$$

and

$$\sqrt{A}_0 = 3.82$$

Eq. (2-16) becomes

$$R_1 \text{(cm)} \cong 3.3 \times 10^{-4} \left(\frac{A_1}{\rho_1 \text{(g/cm}^3\text{)}} \right) R_{air} \text{(cm)} \qquad (2-17)$$

The range computed by Eq. (2-17) may be in error by as much as +15%.

Air is an example of an absorber made up of several components. It is approximately true that the stopping effect of a mixture of atoms or molecules equals the sum of the stopping effects of all the component atoms (Bragg's rule). Letting R_1, R_2, R_3,...indicate the ranges in milligrams per square centimeter of a certain particle in each of several elements, the range R_i of that particle in a compound or in a homogeneous mixture of elements with respective weight fractions w_1, w_2, w_3,... is given approximately by

$$\frac{1}{R_i} = \frac{w_1}{R_1} + \frac{w_2}{R_2} + \frac{w_3}{R_3} + \cdots \qquad (2-18)$$

Example 2-2. Compute the range of 24 MeV α-particles in poly-ethylene, assuming a range in air of 57 mg cm^{-2}.

Solution. Using Eq. (2-13) the range in hydrogen is

$$R_H = 57(0.30 + 0.051 \log \frac{241}{4}) = 19.4 \text{ mg cm}^{-2}$$

while the range in carbon is

$$R_C = 57(1.00 + 0.012 \log \frac{241}{4}) = 57.5 \text{ mg cm}^{-2}$$

Noting that polyethylene contains 85.6% carbon and 14.4% hydrogen by weight, the range in polyethylene becomes by Eq. (2-18)

$$\frac{1}{R_{CH_2}} = \frac{0.856}{57.5} + \frac{0.144}{19.4}$$

$$R_{CH_2} = 44.8 \text{ mg cm}^{-2}$$

IV.   NEUTRONS

Since neutrons are uncharged, they interact almost entirely
with atomic nuclei and produce essentially no ionization directly.
The products of such interactions often cause ionization, and hence
result in typical radiation-chemical reactions. The principal ions
produced are protons and heavier positive ions, and the chemical
reactions caused are similar to those resulting from irradiation
with heavy positive particles. Because of their great penetration,
the results of neutron irradiation are not limited to the surface
layer of the substrate irradiated.

The principal interactions of neutrons with matter include
elastic scattering, inelastic scattering, nuclear reaction, and
capture.

A.   Elastic Scattering by Nuclei

Fast neutrons of high energy may lose considerable amounts of
energy by inelastic collisions, but after they have been slowed
down to intermediate or low energy the loss of further energy
takes place almost entirely through a succession of elastic collisions with nuclei. After the neutrons have been slowed down to
thermal energies, they may either lose or gain energy in further
collisions. This leads to a Maxwell distribution of velocities,
where the fraction of the neutrons with velocities between v and
v + dv is expressed by

$$F(v) \, dv = 4\pi^{-1/2}\left(\frac{M}{2kT}\right)^{3/2} v^2 e^{-MV^2/2kT} \, dv \qquad (2\text{-}19)$$

In this equation the fraction is indicated by F(v)dv, T equals absolute temperature, k equals the Boltzmann constant, and M equals neutron mass. In a distribution of this type the most probable velocity is given by

$$v_m = (\frac{2kT}{M})^{\frac{1}{2}}$$ (2-20)

while the average velocity is

$$\bar{v} = (\frac{8kT}{\pi M})^{\frac{1}{2}} = \frac{2v_m}{\pi^{\frac{1}{2}}}$$ (2-21)

The average kinetic energy of the neutrons is a function of the temperature of the medium,

$$\bar{E} = 3/2kT$$ (2-22)

These results from the Maxwell distribution are only rough approximations at very low temperatures and under conditions where the distribution is substantially altered by neutron leakage or absorption at the surface.

The velocity distribution felt by a sample placed in the medium will be different from that given by Eq. (2-19), because the probability that a particular neutron will strike the sample is proportional to v. This modified distribution is given by

$$F'(v)dv = 2(\frac{M}{2kT})^2 v^3 e^{-Mv^2/2kT} dv$$ (2-23)

and it is this latter distribution that is pertinent in any transmutation or cross-section calculation.

B. Inelastic Collisions with Nuclei

Inelastic scattering is said to take place if a neutron is absorbed by a nucleus, and then a neutron of lower energy given out

by the nucleus. The result is to leave the nucleus in an excited
state, from which it drops back to the ground state by giving off
one or more γ-rays. Evidently, inelastic scattering cannot take
place unless the incident neutron possesses energy at least equal
to that of the lowest excited state of the nucleus, which usually
amounts to a few hundred kiloelectron volts. Inelastic scattering
increases in importance as the energy of the incident neutron
increases.

C. Nuclear Reactions

When a high energy nuetron strikes a target nucleus, it may
become incorporated into the nucleus and another particle may be
emitted. Such nuclear reactions take place when the incident
neutron has an energy above the threshold energy for the reaction,
and this usually implies incident neutron energies of several
megaelectron volts.

An important exception is the reaction of neutrons of thermal
energy with boron-10, ^{10}B (n, α) ^{7}Li. This reaction is commonly
employed in detectors for low energy neutrons. Another important
reaction of thermal energy neutrons is that with nitrogen-14 to
yield carbon-14, ^{14}N (n, p)^{14}C.

D. Neutron Capture

For neutrons of thermal energies, the most likely interaction
with elementary matter is the capture of neutrons by the target
nuclei to give an isotope of the target element. This may lead
first to the formation of a compound nucleus in an excited state,
which returns to the ground state while giving off one or more
γ-rays. Such a reaction is called an (n, γ) reaction.

Neutron capture can be discussed quantitatively in terms of
the microscopic and macroscopic cross section of the absorber
material. If a collimated beam of neutrons strikes perpendicularly
on a very thin slab of absorber material, the microscopic cross
section σ is defined by

$$\frac{-dI}{I} = N\sigma dX \qquad\qquad (2\text{-}24)$$

where I is the intensity of the beam in neutrons per square centimeter per second, dX is the thickness of the slab, and N is the
number of target nuclei per cubic centimeter, so that a square
centimeter of the slab contains N   dX target nuclei.

The macroscopic cross section Σ is defined as follows

$$\Sigma = N\sigma \; cm^{-1} \qquad\qquad (2\text{-}25)$$

Therefore Σ is the total cross section of all nuclei in 1 cm^3 of
the target material. Equation (2-24) can be expressed

$$\frac{-dI}{I} = \Sigma dX \qquad\qquad (2\text{-}26)$$

Hence, Σ is the probability of absorption per unit path length.
Furthermore, it can be shown that Σ is the reciprocal of λ, which
is the average distance traversed by a neutron before capture
takes place.

Although the above discussion was limited to a collimated beam
of neutrons, it can readily be extended to include fields of
neutrons traveling in random directions. Assume that the neutron
moves with an average velocity v. It follows that v/λ equals the
probability that a particular neutron will undergo a reaction per
second. Letting n equal the number of neutrons per cubic centimeter, then R, the reaction rate per cubic centimeter, equals

$$R = \frac{nv}{\lambda} = nv\Sigma = \Phi\Sigma \qquad (2\text{-}27)$$

where nv equals the neutron flux expressed by the symbol Φ. Equation (2-27) is not dependent on the direction that the neutrons are traveling.

Example 2-3. Assume a 150-cm^3 counter tube contains BF_3 gas under normal conditions of pressure and temperature. If the tube is now placed in a neutron flux of 10^{11} neutrons cm^{-2} sec^{-1}, calculate the rate of the resulting (n, α) reaction for neutrons of 0.025 eV average energy.

Solution. It is known that the cross section for this reaction is about 762 b. Computing first the number of boron atoms per cubic centimeter in the tube,

$$\text{Atoms/cm}^3 = \frac{6.02 \times 10^{23}}{22,400} = 2.68 \times 10^{19}$$

The total reaction rate in the volume given is then calculated from Eq. (2-27).

$$\begin{aligned} \text{Rate} = VN\sigma\Phi &= (150)(2.68 \times 10^{19})(762 \times 10^{-24})(10^{11}) \\ &= 3.12 \times 10^{11}/\text{sec} \end{aligned}$$

Generally speaking, the neutrons taking part in a particular reaction possess a range of energies. Letting $\Phi(E)$ equal the neutron flux per unit energy interval, the flux in the energy interval from E to E + dE equals $\Phi(E)dE$. The neutron flux per unit energy interval is equal to the product of v and the neutron density per unit energy interval, n(E).

$$\Phi(E) = n(E)v \qquad (2\text{-}28)$$

The total flux of all energies would then be

$$\Phi = \int_0^\infty \Phi(E) \, dE = \int_0^\infty n(E)v \, dE \qquad (2\text{-}29)$$

where $v = (2E/M)^{\frac{1}{2}}$ and M is the mass of the neutron. The reaction rate R equals

$$R = \int_0^\infty \Sigma(E)\Phi(E) \, dE \qquad (2\text{-}30)$$

where the macroscopic cross section is written $\Sigma(E)$ to indicate that it is a function of energy. Using the average cross section $\overline{\Sigma}$ the reaction rate can be written as

$$R = \Phi\overline{\Sigma} \qquad (2\text{-}31)$$

where the average cross section $\overline{\Sigma}$ is defined as

$$\overline{\Sigma} = \frac{\int_0^\infty \Sigma(E)\Phi(E) \, dE}{\int_0^\infty \Phi(E) \, dE} \qquad (2\text{-}32)$$

The number of atoms (or nuclei) per cubic centimeter of absorber can be expressed as

$$N = \frac{\rho N_0}{W} \qquad (2\text{-}33)$$

where

ρ = density in g/cm^3

W = atomic weight

N_0 = atomic number

By using Eq. (2-33) and the definition of the barn, it can readily be shown that

$$\Sigma \, (\text{cm}^{-1}) \; = \; \frac{0.602 \sigma \rho}{W} \tag{2-34}$$

where σ is in units of barns.

It should be recognized that neutrons can take part in several different types of nuclear reactions. Corresponding to each reaction or process there will be a particular cross section. Noting that the probabilities for the different processes are additive, the total cross section σ_t can be defined as the sum of the cross sections of the individual processes. Letting σ_i equal the cross section of the ith process,

$$\sigma_t \; = \; \Sigma_i \; \sigma_i \tag{2-35}$$

Example 2-4. Assuming a cross section of 2.6×10^3 b for the (n,γ) reaction of neutrons in cadmium, compute the mean free path of 0.025 eV neutrons in cadmium of density equal to 8.6 g/cm^3.

Solution. The mean free path $\lambda = 1/\Sigma$ is computed using Eq. (2-34),

$$\lambda \; = \; \frac{W}{0.602 \sigma \rho} \; = \; \frac{113}{(0.602)(2600)(8.6)} \; = \; 0.0084 \text{ cm}$$

The cross section for capture is usually low for neutrons of high energy, but may rise in a series of resonance peaks at energies in the intermediate range, increasing in the thermal energy region in inverse proportion to neutron velocity. The experimental determination of the cross sections for the various possible nuclear reactions is a complicated and difficult task. Furthermore, there may be large variations in cross section between different elements and between different isotopes of the same element. Experimental measurements are necessary in most cases, because the theory is not adequate to permit the computation of cross sections. An extensive compilation of thermal-neutron cross sections and of curves of cross section versus energy over a large energy range has been published by Hughes and Harvey [10].

V. γ-RAY INTERACTIONS

A. Definition of Absorption Coefficients

When collimated electromagnetic radiation passes through matter in the form of *a narrow beam*, its intensity is reduced. The reduction in intensity *dI* when passing through a small thickness *dX* of matter is expressed by

$$dI = I_i \mu dX \tag{2-36}$$

where I_i is the intensity of incident radiation in ergs per square centimeter per second and μ is the *total linear absorption coefficient* (cm^{-1}) of the absorber. Hence μ represents the fraction of the energy diverted or absorbed from the incident beam by unit thickness of the absorber. Ideally, the radiation intensity measured by the detector under these conditions is due only to that portion of the primary beam, which has not interacted with the absorber in any way. Any scattered radiation is deflected away from the detector and not measured.

When dI and dX are not very small, the correct expression can be obtained by integration of Eq. (2-36) to give

$$I = I_i e^{-\mu X} \tag{2-37}$$

where I represents the intensity of radiation passing through thickness X of absorbing material.

The *mass absorption coefficient* μ/ρ is obtained by dividing the linear absorption coefficient by the density (g/cm^3). The *atomic absorption coefficient* $_a\mu$ and the *electronic absorption coefficient* $_e\mu$ are defined as follows.

$$_a\mu = \frac{\mu A}{\rho N_0} \; cm^2/atom \tag{2-38}$$

$$_e\mu = \frac{\mu A}{\rho N_0 Z} \; cm^2/electron \tag{2-39}$$

where A equals atomic weight and Z equals atomic number of the absorber. Both $_a\mu$ and $_e\mu$ have the units of an area and may be called *cross sections*. Either "cross section" or "absorption coefficient" may be used to indicate the probability of absorption. Values of $_a\mu$ and $_e\mu$ are sometimes given in units of barns/atom or barns/electron, respectively, where 1 b equals 10^{-24} cm^2.

The total absorption coefficient is equal to the sum of various partial absorption coefficients, each of which refers to a particular absorption process. The absorption processes of most importance include the Compton effect, photoelectric effect, and pair production.

B. Photoelectric Effect

In the photoelectric interaction, a photon of energy $h\nu$ ejects a bound electron from an atom or from a molecule, and gives it an energy $h\nu - b$, where b equals the original binding energy of the electron. A quantum of radiation disappears in this reaction, and conservation of energy is possible through the imparting of some momentum to the remainder of the atom. When an incident quantum has an energy greater than the K binding energy of the absorbing element, photoelectric absorption occurs mainly in the

K shell, with the L shell contributing approximately 20% and the outer shells contributing even less. This explains why the probability for photoelectric absorption exhibits sharp discontinuities at energies equal to the binding energies of the K and L shell electrons. The vacancy caused in an inner shell by ejection of an electron is filled by an electron from an outer shell, accompanied by the emission of characteristic x-radiation or of low energy Auger electrons.

The linear absorption coefficient for the photoelectric effect is usually indicated by the symbol τ. Studies of the dependence of τ on the atomic number and the energy of the photon have been made by Heitler [11] and by Bethe and Ashkin [2]. The equation derived which holds for $E_b \ll E \ll mc^2$ (0.51 MeV) can be written as follows

$$\tau \sim NZ^5E^{-3.5}$$

(2-40)

where N is the number of atoms per unit volume, E_b is the binding energy of the orbital electrons, and E is the energy of the incident γ-ray. The γ-ray energy corresponding to a photoelectric absorption of 5% of the total absorption is 0.15 MeV for aluminum, 0.4 MeV for copper, 1.2 MeV for tin, and 4.7 MeV for lead. Except in the case of the heaviest elements, photoelectric absorption is rather insignificant for γ-ray energies greater than 1 MeV. The coefficient τ is correctly referred to as an "absorption" coefficient, because both the secondary x-rays given off as the vacant shell is filled and the photoelectrons rapidly impart their energy to the absorber.

In the photoelectric effect, the photoelectrons are mainly responsible for the ionization produced by low energy photons. The photoelectric effect may be employed to measure γ-ray energies. This is done by determining the total ionization caused by the photoelectrons in a proportional or scintillation counter. Another technique uses a thin foil of high atomic number called a "radiator," which is placed over the γ-emitter, and the energies of the ejected electrons are determined in an electron spectrograph.

C. Compton Effect

In the phenomenon of Compton scattering, the incident photon interacts with one of the orbital electrons in an atom of the absorber. The electrons may be considered to be free electrons, under such conditions that the energy of the incident photon is large compared to the electron binding energy. The Compton effect may be interpreted as an elastic collision between the incident photon and the electron. Following the collision, the energy is shared between the recoil electron and the secondary photon. The secondary photon travels in a different direction from that of the primary photon and is called the scattered photon.

The incident photon has a momentum equal to $h\nu/c$. From a consideration of the conservation of energy and momentum in the photon/electron collision, the following equation can be derived:

$$h\nu' = \frac{h\nu}{1 + (1 - \cos\theta)h\nu/mc^2} \qquad (2\text{-}41)$$

where m is the mass of the electron, $h\nu'$ is the energy of the scattered photon, and θ is the angle between its direction and that of the primary photon. The process can be visualized as shown in Fig. 2-3. The kinetic energy E_{kin} of the recoil electron can be expressed as

$$E_{kin} = h\nu - h\nu' = h\nu \left(\frac{(1 - \cos\theta)h\nu/mc^2}{1 + (1 - \cos\theta)h\nu/mc^2} \right) \qquad (2\text{-}42)$$

Equation (2-42) indicates that when $h\nu \ll mc^2$ (0.51 MeV), the energy received by the recoil electron is insignificant. On the

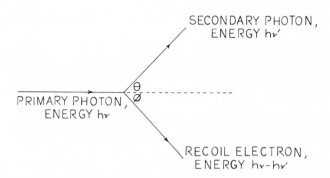

FIG. 2-3.   Diagram of Compton scattering of a γ-ray by an
electron, showing path of secondary photon and recoil electron.

other hand, when $h\nu \gg mc^2$, almost all the incident photon's
energy $h\nu$ is imparted to the recoil electron.

The partial absorption coefficient for the Compton effect is
indicated by the symbol σ.   Since each electron absorbs individu-
ally in the Compton effect, the coefficient σ is proportional to
the number of electrons, that is, proportional to Z.   The influence
of photon energy on the magnitude of σ has been computed by Klein
and Nishina [12], based on quantum mechanical considerations.   When
$h\nu \gg mc^2$, the value of σ can be expressed as

$$\sigma \sim \frac{NZ}{h\nu} \left(\ln\frac{2h\nu}{mc^2} + 1/2 \right) \tag{2-43}$$

Hence this absorption coefficient is approximately inversely proportional to $h\nu$ and gives the fraction of photons of energy $h\nu$
that interact by the Compton process, per electron/cm^2.   A considerable portion of the enrgy of these photons is retained by the
scattered photons.   The fraction of the incident photon energy
retained by the scattered photons is denoted by the Compton
scattering coefficient $_e\sigma_s$.   The fraction of the incident photon

imparted to the recoil electron, per electron/cm^2, is indicated by the *true* or *energy* Compton absorption coefficient $_e\sigma_a$ which is given by

$$_e\sigma_a = (\sigma - _e\sigma_s) \ cm^2/electron \qquad\qquad (2\text{-}44)$$

The ratio of the true to the scatter absorption coefficient is a function of the incident photon energy.

The Compton effect is more important than other types of interaction for photon energies in the range from 1.0 to 5.0 MeV in elements of high atomic number, and over a much more extensive range of energies in elements of low atomic number. In the case of water, the Compton effect is the major type of interaction for photon energies from 30 keV to 20 MeV. Conversion of the Cpmpton electronic absorption coefficient to the atomic, mass, or linear coefficient involves the use of one or more of the constants, atomic number, atomic weight, or density of the absorbing element.

D. Absorption by Pair Production

In this type of absorption, the primary photon disappears and its energy is imparted to the positron and electron pair which is formed. In order for the process to take place, the incident photon must have an energy larger than the total rest mass of the pair, that is, larger than $2mc^2$ = 1.02 MeV. Furthermore, the process must take place in the coulombic field of a nucleus (or less frequently in the field of an electron). The energy over and above that needed to supply the rest-mass energy goes into the kinetic energy E_{kin} of the pair that is formed. Hence,

$$E_{kin} = h\nu - 2mc^2 = h\nu - 1.02 \ MeV \qquad\qquad (2\text{-}45)$$

It may be noted that more than one-half of the energy goes to the positron, which is repelled by the nucleus, while less energy is imparted to the electron.

The probability for the pair-formation event increases in proportion to the square of the nuclear charge. The partial absorption coefficient for pair production is usually given the symbol κ. The value of κ is zero when $h\nu \lesssim 1.02$ MeV, rises linearly at low energies, and increases in proportion to lnE at high energies. The situation can be expressed as follows:

$$\kappa \sim NZ^2(h\nu - 2mc^2) \tag{2-46}$$

in the neighborhood of 1 MeV, and

$$\kappa \sim NZ^2 \ln h\nu \tag{2-47}$$

for photons of very high energy.

Only a portion of the incident photon energy is absorbed in the pair-production event. The positron formed is slowed down and finally combines with an electron, the rest-mass energy of $2mc^2$ appearing in the form of two 0.51 MeV γ-rays emitted in opposite directions (annihilation radiation).

E. Total Absorption Coefficient

The total absorption coefficient μ can be written as

$$\mu = \tau + \sigma + \kappa \tag{2-48}$$

where τ, σ, and κ are the linear absorption coefficients for the photoelectric effect, Compton effect, and pair-production process, respectively. The total mass, total atomic, and total electronic absorption coefficients may be computed similarly by summing the appropriate partial coefficients.

Absorption coefficients for the various elements present in an absorber can be considered separately, and by adding together the absorption contributions for each type of atom the total absorption can be calculated. Electronic absorption coefficients of the elements are weighted in proportion to the number of electrons contributed by each element.

Example 2-5. Compute the electronic absorption coefficient for water, knowing that for 1 MeV γ-rays the total absorption coefficient for oxygen and hydrogen is the same, namely 0.211 b/ electron [28].

Solution. A water molecule contains 10 electrons, 2 from hydrogen and 8 from oxygen. The total absorption coefficient for 1 MeV γ-rays is 0.211 b/electron in oxygen and also 0.211 b/electron in hydrogen. Hence,

$$\left(_e\mu\right)_{H_2O} = 0.2 \ \left(_e\mu\right)_H + 0.8 \ \left(_e\mu\right)_O$$

$$= (0.2)(0.211) + (0.8)(0.211)$$

$$= 0.211 \ b/electron$$

This case is unusual in that the absorption is entirely by the Compton process, and since the Compton electronic absorption coefficient is the same for all materials at a given photon energy, it is the same for water, hydrogen, and oxygen. However, electronic absorption coefficients can also be weighted and combined when they differ for the different elements in an absorber.

On the other hand, atomic absorption coefficients can be combined after weighting them in proportion to the number of each type of atom present in a compound or mixture. By similar reasoning, mass absorption coefficients are weighted in proportion to the mass of each element.

Example 2-6. Compute the atomic and mass absorption coefficients
for 1 MeV γ's in water. Assume atomic absorption coefficients of
0.211 and 1.69 b/atom for hydrogen and oxygen, respectively. Assume
mass absorption coefficients of 0.126 and 0.0636 for hydrogen and
oxygen, respectively [28].

Solution. Each molecule of water contains two atoms of hydrogen
and one atom of oxygen, so that the atomic absorption coefficient
for water can be written

$$(_a\mu)_{H_2O} = 2(_a\mu)_H + (_a\mu)_O$$

$$= (2)(0.211) + 1.69$$

$$= 2.11 \text{ b/molecule}$$

Since water contains 2/18 parts by weight of hydrogen and 16/18
parts by weight of oxygen, the mass absorption coefficient of water
may be expressed

$$(\mu/\rho)_{H_2O} = 2/18(\mu/\rho)_H + 16/18(\mu/\rho)_O$$

$$= 2/18 \times 0.126 + 16/18 \times 0.0636$$

$$= 0.0706 \text{ cm}^2/\text{g}$$

The combination of linear absorption coefficients is more
complicated. It is best to add the elementary electronic, atomic,
or mass absorption coefficients to obtain the corresponding coefficient for the compound and then convert one of these to the
desired linear absorption coefficient for the compound.

The atomic, mass, electronic, and linear absorption coefficients can be converted to each other by employing the equations

$$_a\mu = Z(_e\mu) \tag{2-49}$$

$$\mu/\rho = \frac{N_0 Z}{A}(_e\mu) = \frac{N_0}{A}(_a\mu) \qquad\qquad (2\text{-}50)$$

$$\mu = (\mu/\rho)\rho \qquad\qquad (2\text{-}51)$$

where Z equals the atomic number for an element or the sum of the atomic numbers of the elements present for a compound, ρ equals the density of the absorber, N_0 equals Avogadro's number, and A equals the atomic or the molecular weight as the case may be.

F. Radiation Energy Absorbed by Matter

The quantity of particular interest in radiation chemistry is that portion of energy transferred from the radiation beam to the absorbing material. However, the total absorption coefficient μ refers to the attenuation of a narrow beam of monoenergetic radiation traversing an absorber, but is not a direct measure of the energy transferred to the absorber. Part of the energy of the incident beam, not absorbed by the absorber, may be diverted from the primary beam as scattered or secondary radiation by the following processes

a. Secondary x-ray emission by an atom that has lost an electron through the photoelectric phenomenon.

b. Radiation scattering because of the Compton effect.

c. Production of annihilation radiation by the combination of a positron with an electron.

d. Coherent (Rayleigh) scattering, as contrasted to Compton scattering, which is sometimes called *incoherent*.

e. Bremsstrahlung emission by high energy secondary electrons.

Compton scattering (b) is the only one of these processes that is significant for photons of moderate energy (0.1 to 5 MeV) in materials of low Z values. Processes (a) and (d) become important at very low energies and high Z values, whereas at very high energies (c) and (e) become significant. The energy absorbed by the target material equals the energy loss from the incident beam computed from the total absorption coefficient, less the scattered energy lost by processes (a) through (e).

The energy absorbed from the incident beam is essentially equal to the energy received by the secondary electrons, providing they do not leave the absorbing material. It should be noted that, if the range of the secondary electrons is larger than the size of the sample, then a significant number of these electrons may leave the physical boundaries of the sample and take with them some of the energy that would have remained absorbed. This situation can be prevented by enclosing the sample in similar material of a thickness equal to the maximum range of the secondary electrons. The absorber is then said to be in *electronic equilibrium* with its surroundings.

A variety of absorption coefficients are used to indicate the energy imparted to the secondary electrons (absorbed energy) and the energy scattered by the absorber. The scatter coefficients are indicated by the subscript s, and the *true* or *energy* or *absorption coefficients* are indicated by the subscript a. It is important to note that these subscripts are placed *after* the symbol for the appropriate absorption coefficient. For example, the total linear absorption coefficient equals the sum of a scatter component and an energy absorption component

$$\mu = \mu_s + \mu_a = \tau + \sigma + \kappa \qquad\qquad (2\text{-}52)$$

and in similar fashion the total mass, atomic, and electronic absorption coefficient each equals the sum of a scatter component and an energy absorption component. The scatter component of the total linear absorption coefficient may be written

$$\mu_s = \tau_s + \sigma_s + \kappa_s \qquad (2\text{-}53)$$

and the energy absorption component is

$$\mu_a = \tau_a + \sigma_a + \kappa_a \qquad (2\text{-}54)$$

The Compton linear absorption coefficient can be broken down into two components,

$$\sigma = \sigma_s + \sigma_a \qquad (2\text{-}55)$$

and so can the Compton mass, atomic, and electronic absorption co-efficient. Similar remarks apply to the photoelectric linear, mass, atomic, and electronic absorption coefficients, and to the pair-production linear, mass, atomic, and electronic absorption coefficients.

Absorption coefficients pertaining to linear scatter (μ_s) and linear absorption (μ_a) are defined by the following equations:

$$dI_s = -I'\mu_s dX \qquad (2\text{-}56)$$

$$dI_a = -I'\mu_a dX \qquad (2\text{-}57)$$

where dI_s is the reduction cuased by scattering and dI_a is the reduction resulting from energy absorption when radiation of intensity I' traverses a very small thickness dX of absorbing material. Hence the total reduction in intensity (dI) which is defined by

$$dI = -I'\mu dX \qquad (2\text{-}58)$$

may also be expressed as

$$dI = dI_s + dI_a \qquad (2\text{-}59)$$

In radiation chemistry, the quantity of principal interest is the amount of energy absorbed by a sample located in a radiation beam of known intensity. This can be computed by considering an absorber to be made of a large number of very thin layers, each of thickness dX and perpendicular to the beam of incident rays. The energy absorbed by any one of the layers is given by Eq. (2-57), where I' is the reduced intensity reaching that particular layer. The intensity at any depth is equal to

$$I' = I_i e^{-\mu X} \tag{2-60}$$

where I_i is the incident intensity. The energy absorbed by any thickness dX and depth X according to Eq. (2-57) is

$$dI_a = -I_i \mu_a e^{-\mu X} dX \tag{2-61}$$

The total reduction in intensity resulting from energy absorption as the beam passes through X centimeters can be found by integration of Eq. (2-61) to be

$$I_a = \int_0^X -I_i \mu_a e^{-\mu X} dX \tag{2-62}$$

$$= I_i \mu_a/\mu (1 - e^{-\mu X}) \; ergs/cm^2 \; sec \tag{2-63}$$

By similar reasoning, the reduction in intensity caused by energy scattering is calculated to be

$$I_s = I_i \mu_s/\mu (1 - e^{-\mu X}) \; ergs/cm^2 \; sec \tag{2-64}$$

Hence, the total reduction in intensity in passing through X centimeters is computed by adding Eqs. (2-63) and (2-64) and noting that $\mu = \mu_a + \mu_s$.

$$I_a + I_s = I_i (1 - e^{-\mu X}) \; ergs/cm^2 \; sec \tag{2-65}$$

Integration of Eq. (2-65) with respect to time gives the energy
lost as the beam passes through the absorbing material,

$$E_{lost} = E_i (1 - e^{-\mu X}) \text{ ergs/cm}^2 \qquad (2\text{-}66)$$

where E_i equals incident beam energy in ergs per square centimeter.
The energy absorbed (E_a) can be obtained similarly from Eq. (2-63),

$$E_a = E_i \mu_a / \mu (1 - e^{-\mu X}) \text{ ergs/cm}^2 \qquad (2\text{-}67)$$

If μX is small, the following approximation holds,

$$1 - e^{-\mu X} \sim \mu X$$

so that Eq. (2-67) may be written in the approximate form

$$E_a = E_i \mu_a X \text{ ergs/cm}^2 \qquad (2\text{-}68)$$

Equation (2-68) deviates increasingly from the true value of energy
absorbed as thickness increases. Equation (2-68) can be expressed
in terms of the mass energy absorption coefficient μ_a/ρ as follows:

$$E_a (\text{ergs/g}) = E_i (\text{ergs/cm}^2) \, \mu_a/\rho \qquad (2\text{-}69)$$

Example 2-7. Calculate the energy absorbed by a 5.0 cm thickness
of water from an incident beam of energy 100 ergs/cm^2 of 1.25 MeV
γ-rays from ^{60}Co, assuming $\mu = 0.064 \text{ cm}^{-1}$ and $\mu_a = 0.03 \text{ cm}^{-1}$.
Solution. Using Eq. (2-67), the absorbed energy equals

$$E_a = E_i \mu_a / \mu (1 - e^{-\mu X}) \text{ ergs/cm}^2$$

$$= (100) (\frac{0.03}{0.064}) \left(1 - e^{-(0.064)(5)} \right) \text{ ergs/cm}^2$$

$$= 12.9 \text{ ergs/cm}^2$$

Using Eq. (2-68),

$$E_a = E_i \mu_a X \text{ ergs/cm}^2$$
$$= (100)(0.03)(5) = 15.0 \text{ ergs/cm}^2$$

The value from Eq. (2-68) is only approximate and will deviate more from the true value as the thickness of water increases.

When Eq. (2-69) is used to determine the ratio of energies absorbed by two different absorbers, the errors in the numerator and denominator essentially cancel each other, so that the following holds true over a much wider thickness range.

$$\frac{(E_a)_A}{(E_a)_B} = \frac{E_i(\mu_a/\rho)_A}{E_i(\mu_a/\rho)_B} = \frac{(\mu_a/\rho)_A}{(\mu_a/\rho)_B} \qquad (2\text{-}70)$$

As shown in Eq. (2-54), the total energy absorption coefficient equals the sum of the partial coefficients pertaining to the photoelectric process, the Compton effect, and the pair-production process. Considering first the photoelectric process, it can be shown that

$$\tau_a = \tau\left(\frac{E_0 - fE_s}{E_0}\right) \qquad (2\text{-}71)$$

where E_0 equals the energy of the incident photon and f equals the probability that the binding energy E_s of the ejected electron will be reemitted as a secondary x-ray photon. If $E_0 \gg E_s$, then essentially $\tau_a = \tau$.

The linear energy absorption coefficient for the Compton effect is related to the electronic energy absorption coefficient as follows:

$$\sigma_a = {_e}\sigma_a \left(\frac{\rho N_0 Z}{A} \right) \tag{2-72}$$

and the pair-production linear energy absorption coefficient can be expressed as

$$\kappa_a = \kappa \left(\frac{E_0 - 2m_0 c^2}{E_0} \right) \tag{2-73}$$

where $2m_0 c^2$ (= 1.02 MeV) equals the portion of absorbed energy that appears later in the form of annihilation radiation. Total energy absorption coefficients for several materials have been compiled by Fano [13] and Berger [14].

For materials having a low Z value, there is a range of incident photon energy for which the only absorption process of any importance is the Compton process. In water, for example, these energies range from 0.1 to 4.0 MeV. Under such conditions,

$$\mu = \sigma \tag{2-74}$$

and

$$\mu_a = \sigma_a \tag{2-75}$$

This energy region includes γ-rays from cesium-137 and cobalt-60 when absorbed by monomeric and polymeric systems.

VI. PROCESSES OCCURRING IN THE TRACK OF A
MOVING CHARGED PARTICLE

When a charged particle is slowed down by passage through matter, energy is lost, resulting in the formation of excited states

of the atoms and molecules of the absorber, and in the production
of ions in cases where electrons receive sufficient energy to
eject them from the parent atom. Absorption of γ-rays yields the
same end result, since energy is imparted to electrons and positrons and then dissipated along the paths of these particles.

Hence, the net result of absorbing any type of radiation is
the production of tracks of excited and ionized species, such
species being much the same regardless of the type of radiation
employed. Thus, all radiation will yield similar chemical effects,
but the density of active species along the tracks will depend on
the type and energy of the radiation.

The chemical effects will depend on the density of active
species along the tracks, and hence on specific ionization and
linear energy transfer. If the electrons produced by ionization
have less energy than about 100 eV, their range in liquid or solid
will be small, and any secondary ionization caused by them will
take place in a small cluster or spur close to the path of the
primary particle. If the secondary electron has energy above
100 eV, it is called a δ-ray. Such a δ-ray forms a secondary track
branching off from the track of the primary radiation. It has been
estimated [15] that approximately 50% of the ionizations produced
by a primary particle are in the secondary tracks of δ-rays with
energy over 100 eV, while the remainder of the ionization takes
place along the path of the primary particle. Wilson's study [16]
of cloud-chamber photographs of fast electrons in water vapor
indicated an average of two to three ion pairs in each spur.

When the primary radiation consists of α-particles, a column
of ions and excited species form about the track, a phenomenon
known as *columnar ionization*. Lighter, high velocity particles
give spurs at intervals along the primary track. It has been estimated that γ-rays absorbed in organic liquids produce spurs at an
average interval of 10^4 Å along the track and have an average
diameter of 20 Å [17]. After an electron has slowed down to thermal
energy, it may neutralize a positive ion or add to a neutral molecule to produce a negative ion.

An important quantity in studying energy loss of a charged particle is the *specific ionization*, defined as the total number of ion pairs produced in a gas per unit length of track. A plot of specific ionization for an α-particle versus distance along its path shows that the maximum intensity of ionization is produced near the end of the path where the velocity is low. The meaning of specific ionization is expressed by the relation,

$$\text{Specific ionization} = \frac{dE/dX}{W} \qquad (2\text{-}76)$$

where W is the average energy loss per ion pair produced and dE/dX is called the specific energy loss or stopping power. Values of W for various gases are shown in Table 2-1. Inspection of the table shows that values of W are similar for electrons and α-particles. Actually they are very nearly the same for radiation of all types.

Linear energy transfer (LET) is defined as the linear rate of energy loss absorbed locally by an ionizing particle passing through matter. An approximate average value of the LET may be computed by dividing the total energy of a particle by its path

TABLE 2-1

Values of W

	W (eV/ion pair)	
Gas	α-Particles	Electrons
Air	35.1	34.0
Argon	26.4	26.4
Carbon dioxide	34.4	32.8
Ethane	26.6	24.5
Helium	44.0	42.3
Hydrogen	36.5	36.3
Nitrogen	36.4	34.9
Oxygen	32.4	30.8

length. One reason for the approximate nature of this average is
that the rate of energy loss for a particle changes and passes
through a maximum as the particle slows down. Furthermore, the
energy lost by a primary particle may not be absorbed locally,
but may be transferred to δ-rays or to secondary electromagnetic
radiation.

In computing LET values, the results obtained depend on the
assumptions of the author, and some typical average values are
presented in Table 2-2. It is seen that the LET increases progressively in going from γ-rays to low energy x-rays to α-rays.

TABLE 2-2

Average LET Values in Water

Radiation	Average LET (keV/μm)	
	Gray [19]	Cormack and Johns [20]
25 Mvp betatron x-rays	0.28	0.202
^{60}Co γ-rays, 1.25 MeV	0.36	0.25
200 kvp x-rays	3.25	1.79
^{210}Po α-rays, 5.3 MeV	158	—

VII. RADIATION DOSIMETRY

In order to make quantitative studies of radiation chemistry
reactions, it is necessary to know the amount of radiation absorbed
by a reactant. The term "dose" is often used to describe the
amount of radiation received by a reactant. This term comes from
the field of radiation therapy, where a patient may be given a

"dose" of radiation just as he is given a "dose" of medicine. Measurement of the size of a radiation dose is known as *dosimetry*. Before proceeding it will be necessary to define several terms that are employed in this field.

A. Definition of Radiation Units

A distinction should first be made between *exposure* and *absorbed dose*. The exposure dose is a measure of the amount of radiation incident on a substrate, whereas the absorbed dose is a measure of the radiation actually absorbed by a substrate. If radiations of different types are compared, or if substances of different composition are irradiated in the same radiation field, the amounts of energy absorbed by the substrate can be greatly different for the same exposure dose.

The unit of absorbed dose recommended by the ICRU [21] is the *rad*, which is defined as 100 ergs/g. This unit does not depend on the type of radiation or on the composition of the absorber. Sometimes the absorbed dose may be given in units of electron volts per gram or electron volts per cubic centimeter. One rad equals 6.24×10^{13} eV/g or $6.24 \times 10^{13} \times \rho$ eV/cm^3, where ρ is the density of the absorber in grams per cubic centimeter. The absorbed dose is most useful in radiation chemical studies, since it is a measure of the energy received by the irradiated material.

The *integral absorbed dose* in a given region equals the energy received by the absorber from ionizing radiation in that region. The *gram rad* is the unit of integral absorbed dose and equals 6.24×10^{13} eV or 100 ergs.

The absorbed dose per unit time is called the *absorbed dose rate* and is expressed in *rads per unit time* or electron volts per gram per unit time.

The radiation *exposure dose* is measured in *roentgens* (symbol R) and is a different quantity from the absorbed dose. The ICRU [22]

defines the roentgen as the quantity of x- or γ-radiation such that
the associated corpuscular emission per 0.001293 g of air produces,
in air, ions carrying 1 electrostatic unit of quantity of electricity of either sign. The 0.001293 g is the mass of 1 cm^3 of dry
air at 0°C and 760 mm Hg. While this definition applies only to
x- or γ-radiation, Lea [15] has proposed an extension to charged
particle radiation by defining 1 R of such radiation as that exposure dose which in 0.001293 g of dry air generates ions carrying
1 electrostatic unit of charge of either sign.

The exposure dose per unit time is known as the *exposure dose
rate* and is given in units of *roentgens per unit time.*

The radioactivity of a radioisotope is measured by the number
of nuclear transformations (disintegrations) taking place per unit
time. The unit is the *curie* (symbol Ci), corresponding to exactly
3.7 x 10^{10} transformations per second. This equals the rate of
decay of 1.01 g of ^{226}Ra. A different weight of any other radioisotope would be required to produce the same number of transformations per second.

The *intensity* of radiation at a certain point is defined as
the energy per unit time entering a small sphere at that point per
unit cross-sectional area of the sphere. The units of intensity
are ergs per square centimeter per second, or watts per square
centimeter. Radiation intensity may also be defined as the energy
per unit time crossing a unit area normal to the direction of the
radiation beam. The quantity of radiation entering the unit sphere
is known as the *energy flux*, and it is the time integral of intensity expressed in ergs per square centimeter.

The *flux* of particles or photons at a point equals the number
of particles entering a sphere of unit cross-sectional area at that
point in unit time. Flux is expressed in units of particles per
square centimeter per second.

B. Measurement of Ionizing Radiation

Ionizing radiation can be measured by absolute techniques or secondary methods. One absolute method involves the use of the air-wall ionization chamber. Absolute methods are difficult to employ in routine practice, and secondary methods such as chemical dosimetry or thimble ionization chambers are commonly used in studies of radiation chemistry.

For absolute measurements of radiation exposure in roentgen units, the air-wall or "free-air" ionization chamber is employed. The definition of the roentgen indicates that the ionization that is caused by all secondary electrons ejected from a known mass of air is included. The ejected electrons cause some ionization out- side the volume of space in which they receive their energy. To apply the definition in computing the total exposure in roentgens, this ionization outside the "sensitive volume" must be collected or must be compensated for by an equal amount of ionization that enters the "sensitive volume."

The air-wall chamber provides compensation for the corpuscular emission which leaves the "sensitive volume" before causing all its ionization, as shown in Fig. 2-4. The "walls" of the chamber are composed of air, and any loss of corpuscular emission from the "sensitive volume" is compensated for by the gain of corpuscular emission from the air surrounding the "sensitive volume."

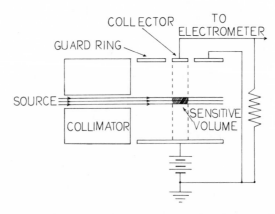

FIG. 2-4. Schematic diagram of a standard free-air chamber.

To insure accurate compensation of this sort, the distance between the walls of the chamber and the nearest edge of the sensitive volume must be greater than the maximum range of the secondary electrons. The distance between the radiation beam and the collecting electrodes must also be greater than this maximum range to insure that the secondary electrons are not collected before they have produced their full yield of ions.

Under such conditions, the charge collected in electrostatic units divided by the sensitive volume in cubic centimeters equals the exposure dose in roentgens. Taking into account corrections to bring the sensitive volume to stp, it can be shown that the collected charge Q due to an exposure dose of X roentgens is expressed by

$$Q\ (C) = \frac{X(R)V(cm^3)273p(mm\ Hg)}{3 \times 10^9 T(^\circ K)(760)} \tag{2-77}$$

where

    V = sensitive volume

    T = absolute temperature

    p = chamber pressure in mm Hg

The current flow i in amperes for an exposure rate of dX/dT can be computed from Eq. (2-77) to give

$$i(A) = \frac{dX/dT(R/h)V(cm^3)273p(mm\ Hg)}{3 \times 10^9 \times 3600T(^\circ K)760}$$

$$= (0.926)(\frac{dX}{dT})(V)(\frac{273}{T})(\frac{p}{700})(10^{-13}) \tag{2-78}$$

The free-air chamber technique is suitable as a standard of exposure dose for x-rays from 10 to 300 kV. For higher energy rays, the very long range of the secondary electrons requires the use of undesirably large chambers, or of inconvenient operation of chambers under several atmospheres of pressure.

By the Bragg-Gray principle, the difficulty of too thin an air wall can be overcome by use of a wall of solid material having a mean atomic number approximately equal to that of air. Such a wall is called an "air-equivalent" wall and may be made of Bakelite,

Lucite, or other plastic. The surface of the plastic is generally
coated with colloidal carbon to give it the required conductive
properties for electrodes of ionization chambers.

Small air-equivalent chambers of this sort are referred to as
thimble chambers. These offer a method that is both sensitive and
convenient for measuring exposure doses of x-ray and γ-radiation.
In the commercial instrument, the small air volume is enclosed in
a plastic cap, and in the center of the cap is an aluminum electrode insulated from the cap. The instrument functions as a small
condenser and before usage is charged to a suitable voltage. Exposure to irradiation causes partial neutralization of the charge,
and the decrease in voltage times the capacitance of the instrument
indicates the amount of ionization resulting from the irradiation.
The observed exposure dose is corrected for temperature and baro-
metric pressure, and an instrument correction is applied which is
determined by comparing the thimble chamber with a standard air-
wall chamber.

At high incident photon energies, the thimble chamber wall
must be surrounded by additional material to insure electronic
equilibrium. An undesirable effect of large wall thicknesses is
to attenuate the radiation, and a correction for this effect must
be applied to the apparent exposure dose at the point of interest.
Up to photon energies of a few megaelectron volts the proper correction may be calculated by using several wall thicknesses and
extrapolating back to zero thickness. Because of wall thickness
difficulties in attaining electronic equilibrium at high photon
energies, an energy of 3 MeV may sometimes be taken as an upper
energy limit for using the roentgen as a unit [18].

C. Calculation of Absorbed Dose for Various
Materials and Mixtures

It is possible to compute the absorbed dose of x-rays or γ-
rays in air, noting 1 R produces in 0.001293 g of air at STP

electrons or positive ions carrying 1 esu of charge. Hence,

$$1 \text{ R} = \frac{1}{0.001293} \frac{(\text{esu})}{(\text{g})} \times 2.082 \times 10^9 \frac{(\text{electrons})}{(\text{esu})}$$

$$\times \ 34 \ \frac{(\text{eV})}{(\text{electron})} \times 1.602 \times 10^{-12} \frac{(\text{ergs})}{(\text{eV})} \times \frac{1}{100} \frac{(\text{rads})}{(\text{ergs/g})}$$

$$= 0.877 \text{ rad (in air)} \tag{2-79}$$

This is equivalent to saying that 1 R equals an energy absorption
of 87.7 ergs in air. This equation holds true for x- and γ-
radiation of all energies greater than 20 keV when absorbed in air.
For other materials the value of the proportionality constant
(0.877) will vary with the compositions of the absorber and the
photon energy.

There are two principal methods for measuring the absorbed
dose in materials other than air. In the first method, the exposure dose in air at a particular point is measured with an ionization (thimble) chamber, and then the other material to be studied
is placed at the same location and exposed to the same radiation.
In this computation it is assumed that neither sample nor thimble
attenuates the radiation significantly, and that the exposure dose
at the point in question has been correctly measured with a properly calibrated thimble chamber. With these assumptions, the
relation between the energy absorbed by air and the sample according
to Eq. (2-70) is given by

$$D_M = D_A \times \frac{(\mu_a/\rho)_M}{(\mu_a/\rho)_A} \text{ rads} \tag{2-80}$$

where $(\mu_a/\rho)_M$ and $(\mu_a/\rho)_A$ are the mass energy absorption coefficients for sample and air, respectively, and D_M and D_A are the
corresponding absorbed doses. By Eq. (2-79),

$$D_A = 0.877R_A \text{ rads} \tag{2-81}$$

where R_A is the exposure dose in roentgens.

Hence,

$$D_M = 0.877 R_A \times \frac{(\mu_a/\rho)_M}{(\mu_a/\rho)_A} = f R_A \text{ rads} \tag{2-82}$$

The value of the factor f is constant for a particular material in the energy range where only Compton absorption takes place, but will vary with energy in the energy range where the pair-production and photoelectric processes are of major importance. Values of f have been compiled for a number of materials and exhibit values of 0.965 and 0.925 for muscle and bone, respectively, using photons of 1.0 MeV energy. The values approximate unity, except for materials of moderate to high atomic number.

The total mass energy absorption coefficient (μ_a/ρ) equals the Compton mass energy absorption coefficient in the Compton absorption region, and the latter is a function of the Compton electronic energy absorption coefficient as shown by the equation

$$\sigma_a/\rho = {}_e\sigma_a (Z/A) N_0 \tag{2-83}$$

As discussed above, the Compton electronic energy absorption coefficient $({}_e\sigma_a)$ is identical for all absorbers and hence

$$\sigma_a/\rho \propto (Z/A) N_0 = \text{no. of electrons per gram} \tag{2-84}$$

This can also be expressed as

$$\sigma_a/\rho \propto (Z/A) \tag{2-85}$$

since N_0 is a constant. Hence, when energy absorption of air and sample takes place mainly by the Compton interaction, Eq. (2-82) may be rewritten

$$D_M = 0.877 R_A \times \frac{(Z/A)_M}{(Z/A)_A} = f R_A \text{ rads} \tag{2-86}$$

Equation (2-86) is useful in computing energy in rads absorbed by monomers or polymers at a point where the exposure dose (R_A) is known. In such computations an average value of Z/A is employed

$$Z/A = \Sigma w_i (Z/A)_i \qquad (2-87)$$

where w_i is the weight fraction of element i in the absorber. For a chemical compound, the average equals the sum of the atomic number of the atoms divided by the molecular weight. Example of values of f computed in this way from Eq. (2-87) are 0.975 for water and 0.99 for methanol.

Calculations of the absorbed dose can be carried out by Eq. (2-82) in energy regions where photoelectric absorption or pair production predominates, but the computation is more complicated. In such cases, the known functional dependence of mass energy absorption coefficients on electronic absorption coefficients is utilized to make suitable substitutions in Eq. (2-82). For mixtures or compounds, an *effective atomic number* is employed in the calculations, which must be computed differently in the photoelectric and pair-production absorption regions.

The other method of measuring absorbed dose involves an ionization measurement made inside the irradiated medium with a gas-filled thimble chamber or similar instrument. The absorbed dose is computed by means of the Bragg-Gray cavity principle [23, 24], which can be written

$$E_M = J_G W_G (S_m)_{gas}^{medium} \qquad (2-88)$$

where E_M is energy absorbed by the medium in ergs per gram; J_G equals ionization produced within the thimble chamber in esu per gram; W_g is the average energy lost by secondary electrons crossing the cavity per ion pair formed in the gas (ergs/ion pair); and $(S_m)_{gas}^{medium}$ is the ratio of the mass stopping power of the medium to that of the gas for such secondary electrons. Equation (2-88) will hold if the cavity is small compared to the range of the ionizing electrons, and if the incident radiation is not significantly attenuated in the medium over a distance equivalent to the dimensions of the cavity. The absorbed dose in the medium is known from the definition of the rad to be given by

$$D_M = 0.01E_M \text{ rads} \tag{2-89}$$

If it is assumed that the cavity gas is air and the ionization caused is expressed in terms of Q, the charge in esu carried by ions of either sign per 0.001293 g of air, then

$$D_M = 0.01 \frac{(\text{rad})}{(\text{erg/g})} \times \frac{Q}{0.001293} \frac{(\text{esu})}{(\text{g air})} \times 2.082 \times 10^9$$

$$\frac{(\text{electrons})}{(\text{esu})} \times 34 \frac{(\text{eV})}{(\text{ion pairs})} \times 1.602 \times 10^{-12}$$

$$\frac{(\text{erg})}{(\text{eV})} \times (S_m)_{air}^{medium}$$

$$= 0.087Q \times (S_m)_{air}^{medium} \text{ rads} \tag{2-90}$$

Equation (2-90) applies when the walls of the ionization chamber are very thin and is used to calculate the absorbed dose rate at a point of interest from the apparent exposure dose rate (Q) at that point measured with a thin-walled cavity ionization chamber.

If the ionization chamber is relatively thick walled and the secondary electrons are formed entirely within the walk of the ionization chamber, then the correct relation is

$$D_M = 0.877Q \times (S_m)_{air}^{wall} \times \frac{(\mu_a/\rho)_M}{(\mu_a/\rho)_{wall}} \text{ rads} \tag{2-91}$$

When the walls of the chamber are made of an air-equivalent material the stopping power ratio of wall to air equals 1, and Eq. (2-91) is written

$$D_M = 0.877Q \times \frac{(\mu_a/\rho)_M}{(\mu_a/\rho)_{wall}} \tag{2-92}$$

In Eq. (2-92), Q equals the exposure dose in roentgens at the point
of interest in the medium, but this is not the same as the exposure
dose in air (R_A) at the same point in the absence of the medium
(because of the attenuating effect of the medium). Values of mass
stopping power ratios relative to air, $(S_m)_{air}^{M}$ or $(S_m)_{air}^{wall}$, have
been tabulated by Whyte [25] and others. For cobalt-60 γ-rays
(1.25 MeV), either of these ratios has a value of approximately
1.1 for polystyrene, Lucite, or Bakelite.

When a mixture is irradiated with γ- or x-rays, the fraction
of the total dose absorbed by each component is proportional to
the weight fraction of that component and also to the mass energy
absorption coefficient for that component. Thus,

$$\frac{D_1}{D_{mix}} = w_1 \times \frac{(\mu_a/\rho)_1}{(\mu_a/\rho)_{mix}} \tag{2-93}$$

where D_1 and D_{mix} are the absorbed doses of the component and mixture, respectively, w_1 equals the weight fraction of component 1,
and

$$(\mu_a/\rho)_{mix} = w_1(\mu_a/\rho)_1 + w_2(\mu_a/\rho)_2 + \cdots + w_n(\mu_a/\rho)_n \tag{2-94}$$

in a mixture of n components. The value of μ_a/ρ is proportional
to Z/A in the Compton region, and in this region Eq. (2-93) becomes

$$\frac{D_1}{D_{mix}} = w_1 \times \frac{(Z/A)_1}{(\overline{Z/A})_{mix}} \tag{2-95}$$

where $(\overline{Z/A})_{mix}$ is an average computed by Eq. (2-87). Equation
(2-95) is employed to compute the fraction of total dose absorbed
by one polymer in a mixture of polymers or by a monomer dissolved
in a solvent. However, Eq. (2-95) holds strictly true only when

the total energy absorbed by component 1 depends entirely on its interaction with the primary radiation and is independent of subsequent interaction with secondary electrons. Since the π-electrons in unsaturated compounds (aromatic compounds) appear to be more readily excited by fast electrons than the σ-electrons of saturated compounds, the portion of energy imparted to unsaturated compounds in a mixture may be higher than expected from Eq. (2-95).

It is sometimes necessary to compute the absorbed dose in one material from the measured absorbed dose in another material. If the radiation is monoenergetic and is absorbed in the Compton region for both materials, Eq. (2-82) becomes

$$D_2 = D_1 \times \frac{(Z/A)_2}{(Z/A)_1} \quad \text{rads} \tag{2-96}$$

where $(Z/A)_1$ and $(Z/A)_2$ are the ratios of atomic number to the atomic weight for materials 1 and 2, respectively, and D_1 and D_2 are the corresponding absorbed doses. If the absorbers are compounds or mixtures, then average values $\overline{Z/A}$ must be used. Absorption of most organic and polymeric materials falls in the Compton region for cobalt-60 and cesium -137 γ-rays, and Eq. (2-96) holds for such systems.

When Compton absorption is unimportant, the corresponding relation between absorbed doses for two materials becomes

$$D_2 = D_1 \times \frac{(\mu_a/\rho)_2}{(\mu_a/\rho)_1} \quad \text{rads} \tag{2-97}$$

where $(\mu_a/\rho)_1$ and $(\mu_a/\rho)_2$ are the mass energy absorption coefficients for materials 1 and 2, respectively.

Equations (2-96) and (2-97) apply only when the absorbed dose is measured in energy absorbed per unit mass. When the absorbed dose is measured in energy absorbed per unit volume, for example eV/cm^3, the absorbed dose is a function of the ratio of the densities of the two absorbers. Equations (2-96) and (2-97) then become

$$(D_v)_2 = (D_v)_1 \times \frac{(Z/A)_2}{(Z/A)_1} \times \frac{\rho_2}{\rho_1} \qquad (2\text{-}98)$$

and

$$(D_v)_2 = (D_v)_1 \times \frac{(\mu_a/\rho)_2}{(\mu_a/\rho)_1} \times \frac{\rho_2}{\rho_1} = (D_v)_1 \times \frac{(\mu_a)_2}{(\mu_a)_1} \qquad (2\text{-}99)$$

where D_v equals absorbed dose per unit volume and μ_a is a linear energy absorption coefficient. Equations (2096) to (2-99) hold strictly only when radiation passing through the absorbers is not significantly attenuated, but they provide useful approximations even when attenuation is appreciable.

Other methods of absorbed dose computation must be used if the radiation source is placed inside the solution or substance being irradiated. For example, α- and β-rays have such short ranges that sources of these rays are often mixed into the material to be irradiated, thus obtaining uniform irradiation of a large sample. Radioisotopic sources used in this way include β-emitting phosphorus-32 and sulfur-35 and, less frequently, α-emitting radon and β-emitting tritium.

Since the sample is generally large compared to the range of the radiation, it can be assumed that the absorbed dose equals the total energy given off by the disintegrating nuclei. If the concentration of the isotope in the absorbing sample equals C millicuries per gram and the average energy released by each disintegration is $\overline{E}$ megaelectron volts, then absorbed dose rate T_M equals

$$T_M = 3.7 \times 10^7 C \; \frac{\text{(disintegrations)}}{\text{(g sec)}} \times \overline{E} \; \frac{\text{(MeV)}}{\text{(disintegration)}}$$

$$\times \; 1.602 \times 10^{-8} \; \frac{\text{(g rad)}}{\text{(MeV)}}$$

$$= 0.593 \times C \times \overline{E} \; \text{rads/sec} \qquad (2\text{-}100)$$

For irradiations lasting longer than the half-life of the isotope, the absorbed dose must be integrated over time to give the actual absorbed dose, D_M,

$$D_M = \int_0^t T_M \, dt \tag{2-101}$$

Using Eq. (2-100) yields

$$D_M = 0.593\overline{E} \int_0^t C \, dt \text{ rads} \tag{2-102}$$

By the law of radioactive decay,

$$C_t = C_0 e^{-\lambda t} \tag{2-103}$$

where the symbols have the usual meaning, and Eq. (2-102) becomes

$$D_M = 0.593 \times \overline{E} \times C_0 \int_0^t e^{-\lambda t} \, dt \text{ rads} \tag{2-104}$$

Integration yields

$$D_M = \frac{0.593 \times \overline{E} \times C_0}{\lambda} (1 - e^{-\lambda t}) \text{ rads} \tag{2-105}$$

If the isotope is allowed to decay completely in the sample, $t \to \infty$, and

$$D_M = (0.593 \times \overline{E} \times C_0)/\lambda \text{ rads} \tag{2-106}$$

D.   Use of Chemical Dosimeters

Chemical dosimetry involves measuring the amount of chemical change caused by irradiation of a selected substrate, followed by

a computation of the absorbed dose. A knowledge of the G value
for the reaction is required, that is, the number of molecules
reacting per 100 eV of absorbed dose. The absorbed dose rate for
the dosimeter can be converted to the absorbed dose for other
materials by use of Eqs. (2-96) through (2-99).

Ideally, the response of a chemical dosimeter should be proportional to the absorbed dose over a wide range of dose, independent of dose rate, independent of the quantum energy of the
radiation, independent of temperature, reproducible to ±5%, and
stable under normal conditions of exposure to light and air.

No known dosimeter fully satisfies this list of requirements,
but the Fricke dosimeter [26] comes close to doing so. The radiation-activated reaction in the Fricke dosimeter is the oxidation
of an acid solution of ferrous sulfate to ferric sulfate in the
presence of oxygen. In preparing the dosimeter solution, Weiss et
al. [27] have recommended dissolving 2 g $FeSO_4 \cdot 7 H_2O$, 0.3 g NaCl,
and 110 ml concentrated H_2SO_4 in enough distilled water to make
5 liters of solution, yielding a solution 0.0014 M in ferrous
sulfate and 0.8 N in sulfuric acid.

Care should be taken to employ pure reagents, and the water
should be distilled from alkaline permanganate and then from acid
dichromate prior to use.

In employing the Fricke dosimeter, a sample of the dosimeter
solution is placed in the radiation field at the desired location
for a measured length of time, and then the concentration of
ferric ions is determined. The usual method for measuring ferric
ion formed is by spectrometric analysis, involving a comparison of
the optical density of the irradiated and nonirradiated solutions
at the wavelength of maximum ferric ion absorption (3040 Å).

The average absorbed dose (D_0) for the volume occupied by
dosimeter solution is then calculated. By definition,

  G (product) = molecules produced/100 eV energy absorbed   (2-107)
and

1 rad = an energy absorption of 100 ergs/g   (2-108)

Combination of Eqs. (2-107) and (2-108) yields

Energy absorbed

$$= 100 \times \frac{\text{(molecules produced/g)}}{G \text{ (product)}} \frac{\text{(eV)}}{\text{(g)}}$$

$$\times 1.602 \times 10^{-12} \frac{\text{(erg)}}{\text{(eV)}} \times \frac{1}{100} \frac{\text{(g rad)}}{\text{(ergs)}}$$

$$= 1.602 \times 10^{-12} \times \frac{\text{(molecules produced/g)}}{G \text{ (product)}} \text{ rads} \qquad (2\text{-}109)$$

Equation (2-109) applies to any dosimeter by definition of its terms. When ferric ions are measured in the Fricke dosimeter,

$$\text{Ferric ion produced (moles/liter)} = \frac{(OD_i - OD_n)}{\varepsilon d} \qquad (2\text{-}110)$$

where OD_i and OD_n are the optical densities of the irradiated and nonirradiated dosimeter solutions, respectively, ε is the molar extinction coefficient at 3040 Å in liters/mole cm, and d is the thickness of the sample in centimeters when optical density is measured. Therefore,

Ferric ion produced (molecules/g)

$$= \frac{(OD_i - OD_n)}{\varepsilon d} \frac{\text{(moles)}}{\text{(liter)}}$$

$$\times \frac{1}{1000\rho} \frac{\text{(liter)}}{\text{(g)}} \times 6.023 \times 10^{23} \frac{\text{(molecules)}}{\text{(mole)}} \qquad (2\text{-}111)$$

Combination of (2-111) and (2-109) gives

$$D_D = \text{energy absorbed} = \frac{OD_i - OD_n}{\varepsilon d G (Fe^{+3})} \times \frac{6.023 \times 10^{20}}{\rho} \times 1.602 \times 10^{-12}$$

$$= \frac{0.965 \times 10^9 (OD_i - OD_n) \text{ rads}}{\varepsilon d \rho G (Fe^{3+})} \qquad (2\text{-}112)$$

The extinction coefficient has a temperature coefficient of +0.7%/°C and can be corrected for temperature if desired. Substitution of $\varepsilon = 2174$, $\rho = 1.024$, and $G(Fe^{+3}) = 15.5$ for cobalt-60 γ-rays in equation (2-112) gives

$$D_D = 2.80 \times 10^4 (OD_i - OD_n)/d \text{ rads} \qquad (2\text{-}113)$$

$$= 1.75 \times 10^{18} (OD_i - OD_n)/d \text{ eV/g} \qquad (2\text{-}114)$$

and for a dosimeter solution of 1.024 density during irradiation,

$$(D_D)_V = 1.79 \times 10^{18} (OD_i - OD_n)/d \text{ eV/cm}^3 \qquad (2\text{-}115)$$

The G value of 15.5 holds well for cobalt-60, but will vary slightly for γ-rays of other energies and for x- and β-rays. The rate of the dosimeter reaction marks an upper limit to the measurable dose rate, and the absorbed dose is not measured accurately above about 50,000 rads. The lower limit of accurate measurement is about 4000 rads in a 1-cm absorption cell.

Other substances that have been employed in dosimeters include ceric sulfate, sodium formate, and calcium benzoate. Glass itself has been used as a dosimeter, by following the degree of darkening as a function of total dose.

REFERENCES

1. Siegbahn, K., *Beta and Gamma Spectroscopy*, North Holland
    Publishing Company, Amsterdam, 1955.

2. Bethe, H. A. and Ashkin, J., *Experimental Nuclear Physics*
    (E. Segre, ed.), Vol. 1, John Wiley, New York, 1953.

3. Evans, R. D., *The Atomic Nucleus*, McGraw-Hill, New York, 1955.

4. Sternheimer, R. M., *Methods of Experimental Physics* (C. L.
    Yuan and C. Wu (eds.), Vol. 5, Pt. A, Academic Press, New York,
    1961.

5. Katz, L. and Penfold, A. S., *Rev. Mod. Phys.*, 24, 28 (1952).

6. Livingston, S. and Bethe, H., *Revs. Mod. Phys.*, 9, 263 (1937).

7. *Report of the International Commission on Radiological Units
    and Measurements (ICRU)*, 1956, Nat. Bur. Std. Handbook 62,
    1957.

8. Price, W. J., *Nuclear Radiation Detection*, 2nd Ed., p. 9,
    McGraw-Hill, New York, 1964.

9. Freidlander, G., Kennedy, J. W., and Miller, J. M., *Nuclear
    and Radiochemistry*, 2nd Ed., p. 95, John Wiley, New York,
    1964.

10. Hughes, D. J. and Harvey, J. A., *Neutron Cross Sections*, U.S.
    At. Energy Comm. Document BNL 325, McGraw-Hill, New York,
    1955.

11. Heitler, W., *The Quantum Theory of Radiation*, Chap. 3, Oxford
    Univ. Press, London, 1944.

12. Klein, O. and Nishina, Y., *Z. Physik*, 52, 853 (1929)

13. Fano, U., *Nucleonics*, 11 (Aug.), 8, (Sept.), 55 (1953).

14. Berger, R. T., *Radiation Res.*, 15, 1 (1961).

15. Lea, D. E., *Actions of Radiations on Living Cells*, Cambridge
    Univ. Press, Cambridge, 1964.

16. Wilson, C. T. R., *Proc. Soc. (London), Ser. A*, 104, 192 (1923).

17. Samuel, A. H. and Magee, J. L., *J. Chem. Phys.*, 21, 1080
    (1953).

18. *Report of the International Commission on Radiological Units and Measurements (ICRU), 1959*, Nat. Bur. Std. Handbook 78, 1961.

19. Gray, L. H., *Brit. J. Radiol. Suppl.*, 1, 7 (1947).

20. Cormack, D. V. and Johns, H. E., *Brit. J. Radiol.*, 25, 369 (1952).

21. *Radiation Quantities and Units, International Commission on Radiological Units and Measurements (ICRU) Report 10a, 1962*, Nat. Bur. Std. Handbook 84, 1962.

22. *Recommendations of the International Commission on Radiological Protection and of the International Commission on Radiological Units*, Nat. Bur. Std. Handbook 47, 1950.

23. Bragg, W. H., *Studies in Radioactivity*, p. 94, MacMillan and Co., London, 1912.

24. Gray, L. H., *Proc. Roy. Soc. (London), Ser. A*, 156, 578 (1936); *Brit. J. Radiol.*, 10, 600, 721 (1937).

25. Whyte, G. N., *Principles of Radiation Dosimetry*, p. 68, John Wiley, New York, 1959.

26. Fricke, H. and Morse, S., *Amer. J. Roentgenol.*, 18, 430 (1927); Fricke, H. and Morse, S., *Phil. Mag.*, 7, 129 (1929).

27. Weiss, J., Allen, A. O., and Schwarz, H. A., *Proc. Intern. Conf. Peaceful Uses At. Energy*, Vol. 14, p. 179, United Nations, New York, 1956.

28. Grodstein, G. W., *X-ray Attenuation, Coefficients from 10 keV to 100 MeV*, Nat. Bur. Std. Circ. 583, 1959.

ILLUSTRATIVE PROBLEMS

Problem 1.   A solution of 50 g of methyl methacrylate in 50 g methanol is irradiated with γ-rays from cobalt-60.   What percentage of the total energy absorbed is absorbed by the methyl methacrylate?

Solution.   In this case, energy is absorbed by the Compton process predominantly, and Eq. (2-95) yields

$$\frac{D_{MMA}}{D_{mix}} = W_{MMA} \times \frac{(\overline{Z/A})_{MMA}}{(\overline{Z/A})_{mix}}$$

Computing first the average values of Z/A for methanol and methyl methacrylate (MMA),

$$\overline{Z/A}_{meth} = \frac{\text{sum of atomic no.}}{\text{molecular wt.}} = \frac{6 + 4 + 8}{32} = 0.562$$

$$\overline{Z/A}_{MMA} = \frac{30 + 8 + 16}{100} = 0.540$$

The average Z/A for the (50/50) mixture is the average of the values for methanol and MMA.

$$(\overline{Z/A})_{mix} = \frac{0.562 + 0.540}{2} = 0.551$$

Hence,

$$\frac{D_{MMA}}{D_{mix}} = \frac{1}{2} \times \frac{0.540}{0.551} = 0.490$$

That is, approximately 49.0% of the energy is absorbed by the methyl methacrylate.

Problem 2. Radiation from cobalt-60 is absorbed by a homogeneous mixture containing 30 and 70% by weight of polyethylene
and polymethyl methacrylate, respectively. What percentage of the
total absorbed energy is absorbed by the polyethylene?

Solution. An equation similar to that in the preceding
problem is used,

$$\frac{D_{PE}}{D_{mix}} = W_{PE} \times \frac{(\overline{Z/A})_{PE}}{(\overline{Z/A})_{mix}}$$

where W_{PE} is the fraction by weight of polyethylene (PE). The
average Z/A for polyethylene, $(CH_2CH_2)_n$, is computed as follows.

$$(\overline{Z/A})_{PE} = \frac{6 + 6 + 4}{28} = 0.572$$

Based on weight fractions, the average Z/A for the mixture becomes

$$(\overline{Z/A})_{mix} = 0.3(\overline{Z/A})_{PE} + 0.7(Z/A)_{MMA}$$

$$= (0.3)(0.572) + (0.7)(0.540) = 0.550$$

Therefore,

$$\frac{D_{PE}}{D_{mix}} = (0.3) \times \frac{(0.572)}{(0.550)} = 0.312$$

indicating that 31.2% of the total energy is absorbed by the
polyethylene.

Problem 3. Assume that the exposure dose rate near a sample
of cobalt-60 was 278 R/min on August 10, 1965. What would be the
exposure dose rate at the same location on August 10, 1966 if the
half-life of cobalt-60 is 5.27 years? Note,

5.27 years = 63.24 months

Solution. For any radioactive decay,

$$\lambda = 0.693/\text{half-life}$$

$$= \frac{0.693}{63.24} = 0.01096 \text{ month}^{-1}$$

Furthermore,

$$I_t = I_0 e^{-\lambda t}$$

$$\log_{10}(I_t/I_0) = -0.43\lambda t$$

$$= -(0.43)(0.01096)(12) = -0.0565$$

Hence,

$$I_t/I_0 = 0.880$$

Hence the exposure dose rate after 12 months would become

New rate = (278) (0.880) = 245 R/min

Problem 4. Assume that the exposure dose rate 1.5 in. from a cobalt source equals 350 R/min. Compute the thickness of concrete needed to reduce the dose rate to 1 mR/h at a point 5 ft from the source. Assume a half-thickness of 2.0 in. for cobalt-60 γ-rays in concrete.

Solution. The effect of distance and shielding must be computed separately. Considering first the inverse square dependence on distance, the dose rate at 60 in. from the source would be

$$350\left(\frac{1.5}{60}\right)^2 = 0.219 \text{ R/min} = 13.1 \text{ R/h}$$

This must be reduced to 0.001 R/h by concrete. The linear absorption coefficient is computed from the half-thickness in concrete,

$$\mu = \frac{0.693}{\text{half-thickness}}$$

$$= \frac{0.693}{2.0} = 0.347 \text{ in.}^{-1}$$

noting that

$$I = I_0 e^{-\mu X}$$

$$\log_{10}(I/I_0) = -(0.43)(0.347)(X) = \log_{10}\left(\frac{0.001}{13.1}\right)$$

$$X = 28.1 \text{ in.}$$

which is the required thickness of concrete.

Problem 5. Compute the mass absorption coefficient for 1 MeV γ-rays in calcium oxide, knowing that the atomic absorption coefficients (barns/atom) for calcium and oxygen are 4.22 and 1.69, respectively.

Solution. Compute first the atomic absorption coefficient for the compound.

$$(_a\mu)_{CaO} = (_a\mu)_{Ca} + (_a\mu)_O$$

$$= 4.22 + 1.69$$

$$= 5.91 \text{ b/molecule}$$

Then compute the mass absorption coefficient.

$$\mu/\rho = \frac{N_0}{A}(_a\mu)$$

$$= 5.91 \times 10^{-24} \times \frac{6.02 \times 10^{23}}{56.1}$$

$$= 0.063 \text{ cm}^2/\text{g}$$

Problem 6. Assume that 1000 μCi of tritium oxide is mixed with 250 mg of acetone. If all the energy is absorbed by the acetone, how much energy (rads) will be absorbed in 1 day, and how much in 1 week?

Solution. The energy of each β-particle from tritium is 5.5 keV. The absorbed dose is computed from Eq. (2-100),

$$T_M = 0.593 \times C \times \overline{E} \text{ rad/sec}$$

where C is the concentration of radioisotope in millicuries per gram, and $\overline{E}$ is in megaelectron volts. Noting that

$$C = \frac{1.0}{0.25} = 4 \text{ mCi/g}$$

the absorbed dose rate is

$$T_M = 0.593 \times 4 \times 5.5 \times 10^{-3}$$

$$= 13.0 \times 10^{-3} \text{ rads/sec}$$

Hence, the absorbed dose in 1 day is

$$T_M = (13.0 \times 10^{-3})(864 \times 10^4) = 1122 \text{ rads}$$

and in 1 week

$$T_M = (7)(13.0 \times 10^{-3})(8.64 \times 10^4) = 7850 \text{ rads}$$

Problem 7. Data obtained with an air-filled cavity ionization chamber indicate an exposure dose rate of 88 R/min inside an aluminum block. Compute the absorbed dose rate at the same location if the chamber walls are very thin compared to the range of the secondary electrons. Note that $(S_m)_{air}^{Al}$ is 0.89 for cobalt-60 γ-rays.

Solution. According to Eq. (2-90), the absorbed dose rate equals

$$D_M = 0.877Q(S_m)_{air}^{medium}$$

$$= 0.877 \times 88 \times 0.89$$

$$= 68.7 \text{ rads/min}$$

Problem 8. A 3 MeV proton has a range of about 14.1 cm in air. From this information, what would be the range in air of 9 MeV tritium nuclei ($_1^3$H)?

Solution. From Eq. (2-14) and (2-15),

$$\text{Range of particle A} = \frac{m_A Z_B^2}{m_B Z_A^2} \times \text{range of particle B}$$

when

$$\text{Energy of particle A} = \frac{m_A}{m_B} \times \text{energy of particle B}$$

Noting that for tritium, m = 3 and Z = 1, while for protons m = 1 and Z = 1, the range of a tritium nucleus equals three times the range of a proton with one third the energy.

Range of 9 MeV tritium = 3 x range 3 MeV proton = 42.3 cm

Problem 9. Assume that a 10-ml sample of $CHCl_3$ is irradiated with cobalt-60 γ-rays for 18 min. Chemical analysis shows that 30×10^{-6} moles of $CHCl_3$ were decomposed. A Fricke dosimeter measurement showed that in 60 min the optical density changed from 0.003 to 0.341 at 3040 Å and at room temperature, when using a 1-cm cell path length.

Compute the G value for $CHCl_3$ disappearance, assuming the following values: $G(Fe^{3+})$ for ^{60}Co radiation is 15.5; density and $\overline{Z/A}$ for the dosimeter solution are 1.024 and 0.553, respectively; the molar extinction coefficient for ferric ion at 3040 Å is 2175; and the density and $\overline{Z/A}$ of chloroform are 1.50 and 0.486, respectively.

Solution. The dose absorbed by the dosimeter is given by Eq. (2-113).

$$D_D = 2.80 \times 10^{-4} (OD_i - OD_n) \text{ rads}$$

$$= 2.80 \times 10^{-4} (0.341 - 0.003) \text{ rads/h}$$

$$= 1577 \text{ rads/10 min}$$

Since the Compton process predominates,

$$D_{CHCl_3} = D_0 \times \frac{(\overline{Z/A})_{CHCl_3}}{(\overline{Z/A})_{dose}}$$

$$= 1577 \times (0.486/0.553)$$

$$= 1387 \text{ rads/10 min}$$

The number of molecules of $CHCl_3$ decomposed is

$$30 \times 10^{-6} \times 6.02 \times 10^{23} \text{ molecules/10 ml}$$

$$= \frac{30 \times 6.02 \times 10^{17}}{10 \times 1.50} \text{ molecules/g}$$

$$= 1.205 \times 10^{18} \text{ molecules/g}$$

and

$$G(-CHCl_3) = \frac{molecules\ decomposed}{100\ eV\ absorbed}$$

$$= \frac{molecules\ decomposed/g}{absorbed\ dose\ (rad)} \times 1.602 \times 10^{-12}$$

$$= \frac{1.205 \times 10^{18}}{1387} \times 1.602 \times 10^{-12}$$

$$= 1390 = G\ value\ for\ CHCl_3\ decomposition$$

Problem 10. A 2030 Ci cobalt-60 source is located at the center of a cubic box 80 cm on a side. Compute the thickness of lead lining in the box needed to give an exposure dose rate on the outside at the center of each side of 0.001 R/h knowing that the half-value thickness for lead is 1.06 cm.

Solution. An unshielded 1 Ci cobalt source gives an exposure dose of 1.32 R/h at 1 m. Hence, at the surface of the box 0.4 m from the source,

$$Exposure/h = 1.32 \times (\frac{2030}{1}) \times (\frac{1}{0.4})^2 = 16,800\ R$$

Therefore, dose rate must be reduced by

$$1.68 \times 10^7\text{-fold} = 2^{24}\text{-fold}$$

Thickness of lead necessary is

$$24 \times 1.06 = 25.4\ cm$$

Problem 11. Probability theory shows that when a reasonably large count is taken, the standard deviation equals the square root of the total count,

$$\sigma = \sqrt{m}$$

If 1000 counts were recorded in 10 min, what would be the standard deviation and how would the counting rate be expressed?

Solution. The standard deviation would be

$$\sigma = \sqrt{1000} = 32$$

and the counting rate would be

$$\frac{(1000 \pm 32)}{10} = 100 \pm 3.2 \text{ counts/min}$$

Problem 12. Assume that a total time of 1 h is allowed for making a counting rate measurement and a background measurement. For the time available, what is the optimum time division between sample and background counting to give the best possible accuracy? Assume sample and background rates of 1000 and 20 cpm, respectively.

Solution. It has been shown that the best accuracy is obtained when

$$\frac{t_b}{t_T} = \left(\frac{r_b}{r_T}\right)^{\frac{1}{2}}$$

where t_b and t_T are the times taken for background and total activity counts, respectively, and r_b and r_T are counting rates for background and sample, respectively.

Therefore,

$$\frac{t_b}{t_T} = \left(\frac{20}{1000}\right)^{\frac{1}{2}} = 0.14$$

Since t_b plus t_T equals 60 min, solving for t_b and t_T yields 7 and 53 min, respectively.

Problem 13. Assume that a 3-ml sample of chloroform was irradiated with γ-rays from cobalt-60 for 20 min. Extraction and titration showed that 18 x 10^{-6} moles of monobasic acid was formed.

Using a Fricke dosimeter, 60 min of irradiation changed the optical density of the dosimeter solution from 0.003 to 0.341 at 3040 Å measured in a 1-cm cell. Compute the G value for the formation of acid from chloroform.

Solution. The density and $\overline{Z/A}$ values for the dosimeter solution are 1.024 and 0.553, respectively. The corresponding values for chloroform are 1.50 and 0.486. The absorbed dose in the dosimeter solution equals

$$D_D = \frac{0.965 \times 10^9 \ (OD_i - OD_n)}{\varepsilon d \rho G(Fe^{3+})} \tag{2-112}$$

Employing the values $\varepsilon = 2175$ at room temperature, $\rho = 1.024$, $d = 1$, and $G(Fe^{3+}) = 15.5$, the absorbed dose becomes

$$D_0 = 2.80 \times 10^4 \ (OD_i - OD_n) \text{ rads}$$

$$= 2.80 \times 10^4 (0.341 - 0.003) \text{ rads/60 min}$$

$$= 3154 \text{ rads/20 min}$$

Since the energy absorption for cobalt-60 γ-rays is mainly by the Compton process in both dosimeter and chloroform, the absorbed dose in the chloroform is given by

$$D_{CHCl_3} = D_D \times \frac{(\overline{Z/A})_{CHCl_3}}{(\overline{Z/A})_D}$$

$$= 3154 \times (0.486/0.553)$$

$$= 2774 \text{ rads/20 min}$$

The number of molecules of acid formed equals

$$18 \times 10^{-6} \times 6.025 \times 10^{23} \text{ molecules/3 ml}$$

$$= \frac{18 \times 6.025 \times 10^{17}}{3 \times 1.50} \text{ molecules/g}$$

$$= 2.410 \times 10^{18} \text{ molecules/g}$$

Now, for any product formed it is known that

$$G \text{ (product)} + \frac{\text{molecules of product/g}}{\text{absorbed dose (rad)}} \times 1.602 \times 10^{-12} \qquad (2\text{-}109)$$

Substituting the above values,

$$G \text{ (acid)} = \frac{2.410 \times 10^{18} \times 1.602 \times 10^{-12}}{2774} = 1390$$

Chapter 3

SHORT-TERM CHEMICAL EFFECTS OF RADIATION ABSORPTION

I. INTRODUCTION

The steps that are intermediate between the initial physical
effect of radiation and the final chemical products are still not
well understood. The basic results of radiation are to ionize and
excite, and both ions and excited molecules can yield free radicals. For most molecules irradiated in the gas phase, roughly
equal numbers of ionized and excited molecules are formed [1].
These primary products may subsequently break down or react with
the substrate to cause further chemical change. Free radicals from
excited and ionized molecules tend to dominate the total mechanism
in many radiation-induced reactions. Ions, excited molecules, and
free radicals will be discussed in turn.

II. IONS

A. Formation of Ions

Ions are generally produced by the loss of electrons from
molecules. Since the electron is removed from the molecule as a

127

whole, the positive charge may exist anywhere in the molecule, but
it will tend to concentrate more at certain positions than at
others. Ionization may theoretically be accompanied by the dissociation of the molecule into ionic and free-radical fragments in
the time needed for one molecular vibration 10^{-13} sec. On the
other hand, data from mass spectroscopic research indicate that for
molecules as large as propane or larger dissociative ionization
rarely occurs, the usual main products being the parent ions.

The ionization of gases has been investigated extensively by
mass spectroscopy. Electrons ejected in irradiated systems may
neutralize positive ions or may be captured by substances having an
electron affinity such as oxygen:

$$e^- + O_2 \rightarrow O_2^- \qquad\qquad (3\text{-}1)$$

Electron capture in the gas phase may also have a dissociative
effect,

$$e^- + H_2O \rightarrow H\cdot + OH^- \qquad\qquad (3\text{-}2)$$

The energy required to produce an ion pair, W, is generally
about twice the ionization potential. The excess energy is employed in producing excitation, so that excitation and ionization
are of similar importance.

Sometimes impurities may reduce the energy required to produce
an ion pair. This can be explained on the ground that energy can
be transferred from excited states of the major constituent to the
minor constituent, causing the latter to ionize,

$$He^* + A \rightarrow He + A^+ + e^- \qquad\qquad (3\text{-}3)$$

For this to happen, the ionization potential of the impurity must
be less than one of the excitation potentials of the major
constituent.

Irradiated liquids appear to form ions just as gases do, but the greater ionization density tends to cause recombination, and the ions cannot be effectively collected even with very high applied fields. Hence, W cannot be precisely measured for liquids, but there are indications that it is of the same order of magnitude as the value determined for gases. The high local field strength in liquids tends to cause ejected electrons to interact with positive ions to produce excited molecules.

Ionization of alkali halides and similar solids raises electrons to the "conduction band," where they are able to move readily through the solid and increase its conductivity. The holes from which the electrons came may also be mobile. Eventually the electrons combine with positive holes or become trapped as lattice defects. Electrons become trapped at negative ion vacancies, while positive holes become trapped at positive ion vacancies, these phenomena being related to certain photo-properties of the crystals.

The ionization properties of high polymers and organic substances appear to be intermediate between those of ionic crystals and of liquids. Organic solids do become conducting during irradiation, perhaps due to the presence of electrons in conduction bands. However, the evidence indicates that trapping of electrons and holes in high polymers probably does not take place except at low temperature.

B. Reactions of Ions

In reactions caused by irradiation, positive ions are generally more important than negative ions. In the first place, there are many systems in which negative ions may not be formed at all. Furthermore, positive ions appear to be more reactive than negative ions. The principal reactions of ions include charge transfer, formation of ion clusters, reaction of ions with molecules, rearrangement or decomposition, neutralization, and electron addition.

One of the simplest reactions is charge transfer (electron transfer), which has been investigated in the gas phase using ion beams. For example,

$$H^+ + CH_4 \rightarrow H\cdot + CH_4^+ \qquad (3\text{-}4)$$

In order for such a reaction to take place, the lowest ionization potential of the molecule whose ion is donating the positive charge must be greater than that of the molecule accepting it. Other examples of charge transfer are

$$He^+ + Ne \rightarrow Ne^+ + He \qquad (3\text{-}5)$$

$$H_2^+ + H_2O \rightarrow H_2O^+ + H_2 \qquad (3\text{-}6)$$

$$Ar^+ + C_2H_4 \rightarrow Ar + C_2H_4^+ \qquad (3\text{-}7)$$

Charge transfer is essentially a resonance process and tends to take place where there is only a small difference in ionization potential between the two species. When the reaction takes place, the energy difference between the two ionization potentials is liberated. Such energy may be imparted to the product ion and cause it to dissociate

$$Ar^+ + CH_4 \rightarrow Ar + CH_3^+ + H \qquad (3\text{-}8)$$

Another reaction of ions is the formation of ion clusters. The idea of an ion cluster was first advanced in 1920 to explain the fact that the number of molecules changed by radiation may be considerably greater than the number of ion pairs formed. According to the hypothesis, several molecules cluster around each ion, and chemical breakdown of the whole cluster takes place on neutralization by an electron, ion, or cluster of opposite charge. Examples of stable ion-molecule clusters that have been studied with the mass spectrometer include H_3O^+ $(H_2O)_7$ and NH_4^+ $(NH_3)_{20}$ [2-6].

Another type of reaction, the reaction of ions with molecules, is a very important class of reactions. Such reactions are believed to take place by the formation of a collision complex, followed by rearrangement of the atoms to give the observed products. The three main categories of ion-molecule reactions are: (1) single particle transfers, (2) condensation reactions, and (3) association reactions.

The following are some examples of single particle transfer reactions:

$$H_2^+ + H_2 \rightarrow H_3^+ + H \tag{3-9}$$

$$CH_4^+ + CH_4 \rightarrow CH_5^+ + CH_3 \tag{3-10}$$

$$Ar^+ + H_2 \rightarrow ArH^+ + H \tag{3-11}$$

$$CH_3OH^+ + CH_3OH \rightarrow CH_3OH_2^+ + CH_3O \tag{3-12}$$

$$H_2O^+ + H_2O \rightarrow H_3O^+ + OH \tag{3-13}$$

Such reactions are essentially hydrogen atom transfers. They are very important in gas-phase radiolysis and usually have very high rate constants.

Another type of single particle transfer is proton transfer, of which the following are examples:

$$H_3^+ + cyclo\text{-}C_3H_6 \rightarrow H_2 + sec\text{-}C_3H_7^+ \tag{3-14}$$

$$C_2H_5^+ + H_2O \rightarrow H_3O^+ + C_2H_4 \tag{3-15}$$

$$CH_5^+ + cyclo\text{-}C_4H_8 \rightarrow CH_4 + sec\text{-}C_4H_9^+ \tag{3-16}$$

$$C_2H_5^+ + NH_3 \rightarrow NH_4^+ + C_2H_4 \tag{3-17}$$

$$CH_5^+ + NH_3 \rightarrow NH_4^+ + CH_4 \tag{3-18}$$

Another reaction in this general category is the transfer of a hydride ion (H^-) to a positive molecule ion. This may be the only reaction of an alkyl ion other than neutralization. Examples include

$$C_3H_5^+ + C_3H_8 \rightarrow C_3H_6 + C_3H_7^+ \qquad (3\text{-}19)$$

$$C_2H_5^+ + C_3H_8 \rightarrow C_2H_6 + C_3H_7^+ \qquad (3\text{-}20)$$

$$C_2D_3H_2^+ + C_6H_{12} \rightarrow C_2D_3H_3 + C_6H_{11}^+ \qquad (3\text{-}21)$$

$$C_4H_8D^+ + (CH_3)_2CDCH_2CH_3 \rightarrow C_4H_8D_2 + C_5H_{11}^+ \qquad (3\text{-}22)$$

Hydride ion transfer is a very efficient reaction and is often the most important reaction for the higher alkanes.

Condensation is another general type of ion-molecule reaction. In this reaction a considerable amount of rearrangement of the collision complex takes place. Examples include

$$C_2H_5I + I^+ \rightarrow C_2H_4 + HI_2^+ \qquad (3\text{-}23)$$

$$C_2H_2^+ + C_2H_2 \rightarrow C_4H_2^+ + H_2 \qquad (3\text{-}24)$$

$$C_4H_6^+ + C_4H_6 \rightarrow C_3H_9^+ + CH_3 \qquad (3\text{-}25)$$

$$C_2H_4^+ + CH_4 \rightarrow C_3H_6^+ + H_2 \qquad (3\text{-}26)$$

$$CH_3^+ + CH_4 \rightarrow C_2H_5^+ + H_2 \qquad (3\text{-}27)$$

$$CH_2^+ + CH_4 \rightarrow C_2H_4^+ + H_2 \qquad (3\text{-}28)$$

In association reactions, the collision complex is the final result.

$$C_2H_4^+ + C_2H_4 \rightarrow C_4H_8^+ \tag{3-29}$$

$$C_3H_7I^+ + C_3H_7I \rightarrow (C_3H_7I)_2^+ \tag{3-30}$$

$$C_2H_5Br^+ + C_2H_5Br \rightarrow (C_4H_{10}Br_2)^+ \tag{3-31}$$

Mass spectroscopic evidence indicates that the lifetimes of such complexes are of the order of 10^{-6} sec. Rate constants are usually very high, up to 2×10^{-9} cm^3/molecule sec. Because of the long life of the ion-molecule complex, the collision producing it is sometimes called a *sticky collision*.

Evidently the whole subject of ion-molecule reactions cannot be reviewed in this discussion, but a few examples of ion-molecule reactions can be given that have been investigated extensively. One such reaction that has been studied by many investigators is the reaction of methane with CH_4^+ [8-11]. It is usually recognized that the main reactions of the ions present in the mass spectrum of methane are

$$CH_4^+ + CH_4 \rightarrow CH_5^+ + CH_3 \tag{3-32}$$

$$CH_3^+ + CH_4 \rightarrow C_2H_5^+ + H_2 \tag{3-33}$$

A small amount of $C_2H_3^+$ is produced by CH_3^+, but the only competing reaction for CH_4^+ is charge transfer, according to Abramson and Futrell [9]. The ions formed, CH_5^+ and $C_2H_5^+$, are essentially unreactive to methane.

The cross sections for CH_4^+ disappearance and CH_5^+ formation have been measured by Giardine-Giudoni and Friedman [10] and compared with those computed theoretically. An experimental arrangement for such studies is the tandem operation of two mass spectrometers, by which a beam of ions of known kinetic energy can be generated, followed by mass analysis of the products of unimolecular or bimolecular reactions of such ions. In the case of CH_4^+

reactions, the cross section for CH_4^+ disappearance remained larger than that for CH_5^+ formation, even after correction for collision-induced dissociation and charge-exchange losses in the analyzer tube. This suggested the possible existence of reactions other than (3-32) for CH_4^+ [10].

Abramson and Futrell [11] used the tandem mass spectrometer to investigate the reaction of CH_4^+ with CD_4 for ion kinetic energies ranging from 0.17 to 6.8 eV. There are several products formed aside from the result of proton transfer (CD_4H^+) at m/e 21. The product observed at m/e 20 is believed to result from charge transfer to CD_4, but may also indicate $CD_3H_2^+$ produced by an exchange reaction in the intermediate complex [12]. Another product was observed at an m/e of 19, possibly $CD_2H_3^+$ or CD_3H^+.

There is a rapid drop in the amount of the m/e 19 product and the CD_4H^+ product at high kinetic energies, indicating that at high kinetic energies an intermediate complex may not exist long enough for exchange reactions to take place. Hence, at higher kinetic energies, only simple proton transfer would be observed. A similar phenomenon of a decrease in the number of exchange reactions at higher energies has been observed for other reactions, which appear to pass through a somewhat loose intermediate complex [13].

Various reactions are capable of forming a product observed at an m/e of 18. At low kinetic energies, the reaction is probably abstraction of D resulting in CH_4D^+. It has been postulated that this abstraction reaction requires a longer-lived intermediate complex than does the reaction of proton transfer. At higher kinetic energies there are two possible reactions leading to an ion having m/e of 18. These reactions, each of which produces CH_3^+, are loss of HD from CD_4H^+ and dissociative charge transfer (loss of D from CD_4^+).

It is quite probable that a product observed at m/e 17 is formed by loss of D_2 from excited CD_4H^+ and is analogous to the production of methyl ions by loss of H_2 from CH_5^+. A strong peak

observed at m/e 15 must be due to CH_3^+ ions and probably results
from dissociation of the CH_4^+ ion on collision. Furthermore, the
evidence indicates that dissociation of CH_4^+ ion on collision with
CH_4 is the most important methyl ion source over the whole range of
kinetic energies studied. Since production of CH_3^+ in this way is
endothermic by about 1.4 eV and since about half of the dissociation reactions take place below kinetic energy of 1.4 eV, it is
evident that methane ions having considerable internal energy must
be reacting to form the observed products.

The reaction of methane with CH_3^+ has also been studied rather
extensively. As noted above, it is commonly agreed that the main
product of this reaction is $C_2H_5^+$. Possibly the reaction takes
place through an intermediate complex, $C_2H_7^+$, although this complex
has not been observed in pure methane up to a pressure of 160 torr
[14]. Evidence for the existence of such an intermediate has been
obtained, however, from a study of the isotope distributions (H and
D) in ethyl ions resulting from the reaction of isotopically
labeled reactants. Randomization of an intermediate $C_2H_nD_{7-n}$ complex has been reported from studies of CH_4-CD_4 mixtures [15, 16].
Identical distributions of isotopic ethyl ions from the dissociation
of $C_2H_3D_4^+$ intermediates prepared from several isotopic reaction
pairs were observed [17].

Hence, there is considerable evidence for an intermediate
complex ($C_2H_7^+$) with a significant lifetime prior to breaking down
the observed products. Evidence for the existence of stable $C_2H_7^+$
has been obtained from studies of ion-molecule reactions of ethane
[18-20]. It has been found that stable $C_2H_7^+$ can be formed in high
yield by proton transfer from CH_3O^+ to ethane. Hence, its stability may be greatly influenced by its energy content. If it is
assumed that single collision stabilization is possible for $C_2H_7^+$,
then the estimated lifetime of the complex must be less than
10^{-10} sec for it *not* to be observed. Such a lifetime would permit
plenty of time for isotopic rearrangement before dissociation.

An investigation was made of the effect of the translational energy of the methyl ion on product distribution. For thermal CH_3^+ the ratio $C_2H_3^+/C_2H_5^+$ was 0.04, and increased to 2.4 for 4.2 eV ion kinetic energy. This indicates that kinetic energy may be converted readily into internal energy. The breakdown sequence of the complex

$$C_2H_7^{+*} \rightarrow C_2H_5^{+*} + H_2 \qquad\qquad (3\text{-}34)$$

$$C_2H_5^+ \rightarrow C_2H_3^+ + H_2 \qquad\qquad (3\text{-}35)$$

is also indicated by the absolute intensities of the product ions. At 4.2 eV the intensity of $C_2H_3^+$ is more than twice its intensity at 0.3 eV, while the $C_2H_5^+$ intensity has greatly decreased.

Reactions of ions with hydrocarbons other than methane have also been studied extensively. Both mass spectrometric and radio-chemical techniques have provided much information on the reaction of ions with hydrocarbons. One of the ion-molecule reactions in this category that was studied very early was the hydride transfer reaction,

$$X^+ + C_nH_{2n+2} \rightarrow XH + C_nH_{2n+1}^+ \qquad\qquad (3\text{-}36)$$

where X^+ is an alkyl ion. This reaction was first investigated mass spectroscopically [21]. More recently, Lias and Ausloos [22] developed a competitive method to determine the relative rates of the reactions, where XH_2 is a perprotonated hydrocarbon added to C_3D_8. The C_3D_8 acts as a source of $C_2D_5^+$ ions. The following kinetic equation is derived from Eqs. (3-34) and (3-35):

$$\frac{k_B}{k_A} = \left(\frac{C_2D_5H}{C_2D_6}_{corr}\right)\left(\frac{C_3D_8}{XH_2}\right) \qquad\qquad (3\text{-}36)$$

The C_2D_6 yield is listed "corr," since it must be corrected for the contribution from the unimolecular breakdown of excited C_3D_8. Such investigations have also been made of hydride transfer reactions for sec-$C_3D_7^+$ and sec-$C_4D_9^+$ with several different hydrocarbons [23, 24]. The relative rates can be converted to absolute rates by using a rate constant of 8×10^{-10} cm^3 $mole^{-1}$ sec^{-1}, which was observed for the reaction

$$C_2H_5^+ + (CH_3)_4C \rightarrow C_2H_6 + C_5H_{11}^+ \qquad (3-37)$$

These investigations show that the rate of H^- transfer for a given ion usually increases as the collision ion cross section increases. On the other hand, the reaction rate increases faster with increase in the molecular weight of XH_2 than does the collision cross section. This probably shows that reaction does not take place at every collision, even though all cases of H^- transfer studied were exothermic. There seems to be a correlation between the probability of the reaction and its exothermicity. For example, the transfer of H^- to $C_2D_5^+$, which is much more exothermic than to sec-$C_3D_7^+$ or sec-$C_4D_9^+$, exhibits a closer correlation between rates and collision cross sections than does the transfer to sec-$C_3D_7^+$ or sec-$C_4D_9^+$.

Generally speaking, mass spectrometric studies of H^- ion transfer do not agree with trends in reactivity shown by the radiolytic data. Several investigators have observed a decrease in effective cross section for these reactions as molecular weight increases [16, 21, 25]. The disagreement between radiolytic and spectroscopic data may come about because radiolytic yields are based on detection of $C_nH_{2n+1}^+$, which may be produced by reactions other than hydride transfer.

Radiochemical investigations of ion-molecule reactions aid in determining the structure of the reacting ion and the reaction complex. For example, Ausloos and co-workers [23, 24] demonstrated that for ethyl, propyl, and butyl ions the relative cross sections of the hydride transfer reactions do not depend on the origin of the carbonium ion. This implies that the reacting ions have the

same structure no matter how they originate. Structure analyses
of the C_3D_7H and C_4D_9H products formed in the reactions

$$C_3D_7^+ + XH_2 \rightarrow C_3D_7H + XH^+ \tag{3-38}$$

$$C_4D_9^+ + XH_2 \rightarrow C_4D_9H + XH^+ \tag{3-39}$$

indicated that the products were made up almost completely of CD_3-
$CDHCD_3$ and $CD_3CDHCD_2CD_3$. These facts indicate that mainly sec-
$C_3D_7^+$ and sec-$C_4H_9^+$ ions are taking part in these H^- transfers.
Hence, H or D atom rearrangements to obtain the most thermodynami-
cally stable arrangements can take place easily in such carbonium
ions. No indication was found of rearrangement of the carbon
skeleton.

Data regarding the reaction complex produced are obtained from
a radiolysis of butane. Results indicated [26] that about 90% of
the propane produced in the radiolysis of $(CH_3)_3CD$ is propane-d_2.
In view of the fact that propane is produced mainly by a hydride
transfer to the propyl ion, the following reactions are required:

$$(CH_3)_3CD^+ \rightarrow C_3H_6D^+ + CH_3 \tag{3-40}$$

$$C_3H_6D^+ + (CH_3)_3CD \rightarrow C_3H_6D_2 + C_4H_9^+ \tag{3-41}$$

If the complex formed in (3-41) underwent rearrangement, the
statistically favored product would be C_3H_7D. Hence it appears
that no randomization of hydrogens in the intermediate complex
occurs in this instance. Similar implications come from a study
of the propane formed in the radiolysis of $CD_3CH_2CH_2CH_3$, which is
mainly propane-d_3 [27].

Since it is known that ion pair yields of products whose precursors are the reactant ions are essentially constant for all
C_nD_{n+2}-XH_2 mixtures, it would appear that hydride ion transfer is
generally the only type of reaction that takes place between an

alkyl ion and a neutral alkane or cycloalkane [23]. Cyclopropane
or XH_2 constitutes an exception to this rule, but in this case
other evidence shows an alternative type of reaction in competition
with hydride transfer. There is also competition with hydride
transfer in the case of alkyl ion reaction with unsaturated hydrocarbons, where proton transfer and condensation reactions may
possibly take place.

There are indications that H^- transfer takes place from alkanes
to smaller olefinic ions [28, 29],

$$C_nH_m^+ + XH \rightarrow C_nH_{m+1} + X^+ \tag{3-42}$$

and also to carbonium ions of the general formula $C_nH_{2n-1}^+$ [30],

$$C_nH_{2n-1}^+ + XH \rightarrow C_nH_{2n} + X^+ \tag{3-43}$$

Studies of H^- transfer to olefinic ions by radiolytic experiments
is difficult, because the neutral product formed is a free radical.
One technique that has been successful in such work is to emply
H_2S as a free radical scavenger. In such investigations it has
been discovered that the transfer of H^- from various alkanes to
$C_3D_6^+$ is more probable when the H^- donor is a branched alkane
containing one or more tertiary H atoms.

An interesting additional reaction of considerable importance
is the transfer of a proton from various donor ions to hydrocarbons. For example, the formation of protonated methane has been
studied extensively. One of the early reactions of this type to be
observed was the transfer of a proton from H_3^+ or CHO^+ to an alkane
followed by the breakdown of the protonated alkane [31, 32] as
shown by work with a tandem mass spectrometer. Radiochemical
studies of CH_4 or H_2 containing traces of deuterated alkanes indicated proton transfer from CH_5^+ or H_3^+ to the alkanes, as shown
by the analysis of the products formed [33, 35].

As an example, when $n-C_5D_{12}$ is mixed with CH_4 and irradiated, the ion $C_5D_{12}H^+$ is formed and then decomposes as follows:

$$C_5D_{12}H^+ \rightarrow CD_3H + C_4D_9^+ \qquad\qquad (3-44)$$

$$\rightarrow C_2D_5H + C_3D_7^+ \qquad\qquad (3-45)$$

$$\rightarrow C_3D_7H + C_2D_5^+ \qquad\qquad (3-46)$$

It was postulated that the perdeutero-alkanes were formed by alkyl ions undergoing deuteride transfer reactions with the $n-C_5D_{12}$. Similar results were obtained when a mixture of H_2 and $n-C_5D_{12}$ was irradiated. In the latter case, more than 95% of the methane formed was CD_3H, strongly suggesting that the protonated precursor ion was $CD_3CD_2CD_2CD_2CD_3H^+$. This indicates very little likelihood of a rearrangement of hydrogen atoms in the protonated alkane. In related work, Ausloos and Lias [36] found that proton transfer to cyclopropane gives mainly a stable $C_3H_7^+$ ion, which acquires the secondary structure $(CH_3CHCH_3^+)$ before reacting with an interceptor molecule. Ausloos also observed that proton transfer to ethylene, propylene, and 2-butene forms principally $C_2H_5^+$, $sec-C_3H_7^+$, and $sec-C_4H_9^+$, respectively.

Aquilanti and Volpi [37] have obtained mass spectrometric evidence for the following reaction at high pressure,

$$H_3^+ + C_2H_6 \rightarrow (C_2H_7^+)^* + H_2 \qquad\qquad (3-47)$$

$$(C_2H_7^+)^* \rightarrow C_2H_5^+ + H_2 \qquad\qquad (3-48)$$

$$(C_2H_7^+)^* + M \rightarrow C_2H_7^+ + M \qquad\qquad (3-49)$$

suggesting that the excited protonated ion undergoes collisional deactivation. There is considerable evidence for such collisional deactivation of excited ions, including the fact that above 100 torr there are essentially no differences in the modes of decomposition of protonated alkanes formed by H_3^+ and CH_5^+, although the H_3^+ ion forms an excited protonated ion having about 2 eV more energy.

Extensive work has been done on the transfer of protons by ethyl ions to polar compounds such as CH_3NO_2, $(CH_3)_2N_2$, CH_3OCH_3, and CH_3OH [38]. The rate of proton transfer for these materials is about 10 to 100 times as great as for the competing hydride transfer reaction. Ausloos [24] also investigated the proton transfer reaction from $C_3H_7^+$,

$$C_3H_7^+ + X \rightarrow C_3H_6 + XH^+ \qquad (3\text{-}50)$$

where X is a molecule such as 1-pentene, CH_3NO_2, or 4-methyl-cis-pentene. When X is an olefin, proton transfer is in competition with hydride transfer and condensation reactions with carbonium ions.

In general terms, when an unsaturated hydrocarbon ion reacts with a saturated hydrocarbon molecule, there is a competition between several possible reactions. An example of one competing reaction is the H_2^- transfer reaction, which can be written

$$C_nH_m^+ + XH_2 \rightarrow C_nH_{m+2} + X^+ \qquad (3\text{-}51)$$

A reaction of this type that was observed very early was [16]

$$C_2D_4^+ + C_3D_8 \rightarrow C_2D_6 + C_3D_6^+ \qquad (3\text{-}52)$$

It has recently been found that the breakdown of cycloalkane ions generally gives large yields of olefinic ions. Ausloos *et al.* [23] used the radiolysis of cycloalkanes to form such ions for the study of relative rates of H_2^- transfer reactions.

One investigation covered the reactions of $C_3D_6^+$ ions with several hydrocarbons for which the rate of the H_2^- transfer reaction

$$C_3D_6^+ + XH_2 \rightarrow C_3D_6H_2 + X^+ \tag{3-53}$$

was measured relative to the reaction

$$C_3D_6^+ + C_5D_{10} \rightarrow C_3D_8 + C_5D_8^+ \tag{3-54}$$

by use of the equation

$$\frac{k_{54}}{k_{53}} = \frac{C_3D_6H_2}{C_3D_8} \times \frac{C_5D_{10}}{XH_2} \tag{3-55}$$

Relative rate constants computed in this way showed that there is
an increase in the rate of the H_2^- transfer with increasing molecular weight of the neutral molecule when the latter is a normal
alkane. The rate appeared to decrease with increasing degree of
branching of the neutral molecule in a sequence of isomers. Such
a drop in the H_2^- transfer rate was usually balanced by an increase in the rate of the H^- transfer reaction, so that the total
reactivity for different isomers is essentially constant.

The general conclusion is that the H_2^- transfer reaction involves a loosely bound complex, and that no rearrangement of H
atoms takes place during the lifetime of the reaction complex.
Furthermore, the H_2^- transfer reaction is quite stereospecific in
character, as shown by the fact that the terminal hydrogen of n-
butane or isobutane is transferred almost solely to the center
atom of $C_3D_6^+$. It has also been found that the reaction involves
only the transfer of hydrogen atoms on adjacent carbons of the
XH_2 molecule.

Another sort of ion-molecule reaction has been observed
recently, which can be expressed as

$$C_nH_m + XH_2^+ \rightarrow C_nH_{m+2} + X^+ \tag{3-56}$$

where C_nH_m represents cyclopropane or an unsaturated hydrocarbon having fewer carbons than the parent ion, XH_2^+. A reaction in competition with reaction (3-56) would be the H transfer reaction

$$C_nH_m + XH_2^+ \rightarrow C_nH_{m+1} + XH^+ \qquad\qquad (3-57)$$

Transfers of H_2 have been studied in the vapor phase [39, 40] and also in the condensed phase [41], employing radiolytic methods. Recent rate determinations of H_2 transfer from cyclohexane ions by radiolytic and mass spectrometric techniques were in good agreement, showing a definite correlation between ion-molecule investigations made by the two methods. The evidence indicated that the collision complex produced in the H and H_2 transfers allows no randomization of hydrogens. Hence, such reactions show a definite stereospecificity. In the mass spectrometric study of butane ions, it was found that the hydrogens are lost preferentially from the 2,3 positions during H_2 transfer. Radiolytic investigations indicated that n-butane produced in cyclohexane containing 1-butene-d_8 consisted entirely of $CD_2HCDHCD_2CD_3$.

The relative likelihood of H_2 transfer in the vapor phase was found quite different from that in the liquid phase. This indicates that the probability of H_2 transfer relative to H transfer varies with the density of the reaction medium. This implies that collisional deactivation of the active complex is important in determining the observed course of the reaction.

Once an ion has been formed radiolytically, it can undergo *rearrangement or decomposition*. Ion decomposition can be studied in mass spectrometers, where it is responsible for most of the ions observed other than parent ions. Such decompositions take place in the 10^{-5} sec between the formation of the parent ion and the collection of all ions. On the other hand, when a gas is irradiated at atmospheric pressure, each molecule has one collision every 10^{-9} sec, so that collisional deactivation is likely to occur prior to unimolecular decomposition. Hence, the fragmentation pattern on

irradiation will differ from that found in the mass spectrometer.
Collisions take place even more rapidly in solids and liquids. For
these reasons, there is not much direct relation between mass
spectra and radiation chemistry, although mass spectra have use as
a fertile source of ideas.

The following are examples of ion decompositions that may
occur during irradiation:

$$CH_3CH_2CH_2NH_2^+ \rightarrow C_2H_5 + CH_2NH_2^+ \tag{3-58}$$

$$CH_4^+ \rightarrow CH_3 + H^+ \tag{3-59}$$

$$C_8H_{16}^+ \rightarrow C_7H_{13}^+ + CH_3 \tag{3-60}$$

$$C_4H_8^+ \rightarrow C_3H_5^+ + CH_3 \tag{3-61}$$

$$CH_3OH^+ \rightarrow CH_3O^+ + H \tag{3-62}$$

The fragments formed by dissociation may both be molecular or
both free radicals, but one or the other must have a positive
charge. The charge usually stays with the fragment having the
lowest ionization potential, which is normally the larger of the
two fragments. For example, excited ionized ethane yields hydrogen
and ethylene,

$$C_2H_6^+ \rightarrow C_2H_4^+ + H_2 \tag{3-63}$$

In this case the charge stays with the ethylene (ionization
potential 10.5 eV) rather than with hydrogen (ionization potential
15.4 eV).

Positive ions formed by ionizing radiation are almost always
vibrationally excited, because they are formed by vertical Franck-
Condon transitions from the ground state of the neutral molecule,

and the internuclear distance in the ion is commonly different from that in the neutral molecule. Furthermore, it is possible for ions to be electronically excited if the radiation causing ionization has an energy greater than the lowest ionization potential of the molecule, a situation that holds true for a large fraction of the ionizing events caused by ionizing radiation.

The bond that breaks in the dissociation of an ion is not necessarily at the site of the original interaction with the ionizing radiation. The electrons in the ion readily shift position, corresponding to a migration of charge within the ion. Also, the excitation energy becomes distributed within the ion in a random manner among the various vibrational and electronic excited states. If enough energy becomes concentrated in a given bond, the ion can dissociate at that bond. Dissociation tends to take place at relatively weak bonds, which require less than the average amont of energy for breakage. For example, bonds of aliphatic halogen compounds usually break at the weak C-H bond during irradiation. Dissociation is not immediate, and there is usually time for other processes to compete with dissociation.

Rearrangement of ions before or after fragmentation may lead to products not obtainable by breakage of a single bond of the parent ion. For example, the production of ethylene from $C_2H_6^+$ results from the migration of a hydrogen atom. As another example, isobutane shows a peak at m 29, caused by the $C_2H_5^+$ ion, which is explained by hydrogen migration followed by the breaking of two C-C bonds. Alcohols often show a peak at m 19, probably caused by the hydronium ion H_3O^+, which results from a rearrangement involving two hydrogen atoms. Furthermore, rearrangements are often indicated by the mass spectra of unsaturated hydrocarbons. The mass spectra of the following substances can hardly be distinguished, possibly owing [7] to a loss of molecular·structure on ionization and a redistribution of bonds in a random fashion:

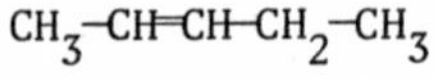

$$CH_3-CH=CH-CH_2-CH_3$$

2-Pentene

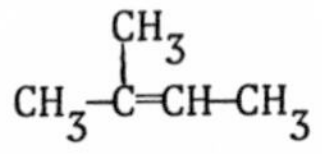

$$CH_3-\overset{\overset{\displaystyle CH_3}{|}}{C}=CH-CH_3$$

2-Methyl-2-butene

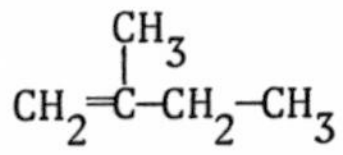

$$CH_2=\overset{\overset{\displaystyle CH_3}{|}}{C}-CH_2-CH_3$$

2-Methyl-1-butene

$$CH_2=CH-\overset{\overset{\displaystyle CH_3}{|}}{C}H-CH$$

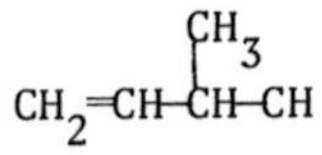

3-Methyl-1-butene

One of the most important reactions that ions can undergo is *neutralization*. A positive ion recombines with an electron (or negative ion to yield an excited molecule,

$$X^+ + e^- \rightarrow X^{**} \tag{3-64}$$

or

$$X^+ + X^- \rightarrow X^* + X \tag{3-65}$$

The excited molecule may then dissociate to give molecular products or free radicals,

$$X^* \ (\text{or } X^{**}) \rightarrow M^* + N \tag{3-66}$$

$$X^* \ (\text{or } X^{**}) \rightarrow R\cdot^* + S\cdot \tag{3-67}$$

Excited radicals are produced by neutralization of an ion which is also a free radical,

$$R\cdot^+ + e^- \rightarrow R\cdot^{**} \tag{3-68}$$

Translationally or vibrationally excited molecules or radicals
formed in this manner may be very reactive. Such highly excited
radicals cannot be removed by normal scavenging techniques employed
for removing thermal radicals by reactions with low concentrations
of chemically reactive substances. These highly reactive radicals
are referred to as "hot radicals."

The ionization potential of water is 12.61 eV. Therefore, the
reaction $H_2O^+ + e^- \rightarrow H_2O$ forms water molecules in a highly excited
state. Hence, this neutralization reaction has been postulated as
a source of H and OH radicals. The H and OH radicals would be
examples of "hot radicals."

Recombination (neutralization) reactions may be modified by
secondary reactions taking place before, and in competition with,
neutralization by combination with an electron. The capture of
electrons by molecules having high electron affinities is an exothermic reaction, thus the energy given off on subsequent recombination is decreased by an amount equivalent to the electron
affinity. This may result in an energy release lower than the
minimum dissociation energy of the product, in which case no bond
scission would take place, but the formation of excited molecules
may lead to further reaction on subsequent collision.

Solvation of the positive ion may also change the neutralization reaction if the solvation is strongly exothermic, as in the
case of ammonia [6] or water vapor [5]. Evidence has accumulated
that charge recombination reactions may not be dissociative. Such
evidence is based on product yields and reaction mechanisms in the
radiolysis of oxygen [42], carbon monoxide [43, 44], carbon dioxide
[45], and nitrogen/oxygen mixtures [46].

One possible fate of slow electrons is capture by molecules
having a strong electron affinity such as osygen or halogens, an
attachment that may be either associative or nondissociative,

$$X + e^- \rightarrow X^- \qquad\qquad\qquad\qquad (3\text{-}69)$$

$$X + e^- \rightarrow A^- + B \qquad\qquad (3\text{-}70)$$

Various ion-molecule reactions involving negative ions such as O^-, Cl^-, or OH^- have been observed in the mass spectrometer [47, 48]. Such a capture step would be followed by a charge-recombination step involving a negative ion rather than an electron.

Examples of compounds having high electron affinity include halogens, organic halogen compounds, and oxygen, and these may react as follows:

$$Cl_2 + e^- \rightarrow Cl^- + Cl\cdot \qquad\qquad (3\text{-}71)$$

$$C_2H_5I + e^- \rightarrow C_2H_5\cdot + I^- \qquad\qquad (3\text{-}72)$$

$$O_2 + e^- \rightarrow O_2^- \qquad\qquad (3\text{-}73)$$

In liquid water the reaction is

$$H_2O_{aq} + e^- \rightarrow OH_{aq}^- + H\cdot \qquad\qquad (3\text{-}74)$$

the latter reaction being endothermic, with the necessary energy furnished by solvation of the ions.

In reaction (3-69) there is no way of dissipating any excess energy of the electron. Hence, the captured electron can only have energies that accomodate to the energy levels in the negative ion formed, implying a resonance process. Reaction (3-70) takes place when the capture of an electron by X liberates enough energy to break one of the bonds of the molecule. Dissociation generally follows the capture of electrons having kinetic energy appreciably greater than thermal energy.

There is mass spectrometric evidence for another type of re-action, which causes the formation of an ion pair,

$$X + e^- \rightarrow A^+ + B^- + e^- \qquad (3\text{-}75)$$

This reaction is not a resonance process and can take place over a much wider range of electron energy than electron capture. Organic halogen compounds undergo this reaction, probably because of the high electron affinity of the halogens. For example, ethyl chloride undergoes the following reaction at electron energies of about 9 eV:

$$C_2H_5Cl + e^- \rightarrow C_2H_5^+ + Cl^- + e^- \qquad (3\text{-}76)$$

even though the ionization potential of ethyl chloride is 11.2 eV, which is the lowest electron energy for the reaction,

$$C_2H_5Cl + e^- \rightarrow C_2H_5Cl^+ + 2\ e^- \qquad (3\text{-}77)$$

Magee and Burton [49] have noted that electron capture by a neutral molecule is in competition with capture by a positive ion. The probability of these two reactions in the path of a densely ionizing particle will change as the track ages and the ions diffuse apart. When the track is first formed, ions and electrons will be abundant near the track and electrons will be captured mainly by positive ions. As the track ages the ions will diffuse apart and capture of electrons by neutral molecules will become more probable.

Any negative ion present will eventually [50] neutralize a positive ion,

$$X^- + X^+ \rightarrow X^* + X \qquad (3\text{-}78)$$

One or both of the resulting neutral molecules will be excited. The excitation energy will be less than that obtained by a positive ion neutralized by an electron.

Hamill *et al.* [51] have carried out electron-capture experiments by irradiating frozen solutions of biphenyl or naphthalene in organic solvents. The ions $C_{12}H_{10}^-$ and $C_{10}H_8^-$ from biphenyl and naphthalene produced characteristic ultraviolet absorption spectra, which were observed and studied. When a second solute was present competing with biphenyl or naphthalene for electrons, the decrease in $G(C_{12}H_{10}^-)$ or $G(C_{10}H_8^-)$ gave a measure of the electron affinity of the second solute.

Electron attachment by a minor component of an irradiated solution may explain the disproportionately high decomposition of the minor component sometimes observed. For example, irradiation of a dilute solution of CCl_4 in benzene leads to greater decomposition of the CCl_4 than would be expected from the separate radiolysis of the two components. The result can be explained by assuming that CCl_4 captures the electrons formed by ionization of both components [52].

Electrons lose energy much more slowly in a medium once they have dropped to an energy (0.5 to 4 eV) below that of the lowest excitation energy of the medium [53]. The rates of energy loss $(-dE/dt)$ for an electron with energy of 10 to 20 eV and one with an energy less than E_x, the minimum excitation energy of the medium, are about 10^{16} and 10^{13} eV/sec, respectively, in liquid water. These low energy long-life electrons are sometimes referred to as *subexcitation electrons.*

They have particular significance in systems where a minor component has a lower excitation energy (E_m) than that of the major component (E_x). In such a case, almost all the subexcitation electrons with energies between E_m and E_x may cause excitation of the minor component, which would therefore play a more major role in the radiolysis reaction than would be expected from its concentration. In such a case, and providing electron capture by the minor component takes place, the fraction of the total energy absorbed by the minor component would be greater than that predicted

by Eqs. (2-93) and (2-95). Platzman [53] computed that for high
energy radiations about 15 to 20% of the absorbed energy might be
dissipated by subexcitation electrons.

III. EXCITED MOLECULES

A. Formation of Excited Molecules

Electronically excited neutral molecules are formed in two
ways by the action of ionizing radiation: (1) by direct excitation
to a specific electronic level above or below the ionization
potential, and (2) as products in charge recombination reactions.
These reactions are

$$X \rightarrow X^{*} \tag{3-79}$$

$$X \rightarrow X^{+} + e^{-} \; [\text{or} \; (X^{+})^{*} + e^{-}] \tag{3-80}$$

$$X^{+} + e^{-} \rightarrow X^{**} \rightarrow X^{*} \tag{3-81}$$

The highly excited molecules formed on neutralization (X^{**}) may
lose part of their energy through collisions with other molecules
and drop to lower excited states (X^{*}) similar to those formed
photochemically.

It should be noted that there is a definite probability for a
molecule to receive energy in excess of its lowest ionization
potential without immediate loss of an electron, thus forming an
electrically neutral molecule having energy greater than the ionization energy ("superexcited molecules"). Instead of ionizing,
such a molecule may dissociate to form smaller molecules or free
radicals, one or both of which may be electronically excited. The

typical molecular excitation levels reached by the absorption of
ionizing radiation are known to be very high [54].

The cross section for excitation to the various energy levels
is proportional to the energy ratio of oscillator strength/energy
transition. For low energy electrons, kinetic energies <100 eV,
optically forbidden transitions can be induced by electron exchange. Also, the absorption of energy from ionizing radiation is
quite nonspecific and can excite any part of the molecule. These
data, in addition to the high quantum energies of ionizing radiation, indicate that photochemical information must be used with
great care in the interpretation of radiation chemical processes.

Energy of excitation can be transferred from one molecule to
another,

$$X^* + Y \rightarrow X + Y^* \tag{3-82}$$

There is still a considerable amount of controversy concerning the
mechanism of energy transfer. There are four possible mechanisms
by which the transfer of energy can take place. First, there are
"collisions of the second kind," by which energy is transferred
from an electronically excited molecule to a molecule in its ground
state, for example,

$$Xe^* + H_2 \rightarrow Xe + H_2^* \tag{3-83}$$

A well-known example is the irradiation of a mixture of mercury and
hydrogen with light of 2537 Å, which raises the mercury to a triplet excited state, followed by collisional transfer of energy to
hydrogen, resulting in the dissociation of the latter molecule.
An example of energy transfer in liquid media involves energy absorption by a solvent followed by transfer to a small concentration
of organic scintillator, which subsequently fluoresces.

A second mechanism is by fluorescence of the excited molecule
X^*, the fluorescence being absorbed by the second molecule Y and
raising it to the excited state Y^*.

A third method is known as inductive resonance, or nonradiative resonance, by which energy is transferred by quantum-
mechanical resonance from a solvent molecule to a distant molecule,
molecules intermediate between the two playing no part in the
transfer. The requirements are similar to those for emission and
reabsorption of fluorescent radiation, that is, that there be maximum overlap of the emission spectrum of the sensitizing molecule
(X) and the absorption spectrum of the accepting molecule (Y), but
the process is believed to involve both molecules simultaneously.
The process can take place between molecules separated by distances
much larger than molecular dimensions, that is, by distances of the
order of 50-100 Å. The process has been observed in gases and is
found to agree well with experimental results on dye solutions.
This mechanism may explain why anthracene crystals containing 1/10
of 1% naphthacene do not show the blue-violet fluorescence of
anthracene, but instead exhibit the yellow-green fluorescence of
naphthacene.

The fourth method of energy transfer is exciton transfer,
which is important in ionic crystals and large polymeric molecules.
A strong coupling of the molecules is required for this method.
It might appear that exciton transfer would explain the observations of anthracene/naphthacene, but measurements have shown that
the molecules in anthracene crystals interact so weakly that
exciton transfer is improbable.

The evidence indicates that the exciton moves rapidly from
one molecule to another, remaining for a very brief time (less than
the period of one vibration) in each. The energy of the exiton is
equal to that of the vibrationless lowest excited state, since any
additional electronic or vibrational energy will be lost very
rapidly. The lifetime of the exciton is about 10^{-8} sec, that is,
approximately the same as if it were located on a single molecule.
If the crystal contains an impurity molecule having an excited
state lower than the host molecule, the exciton may become located
on the impurity molecule and raise it to an excited state having

both electronic and vibrational energy. If the impurity loses
vibrational energy, the exciton can no longer return to the host
molecule but will remain on the impurity. If the impurity than
fluoresces, the fluorescence will be typical of the impurity
instead of the host (major component) molecule. Such trapping of
absorbed radiation energy followed by emission of fluorescence
from a minor component in a solid is often noted with the solids
employed as organic scintillators.

In order to explain the high rate of exciton transfer in
-cintillator solutions, Lipsky and Burton [55] have postulated that
in liquid hydrocarbons small groups of ordered molecules (domains)
may be present. Exciton transfer can take place within these
groups of 10 or 15 molecules, so that collision with any molecule
of a domain containing an excited molecule has essentially the
same result as collision with the excited molecule itself.

B.   Reactions of Excited Molecules

When excited molecules do not fluoresce, they can lose their
energy by internal conversion to yield a strongly vibrating lower
electronic state. The energy of vibration may then be removed by
collision. Excited molecules may reach a stable state by some
type of molecular rearrangement [56], such as the conversion of a
trans compound into an equilibrium mixture of *cis* and *trans* forms,,

$$
\begin{array}{c}
\diagdown \\
\diagup
\end{array}
C=C
\begin{array}{c}
\diagup H \\
\diagdown
\end{array}
\quad \rightarrow \quad
\begin{array}{c}
\diagdown \\
\diagup
\end{array}
C=C
\begin{array}{c}
\diagdown H \\
\diagup
\end{array}
$$
(3-84)

Excited molecules can also decompose into free radicals,

$$(X : Y)^{*} \rightarrow X\cdot + Y\cdot \tag{3-85}$$

but unless the decomposition is an energetic one there is a

possibility that the molecules may recombine within the solvent
cage (Franch-Rabinowitch [57] effect), resulting in no net reaction.
Another mode of breakdown gives molecular products directly, as in
the unimolecular decomposition of methyl butyl ketone [58] when
irradiated by ultraviolet light,

$$CH_3(CH_2)_3COCH_3 \rightarrow CH_3CH\!=\!CH_2 + CH_3COCH_3 \qquad (3\text{-}86)$$

Such dissociation into molecular products is less common than dissociation into free radicals.

An example of an excited molecule decomposition that has been
extensively studied is the decomposition of water vapor. The main
decomposition reactions of excited water molecules as indicated
by photochemical studies are [59]

$$H_2O \rightarrow H(^2S_{\frac{1}{2}}) + OH(^2\pi) \qquad (3\text{-}87)$$

$$H_2O \rightarrow H(^2S_{\frac{1}{2}}) + OH(^2\Sigma^+) \qquad (3\text{-}88)$$

$$H_2O \rightarrow H_2 + O(^1D) \qquad (3\text{-}89)$$

Such reactions may tive only a rough indication of excited neutral
species formed by ionizing radiation, but it is of interest that
residual yield of hydrogen unaffected by H atom scavenger or
electron scavengers and which may result from Eq. (3-89) has been
reported [60].

The principal positive ions produced from water vapor in the
mass spectrometer [61] include H_2O^+ (77%) and OH^+ (18%). Other
positive ions are of lesser importance. Negative ions such as H^-,
O^-, and OH^- have been found [61] in very small yield. In the
presence of O_2, the ion H_3O^+ has been observed in addition to complex ions such as $(H_2O_2)_nO_2^-$, $(H_2O_2)_nO^-$, and $(H_2O)_nOH^-$ with $n \lesssim 5$
[62]. Both of the main positive ions undergo ion-molecule reac-
tions [63] with water molecules, so that at pressures generally

used in radiation chemistry, it is generally the hydronium ion
H_3O^+ which is eventually neutralized,

$$H_2O^+ + H_2O \rightarrow H_3O^+ + OH \tag{3-90}$$

$$OH^+ + H_2O \rightarrow H_3O^+ + O \tag{3-91}$$

$$D_2O^+ + D_2O \rightarrow D_3O^+ + OD \tag{3-92}$$

$$OD^+ + D_2O \rightarrow D_3O^+ + O \tag{3-93}$$

The rate constants for these reactions are quite high [64],

$$k_{90} = 4.9 \times 10^{-10} \text{ cm}^3 \text{ molecule}^{-1} \text{ sec}^{-1} \tag{3-94}$$

$$k_{91} = 4.7 \times 10^{-10} \text{ cm}^3 \text{ molecule}^{-1} \text{ sec}^{-1} \tag{3-95}$$

$$k_{92} = 3.7 \times 10^{-10} \text{ cm}^3 \text{ molecule}^{-1} \text{ sec}^{-1} \tag{3-96}$$

$$k_{93} = 2.2 \times 10^{-10} \text{ cm}^3 \text{ molecule}^{-1} \text{ sec}^{-1} \tag{3-97}$$

Reactions (3-91) and (3-93) are significant because they could lead
to high concentrations of O atoms, although no definite chemical
evidence has been obtained to show what part O atoms may play. The
O atoms may possibly react with water molecules to produce hydroxyl
radicals, although such a reaction as (3-98) is endothermic to the
extent of 19 kcal for ground state O atoms,

$$O + H_2O \rightarrow 2 \text{ OH} - 19 \text{ kcal} \tag{3-98}$$

Therefore, if scavengers are present such as olefins or aldehydes,
which are known to react rapidly with O atoms, some of the products
may be attributable to O atom reactions.

The hydronium ion (H_3O^+) is a major product of water vapor radiolysis, but it probably reacts in hydrated form $H_3O^+(H_2O)_n$. The formation of such complex hydrates has been shown mass spectrometrically at pressure less than 1 Torr in irradiated water vapor by Kebarle and Hogg [5], and some information is available concerning the effect of temperature on the equilibrium,

$$H_3O^+ + n\text{-}H_2O \rightleftarrows H_3O^+ (H_2O)_n \qquad\qquad (3\text{-}99)$$

For hydrates with values of n = 7, there is an upper limit [65] of about 17 kcal/mole for the heat of reaction of an average hydration step such as

$$H_3O^+ + H_2O \rightarrow H_3O^+ (H_2O) \qquad\qquad (3\text{-}100)$$

$$H_3O^+ (H_2O) + H_2O \rightarrow H_3O^+ (H_2O)_2 \qquad\qquad (3\text{-}101)$$

The products of charge neutralization depend on the form of the hydronium ion. When the simple hydronium ion is neutralized, the overall charge neutralization reaction (3-102) to yield two H atoms and one OH radical is exothermic by about 22-26 kcal/mole.

$$H_3O^+ + e^- \rightarrow H_2O^* + H + 142\text{-}146 \text{ kcal/mole} \qquad\qquad (3\text{-}102a)$$
$$\downarrow$$
$$H + OH + 22\text{-}26 \text{ kcal/mole} \qquad\qquad (3\text{-}102b)$$

However, in view of the exothermicity of the hydration steps, in the charge neutralization of a hydrate with n = 3 or 4, the excess energy released is not enough to dissociate a water molecule so that only one H will be produced for each ion originally present. If the complete solvation energy of the hydronium ion (about 100 kcal/mole) is realized in the vapor phase, then charge neutralization by an electron is exothermic to an extent of only 42-46 kcal/mole.

At high temperature and low pressure when the simple hydronium ion predominates, the charge neutralization reaction will change when neutralization is by a negative ion rather than an electron. With nitrous oxide as an electron scavenger, the negative ion taking part in neutralization will be O^- or OH^-.

$$N_2O + e^- \rightarrow N_2 + O^- \qquad\qquad (3\text{-}103)$$

$$O^- + H_2O \rightarrow OH + OH^- \qquad\qquad (3\text{-}103a)$$

Neutralization of H_3O^+ by OH^- is exothermic by more than 200 kcal, and since the dissociation energy of the OH bond in water is about 119 kcal/mole, there is enough energy available to dissociate a water molecule.

$$H_3O^+ + OH^- \rightarrow 2\ H_2O^* + 211\text{-}215\ \text{kcal/mole} \qquad (3\text{-}104)$$
$$\downarrow$$
$$H + OH + H_2O + 92\text{-}96\ \text{kcal/mole} \qquad (3\text{-}105)$$

Similar remarks apply to neutralization by the O^- ion,

$$H_3O^+ + O^- \rightarrow H_2O^* + OH + 213\ \text{kcal/mole} \qquad (3\text{-}106)$$
$$\downarrow$$
$$H + OH + 94\ \text{kcal/mole} \qquad (3\text{-}107)$$

For both O^- and OH^-, when the positive and negative ions are solvated there is not enough energy released in the charge recombination to dissociate a water molecule. Recently, the H_3O^- ion has been observed by Milton [66]. Its role in charge neutralization cannot be defined until more information is available on its abundance and stability.

If scavengers having a higher proton affinity than water are present, such as CH_3OH or NH_3, proton transfers such as the following can take place.

$$H_3O^+ + CH_3OH \rightarrow H_2O + CH_3OH_2^+ \tag{3-108}$$

$$H_3O^+ + NH_3 \rightarrow H_2O + NH_4^+ \tag{3-109}$$

As a result of such reactions, the central ion in the ion-molecule cluster would no longer be the hydronium ion. Furthermore, the number distribution of molecules in the cluster would not neces- sarily be proportional to their concentrations in the total mixture [65]. Hence, the products of the charge neutralization reaction, and the subsequent reactions of such products, could be altered significantly by the presence of a trace of substance of high proton affinity.

Published data on radiolysis yields from water vapor are rather scarce, and most new data of significance have been published since 1963. In the absence of added scavengers, published yields of hydrogen have shown a wide variation from $G(H_2) = 0.015$ [69] to $G(H_2) = 5.9$ [67]. The following experimental factors were perhaps involved in the wide variation in yields: (1) lack of careful cleaning of irradiation vessels, (2) variations in the condition of vessel walls, (3) taking up of traces of contamination by the water in transfer prior to irradiation, and (4) the determination of integral G values rather than studying G as a function of dose rate. A study [68] of yield versus dose using ^{60}Co γ-radiation has in- dicated two main conclusions: (1) after a steady-state hydrogen concentration has been formed, no further radiolysis takes place, in agreement with results obtained for liquid water, and (2) the previous treatment of the irradiation cell has a strong influence on the steady-state hydrogen concentration. It is worth noting that the lowest yield quoted by Firestone [69], $G(H_2) = 0.015$ for a measurement at a single dose, is in close agreement with the value that can be computed from Anderson's data at about the same total dose (10^{21} eV g^{-1}).

The fundamental stability of water vapor to ionizing radiation has been shown by Anderson [70] using 1.5 MeV protons. The same silica cell was used repeatedly for successive irradiations, and samples were taken through an array of break seals, resulting in minimum changes in the surface condition of the cell. At high dose rates of 7×10^{19} to 10^{22} eV g^{-1} sec^{-1}, reproducible steady-state concentrations of hydrogen and oxygen were observed (Fig. 3-1), which showed a linear dependence on (dose rate)$^{\frac{1}{2}}$. At the highest dose rate employed, the ratio of $[H_2]/[O_2]$ was about 2, but the ratio fell as the dose rate decreased. The evidence suggested the loss of an almost constant amount of oxygen to the walls of the irradiation cell, since only trace amounts of hydrogen peroxide were found at 125°C.

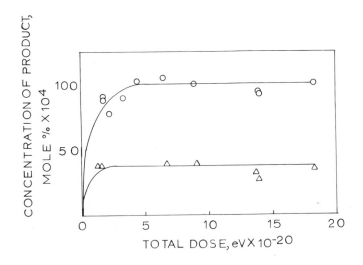

FIG. 3-1.   Steady state concentration of hydrogen and oxygen in the proton radiolysis of water vapor; temperature, 125°C; pressure, one atmosphere; dose rate, 5×10^{20} eV g^{-1} sec^{-1}. [From A. P. Anderson, in *Fundamental Processes in Radiation Chemistry* (P. Ausloos, ed.), John Wiley and Sons, New York, 1968, p. 302.]

Results of greater importance to an understanding of the role of ions and excited states in the radiolysis have only been attained through the addition of reactive scavengers. This approach has been used in studying the nonscavengeable yield of hydrogen, the radical and electron yields, isotope effects, and the effects of temperature and pressure.

As a result of the work of several investigators [71-73], it is known that a nonscavengeable yield of hydrogen is produced in the presence of both electron and H atom scavengers in the radiolysis of water vapor with a G value of about 0.5. Yields of electrons have G values of the order of 3.0, as determined in the same investigations. Anderson *et al.* [60] found no change in hydrogen yield at temperatures up to 700°C. Johnson and Simic [71] found the yield constant up to 400°C, and noted a slight increase in the yield in the presence of high concentrations of electron scavengers (over 20% N_2O and SF_6) at 140°C. This increase suggests that the nonscavengeable hydrogen comes from excited water molecules, as in the photolysis reaction where molecular detachment of hydrogen becomes significant for light of 1236 Å wavelength [74].

When water vapor is irradiated with ionizing radiation, other contributions to the hydrogen yield could come from charge neutralization reactions with a specific orientation of water molecules around the hydronium ion, or from the following reactions representing resonance capture of electrons by water molecules [73]:

$$e^- + H_2O \rightarrow H^- + OH \tag{3-110}$$

$$H^- + H_2O \rightarrow H_2 + OH^- \tag{3-111}$$

A study of radical and ion yields is also important to an understanding of the raiolysis mechanism of water vapor. The first important studies of radical yields were made by Firestone [75], who investigated the tritium β-radiation-induced exchange reaction between D_2 and H_2O over a wide temperature range.

$$D_2 + H_2O \rightarrow HD + HDO \qquad\qquad\qquad (3\text{-}112)$$

At temperatures greater than 150°C the exchange takes place by a chain reaction. At low temperatures the yield of HD becomes constant at $G(HD) = 11.7 \pm 0.6$, which was postulated to be equal to $G(-H_2O)$.

Several independent studies [71, 73, 76] in the presence of H scavengers have given values of $G(-H_2O)$ of about 7 under conditions of temperature and pressure similar to those of Firestone. Good agreement was attained in those investigations on the total yield of reducing species, for irradiations at less than 200°C and pressures of about 1 atm. The only reported value for $G(OH)$ from an $H_2O/NH_3/O_2$ system [68] is consistent with the total yield of reducing species.

The measurement of electron yields by the suppression of hydrogen production in the presence of electron scavengers [71, 76, 77] has given values of $G(e^-) = 3.0 \pm 0.3$. Values for the yields of H atoms from neutral excited molecules have also been obtained from these investigations, indicating $G(H) = 3.5$, which is much higher than the postulated maximum value for liquid water, $G(H) = 0.6$. The lower yield in the liquid may be due to collisional deactivation of excited molecules or to rapid recombination of radicals within the solvent cage.

While the use of isotopes is a valuable technique for providing additional information, it is recognized that different isotopes of the hydrogen atom may be formed at different rates. Yang and Marcus [72] studied isotope effects in the γ-radiolysis of tritiated water vapor and in H_2O/D_2O mixtures, each containing cyclopentane. Analysis of the hydrogen formed indicated relative formation rates of $k_H/k_T = 1.7 \pm 0.1$ and $k_H/k_D = 1.2 \pm 0.1$. Johnson and Simic [71] have questioned the accuracy of these results. Employing electron scavenger studies, Johnson found a ratio of $k_H/k_D = 1.1$ in processes other than neutralization of the hydronium ion. By analogy with this work, Johnson postulated that the (similar) low ratios obtained by Yang and Marcus were due to the presence of a low concentration of an electron scavenger.

Not much information is available concerning the effect of temperature and pressure on the decomposition of water vapor during irradiation. However, Anderson and Winter [76] reported a stepwise increase in $G(H_2)$ with increasing temperature in the presence of scavengers such as NH_3, cyclo-C_6H_{12}, and C_2H_5OH. The change was attributed to a change in the ion-molecule equilibrium (Eq. 3-99), causing an increased formation of H atoms on neutralization of the ion H_3O^+ $(H_2O)_n$, $0 \lesssim n \lesssim 2$. The increased hydrogen yield $G(H_2) \cong 3.5 \cong G(e)$ is in agreement with this postulate, and by computing relative equilibrium constants for the ion-molecule equilibrium, a value of 78-100 kcal/mole for the solvation energy of the hydronium ion in the vapor phase was computed, which is about equal to the solvation value in liquid water ($\sim$ 100 kcal/mole). On the other hand, work by Johnson and Simic [78] on the radiolysis of D_2O/C_6H_{12} showed no evidence for a stepwise increase in hydrogen yield, indicating a conflict in experimental data and in interpretation between Anderson and Johnson.

Work on the effect of pressure on water vapor radiolysis has been very limited. Little change in yield has been observed at pressures from 1 to 3 atm, but Anderson and Winter [76] have reported increased yields of hydrogen at lower pressures. Investigation has shown that at least part of this increase is associated with a wall effect, since the yield is a function of surface/volume ratio at a given pressure. This evidence invalidates the simple postulate of an excited water molecule readily deactivated by collision.

Ammonia is another example of a molecule whose radiolytic decomposition has been studied extensively. There are several similarities between the primary processes that take place when ionizing radiation is absorbed by ammonia or by water vapor, but the subsequent radiation chemistry of the two molecules shows quite a contrast in the absence of scavengers. Whereas water vapor is

somewhat stable to radiation, the equilibrium between ammonia and
its radiolysis products, mainly nitrogen and hydrogen, is not
reached until about 85-97% of the ammonia is decomposed.

The principal ions formed in the mass spectrometric decompo-
sition pattern of NH_3 are NH_3^+ and NH_2^+, with NH^+ and N^+ present in
very small amounts [79]. Traces of other hydrides such as N_2H^+,
$N_2H_2^+$, $N_2H_3^+$, $N_2H_4^+$, and $N_2H_5^+$ have also been found [80]. However,
mass spectrometric measurements [65] at higher pressures have shown
that the principal and probably only ion present in the radiolysis
of ammonia is the solvated ammonia ion NH^+ $(NH_3)_n$. The primary
ions are probably converted to the ammonium ion by ion-molecule
reactions (3-113) and (3-114), folloed by electrostatic attraction
of NH_3 molecules [reaction (3-115)] to produce the ammoniated complex with $n \lesssim 20$ at pressures less than 200 Torr.

$$NH_3^+ + NH_3 \rightarrow NH_4^+ + NH_2 \qquad\qquad (3\text{-}113)$$

$$NH_2^+ + NH_3 \rightarrow NH_4^+ + NH \qquad\qquad (3\text{-}114)$$

$$n\ NH_3 + NH_4^+ \rightleftarrows NH_4^+\ (NH_3)_n \qquad\qquad (3\text{-}115)$$

Charge neutralization of the ammonium ion by electrons can
take place by the following exothermic reactions:

$$NH_4^+ + e^- \rightarrow NH_3 + H + 111\ \text{kcal/mole} \qquad\qquad (3\text{-}116)$$

$$NH_4^+ + e^- \rightarrow NH_2 + 2\ H + 7\ \text{kcal/mole} \qquad\qquad (3\text{-}117)$$

$$NH_4^+ + e^- \rightarrow NH_2 + H_2 + 111\ \text{kcal/mole} \qquad\qquad (3\text{-}118)$$

If the complete solvation energy of the ammonium ion is realized in
the vapor phase (about 84 kcal/mole), then the reaction equivalent
to reaction (3-117) is endothermic by about 77 kcal/mole.

Photochemical studies indicate that excited ammonia molecules can break down to give H, NH_2, NH, and H_2. At wavelengths less than 1500 Å the main decomposition is by reaction (3-119), but at shorter wavelengths, reactions (3-120) and (3-121) appear to be of some importance [74].

$$NH_3^* \rightarrow H(^2S_{\frac{1}{2}}) + NH_2 \tag{3-119}$$

$$NH_3^* \rightarrow H_2 + NH(^1\pi) \tag{3-120}$$

$$NH_3^* \rightarrow H + H + NH(^3\Sigma) \tag{3-121}$$

It is recognized that photolysis results can only give a rough guide to some of the reactions caused by ionizing radiation, but it is worth noting that Jones and Sevorski [63] have observed a yield of nonscavengeable hydrogen (G = 0.75) and Anderson [70] has observed the NH radical, which appears to react with second-order kinetics at the high dose rates used in pulse radiolysis (10^{23} eV cm^{-3} sec^{-1}).

An interesting device for studying NH_3 radiolysis was described by Milton [81], who employed a three-compartment source so that reactions could be studied over a pressure range from 10^{-9} to 10 Torr. Each compartment used a separate electron source, and with such equipment he measured interalia ionization cross sections, ion-molecule reactions, cross sections for the production of free radicals, and G values. He observed that the principal transient species were NH_2, H, and NH_4^+, and that all the primary ions eventually reacted with NH_3 to form NH_4^+ and H or NH_2. Furthermore, the following secondary radicals were found: NH_4, N_2H_2, N_2H_3, and N_2H_5. The primary products of ionizing radiation were 58.8% positive ions (0.4% negative ions) and 40.8% free radicals, whereas positive ion reactions were responsible for 54% of the hydrogen produced and for 65% of the nitrogen.

The complete tentative mechanism for the radiolysis of ammonia
includes the primary processes (3-113) through (3-121) plus the
following sequence of reactions.

$$H + H + M \rightarrow H_2 + M \qquad\qquad (3\text{-}122)$$

$$H + N_2 \rightarrow NH_3 \qquad\qquad (3\text{-}123)$$

$$NH_2 + NH_2 \rightarrow N_2H_4 \qquad\qquad (3\text{-}124)$$

$$NH_2 + NH_2 \rightarrow NH_3 + NH \qquad\qquad (3\text{-}125)$$

$$H + N_2H_4 \rightarrow NH_3 + NH_2 \qquad\qquad (3\text{-}126)$$

$$H + N_2H_4 \rightarrow H_2 + N_2H_3 \qquad\qquad (3\text{-}127)$$

$$NH_2 + N_2H_4 \rightarrow NH_3 + N_2H_3 \qquad\qquad (3\text{-}128)$$

$$NH + NH_3 \rightarrow 2\ NH_2 \qquad\qquad (3\text{-}129)$$

$$NH + NH \rightarrow N_2 + H_2 \qquad\qquad (3\text{-}130)$$

$$2\ N_2H_3 \rightarrow N_2H_4 + N_2 + H_2 \qquad\qquad (3\text{-}131)$$

$$H + NH_3 \rightarrow H_2 + NH_2 \qquad\qquad (3\text{-}132)$$

Reaction (3-131) is merely a stoichiometric summary of products
from N_2H_3 and is probably made up of a series of obscure steps.
Hydrazine is not formed as a product probably because of its quick
destruction in reactions (3-126) through (3-128). Jones and
Sevorski [63] have shown that (3-127) predominates over (3-126).
However, hydrazine is formed [63] as a product in a gas-flow system
where it is removed quickly from the irradiated area.

The radiolysis of liquid ammonia [82] shows two main differences from the vapor-phase radiolysis: (1) the overall yield is lower and (2) hydrazine is formed in small amounts in static systems. These differences are possibly caused by collisional deactivation of excited molecules in the liquid phase.

The only products of gas-phase radiolytic ammonia decomposition are hydrogen and nitrogen in a 3:1 ratio. Several investigators have obtained results in agreement within ± 10% yields at room temperature: $(G(H_2) = 6; G(N_2) = 2;$ and $G(-NH_3) = 4$. The yields increase as temperature increases, and there is general agreement on $G(H_2) = 14.5$-15.0 as a limiting value at high temperature.

Yields are known to decrease with increasing dose rate [83]. Jones and Sevorski [63] observed that the dose rate effect is influenced by pressure and defined two pressure regions: less than 400 Torr where the decomposition yield is independent of pressure and over 400 Torr where the yield increases with decreasing pressure. At 700 Torr, he found the yield $G(-NH_3)$ dropped from 3.6 to 2.6 with a 100-fold increase in dose rate. This may have resulted from the quenching of excited NH_3^* molecules formed by the $H + NH_2$ recombination.

There was an observed decrease [83] in $G(-NH_3)$ from 4.0 to 1.6 over an intensity range from 10^{14} to 10^{19} eV ml^{-1} sec^{-1}. The explanation postulated was based on collisional deactivation of excited molecules.

An extensive investigation of the effect of temperature on yields was made by Jones and Sevorski [63], who gave particular attention to the primary H atom yields. At 23°C, $G(H)$ increased from 10.0 ± 1.5 at 200 Torr to 11.7 ± 0.5 at 600 Torr, based on determinations with added deuterium and added hydrazine, with a maximum primary decomposition yield of $G(-NH_3) = 13.2$. At higher temperatures the hydrogen yield increased to $G(H_2) = 15.1 ± 0.5$ in a static system and 15.9 ± 1.0 in a flow system. These are larger yields than the value of 12.5 computed on the basis of a primary

yield of H atoms of 11.7 and of molecular hydrogen of 0.7, with the
further assumption that the H+NH$_3$ reaction predominates over the
H + H combination.   Hence the measured $G(H_2)$ at high temperatures
may be as much as 4.4 greater than the computed value, taking into
account the uncertainty limits of the experimental values.   Jones
postulated that the increase is caused by a more efficient decomposition of NH$_3^*$ at higher temperature or by a decrease in the W
value.

The addition of xenon to ammonia increases the decomposition
yield [83], but added nitrogen has little effect, and hydrogen
seems to inhibit the decomposition.   The increase in yield due to
xenon is probably not caused by charge transfers such as (3-133)
and (3-134),

$$Xe^+ + NH_3 \rightarrow NH_3^+ + Xe \qquad\qquad (3-133)$$

$$Xe^+ + NH_3 \rightarrow NH_2^+ + H + Xe \qquad\qquad (3-134)$$

since the ionization potentials of both nitrogen (15.6 eV) and
hydrogen (15.4 eV) are greater than that for NH$_3$ (10.15 eV), and it
is therefore concluded that reactions of electrically excited
neutral states of Xe must be significant in the decomposition
mechanism.

The effects of added nitrogen and hydrogen are not evident,
but it has been observed that ammonia can be produced radiolytically from its elements, and hence any possible increase in NH$_3$
decomposition by charge or enrgy transfer must be compensated by
its reformation.   It is thought that hydrogen may repress the decomposition through the reaction

$$NH_2 + H_2 \rightarrow NH_3 + H \qquad\qquad (3-135)$$

The decomposition yield decreased by 30% when an electric
field was applied at a gas pressure of 62 cm [84], although no

effect was noted at 20 cm. The decreased yield was explained by
ion neutralization without decomposition at the electrodes. Other
investigators [85] found no decrease in yield at either 62 or 20 cm.
At high field strengths, increased decomposition is found, probably
caused by acceleration of secondary electrons.

Jones and Sevorski [63] have shown qualitative resemblances
between the raiolysis and photolysis of ammonia in an investigation
of the effects of gas-flow rate, pressure, and dose rate in the
decomposition caused by electron radiation. It was found that
yields of hydrazine may ran as high as $G(N_2H_4) = 4$ in a flow system
at 300°C and concluded that the limiting yield of hydrazine can be
expressed as

$$G(N_2H_4) = 15.9e^{-1930/RT} \tag{3-136}$$

While the above discussion applies to gaseous ammonia, it is
of interest to examine comparable radiolytic decomposition data for
liquid ammonia. When liquid ammonia was decomposed by γ-rays from
cobalt-60 at 20°C, Dainton [86] obtained the following initial
yield values: $G(H_2) = 0.84 \pm 0.02$, $G(N_2) = 0.23 \pm 0.01$, and
$G(N_2H_4) = 0.18$, and computed the following primary yields from
these and other data: $G(H_2) = 0.84 \pm 0.25$, $G(N_2H_4) = 0.9 \pm 0.5$,
$G(-NH_3) = 2.1$, and $G(e) = G(NH_2) = 0.35$. Sutherland and Kramer
[87] found no nitrogen at doses of less than 30×10^3 rads, but
above that level confirmed yields of $G(H_2) = 0.96$, $G(N_2) = 0.21$,
and $G(N_2H_4) = 0.18$ in good agreement with the results of Dainton
et al. [86]. From a study of the material balance, Sutherland
concluded that additional products are formed below 30×10^3 rads
and above 40×10^3 rads. Infrared and ultraviolet studies showed
that HN_3 is formed in the high dose region, and the low dose
region produced an unknown species containing an -N = N-group and
was identified as one of a mixture of the products N_2H_2 (diazene),
N_3H_3 (triazine), or N_4H_4 (tetrazine).

IV. FREE RADICALS

A. Introduction

The reactions of ions and excited molecules in radiation
chemistry are fairly well known in general terms, but the details
of reaction mechanisms, especially in the irradiation of liquids
and solids, are difficult to obtain. It is recognized that both
ions and excited molecules can produce free radicals as well as
stable molecules. Free radicals result when a covalent bond is
broken in such a way as to leave one bonding electron with each
fragment.

$$X : Y \; \rightleftarrows \; X\cdot + Y\cdot \tag{3-137}$$

The reverse reaction can take place, resulting in the formation
again of the original stable molecule. Free radicals are usually
electrically neutral, but not always. Examples of charged radicals
include $\cdot CH_3^+$ and O_2^-. Several reviews have been published on the
subject of free radicals [88, 89].

One of the best ways of investigating free radicals is by
electron spin resonance (ESR), often referred to as paramagnetic
resonance. This method is based on the electronic angular momentum
of free radicals and affords an unambiguous technique of detection.
It is employed to measure radical concentrations and yields and has
furnished some information about the nature and structure of free
radicals. The ESR approach has been employed mainly for solids,
because the rigid structure of the solid tends to keep the radicals
from reacting with each other or with other substances present.
Ultraviolet absorption spectroscopy is also helpful in studying
radicals and has been utilized to compare free radicals produced
by radiation with similar radicals produced by other means.

It is significant in the irradiation of dilute solutions that
the solute, being present at low concentration, is much less likely
to be altered directly by the radiation than by the attack of
radicals produced from the solvent, an effect known as "indirect
action." When a substance is irradiated in bulk and in solution,
the number of molecules altered for a given dosage may be much the
same in the two cases, but the proportion of molecules altered is
much greater in the solution. Hence, measurable changes often take
place with solutions at much lower dose levels than would be
required to affect the bulk material.

Dilute solutions may be employed to measure the uield of free
radicals from a liquid. The free radicals produced from the
solvent have, essentially, the possibility of reacting with each
other or with the solute. They tend to react with each other at
very low solute concentrations, but as solute concentration increases a point is reached where nearly all the radicals react with
the solute. In the latter situation, the number of solute molecules altered per 100 eV of energy absorbed by the total system is
independent of solute concentration. With certain solutes, called
radical traps or scavengers, the number of solute molecules altered
is a measure of the number of free radicals produced from the
solvent.

Scavengers that have been used in this way include vinyl
monomers, diphenylpicrylhydrazyl (DPPH), and iodine. This method
has mainly been used for radiations of LET less than 3 keV/μm. The
method has certain deficiencies, for example, there may be a chance
of charge or energy transfer form solvent to solute, although this
chance can be minimized by using solute concentrations less than
10^{-3} M. Another problem is that there is often some uncertainty
about the reaction mechanism. Generally speaking, there is fair
agreement between results in some cases and poor agreement in
others.

The character of free radicals produced by radiation can often
be postulated from mass spectrometric or photochemical experiments,
but there is generally some uncertainty resulting from cage effects,
collisional deactivation, and other phenomena. Some of the better
evidence to date has come from a knowledge of the final products of
irradiation, with or without the presence of a scavenger. It is
probable that free radicals are most commonly formed by abstraction
of a hydrogen atom from a molecule. Radicals may be generated
deliberately in an irradiation system to permit the effect on the
system of a single radical species to be studied. Methods of radical preparation, other than by ionizing radiation, include the
transfer of enough energy to a molecule to cause dissociation
(thermal or photo) or an oxidation-reduction reaction. The latter
involves the transfer of a single electron, or an atom or group
with an unpaired electron, from one molecule to another.

Perhaps the most unique property of free radicals is their instability, related to the presence of the unpaired electron. Radicals may be quite reactive, reacting so that the odd electron is
paired with a similar electron in another radical or lost through
an electron transfer reaction. On the other hand, the radical may
react in such a way as to produce a second and more stable radical.
The reactivity and also the stability of free radicals is much influenced by their structure. Radicals may either be quite reactive
under normal conditions or elese rather unreactive and stable. The
following are examples of the two types of radicals.

Reactive Radicals

$H\cdot, \cdot OH, Cl\cdot, \cdot CH_3, Ph\cdot$

Stable Radicals

$Ph_3C\cdot$ and substituted triphenylmethyl radicals

(semiquinone)

(DPPH)

NO, NO_2, O_2

The radicals listed in the first group are small and have a very
brief life under normal conditions. Such radicals can be trapped
and preserved [90] in solids at low temperature, where they cannot
diffuse to each other and react. The second group of radicals is
more stable and larger, so that the odd electron is spread over a
larger volume. Nitric oxide and nitrogen dioxide are listed in the
second group because each has one unpaired electron and performs in
an unreactive free radical. Oxygen has a triplet ground state and
acts as a diradical.

The stability of a radical is increased when the hydrogen on
the carbon carrying the electron is replaced by a halogen, the
effectiveness increasing in the order $F < Cl < Br < I$. Unsaturated
substituents increase the stability when located on carbon atoms
next to the one carrying the odd electron, but not when the carbon
having the electron is part of the unsaturated grouping.

Hydrogen atoms, and also -OH radicals, tend to react with the
first molecule with which they collide. Less reactive radicals are
more selective in what substrate they may attack. Generally speaking, radicals are less selective at high temperature, but even very
reactive radicals may be selective at low temperatures.

If a radical attacks a substrate as follows,

$$X \cdot + Y - Z \rightarrow X - Y + Z \cdot \tag{3-138}$$

the newly formed radical is usually less reactive than the attacking radical. For example, hydrogen atoms react with iodine to
produce hydrogen iodide,

$$H \cdot + I_2 \rightarrow HI + I \cdot \tag{3-139}$$

but the reverse reaction,

$$I \cdot + HI \rightarrow H \cdot + I_2 \tag{3-140}$$

which replaces a stable iodine atom by a reactive hydrogen atom, is very improbable. The total energy change in (3-139) is the difference between the energy released in forming a H-I bond (70.5 kcal) and that absorbed in breaking an I-I bond (35.6 kcal):

$$D(H\!-\!I) - D(I\!-\!I) = 70.5 - 35.6 = 34.9 \text{ kcal/mole} \qquad (3\text{-}141)$$

Energy is released, and the reaction is favored in the direction indicated. The reverse reaction between iodine atoms and hydrogen atoms is much less probable since 35 kcal/mole must be supplied in order for the reaction to take place. Reactions such as (3-139) may require a certain amount of energy of activation. Energy of activation is especially important if the attacking radical can react with the substrate in several ways, in which case the reaction with the lowest energy of activation will be the most probable.

Radicals can be classified as electron acceptors or electron donors, in line with their tendency to gain or lose electrons. Halogen radicals tend to gain electrons, and preferentially attack centers of high electron density in the substrate. Methyl radicals are nucleophilic and tend to attack electron-deficient centers in the substrate.

A special characteristic of radiation-formed free radicals is that they are produced in high local concentrations along the tracks of the particles, and hence tend to react with each other more than if the distribution were uniform. This aspect is notable with densely ionizing radiation such as α-particles.

Radicals produced by radiation sometimes possess energy in excess of thermal energy ("hot radicals") and undergo reactions that might not normally occur owing to a high energy of activation. Such "hot radicals" should be distinguished from hot atoms, such as tritium formed by the nuclear reaction ^{3}He (n, p) ^{3}H.

"Hot" or energetic radicals exceed thermal energy by only a few electron volts, and may be produced by the neutralization of molecular or radical ions,

$$A^+ + e^- \rightarrow A^{**} \rightarrow R\cdot^* + S\cdot \tag{3-142}$$

$$R\cdot^+ + e^- \rightarrow R\cdot^* \tag{3-143}$$

They may also be produced by the dissociation of excited molecules, providing the excitation energy exceeds the dissociation energy of the broken bond,

$$A^* \rightarrow R\cdot^* + S\cdot \tag{3-144}$$

Energetic radicals formed by reaction (3-142) and (3-144) have less tendency to be trapped in a solvent cage than similar radicals produced with only thermal energy, resulting in a greater probability that such radicals will react with the substrate. The extra energy of the hot radical also favors reaction with the substrate. Also, the reactions of hot radicals are less likely to be influenced by temperature than those of thermal energy radicals. It is known that hot radicals formed when a frozen substrate is irradiated at liquid nitrogen temperature may react with the substrate. The same radicals formed under such conditions with only thermal energy would not react because they lack the required energy of activation.

B. Reactions of Free Radicals

The principal reactions of free radicals are abstraction, combination, disproportionation, addition, decomposition, and rearrangement.

1.   Abstraction

The abstraction reaction is often called a transfer reaction and in general can be written

$$R\cdot + XY \rightarrow RX + Y\cdot \qquad (3\text{-}145)$$

The species abstracted, X, is usually a hydrogen or halogen atom. Specific examples include the following:

$$CH_3\cdot + C_2H_6 \rightarrow CH_4 + C_2H_5\cdot \qquad (3\text{-}146)$$

$$n\text{-}C_3F_7\cdot + H_2 \rightarrow n\text{-}C_3F_7H + H\cdot \qquad (3\text{-}147)$$

$$H\cdot + C_2H_6 \rightarrow H_2 + C_2H_5\cdot \qquad (3\text{-}148)$$

$$C_2H_5\cdot + C_6H_{14} \rightarrow C_2H_6 + C_6H_{13}\cdot \qquad (3\text{-}149)$$

2.   Combination

Energies of activation for radical combinations are very small and are generally considered to be zero. The following are examples of radical combination:

$$I\cdot + I\cdot + M \rightarrow I_2 + M \qquad (3\text{-}150)$$

$$Br\cdot + Br\cdot + M \rightarrow Br_2 + M \qquad (3\text{-}151)$$

$$CH_3\cdot + CH_3\cdot \rightarrow C_2H_6 \qquad (3\text{-}152)$$

$$C_2H_5\cdot + C_2H_5\cdot \rightarrow C_4H_{10} \qquad (3\text{-}153)$$

$$CH_3\cdot + C_2H_5\cdot \rightarrow C_3H_8 \qquad (3\text{-}154)$$

$$CF_3\cdot + CF_3\cdot \rightarrow C_2F_6 \qquad (3\text{-}155)$$

Combination of atoms usually requires the presence of a third body,
M, to remove the excess energy. If this is not done, the atoms
dissociate again. For polyatomic radicals, a third body is usually
not necessary, because the excess energy can be taken up by the
various molecular vibrational degrees of freedom.

In a gaseous system, the wall of the vessel can serve as the
third body in the combination of atoms to form a diatomic molecule.
For this reason, surfaces may affect gas-phase radiolyses in which
free atoms are intermediates, but will have little influence on the
radiolysis of liquids, where there is considerable opportunity for
the rapid loss of energy by collisions.

The energy available from radical combination may be retained
long enough by the product to affect its reactions. For example,
the gas-phase combination of hydrogen atom with ethyl radical
produces an excited ethane molecule, which may break down into two
methyl radicals [91, 92].

$$H\cdot + C_2H_5\cdot \rightarrow C_2H_6^* \rightarrow 2\ CH_3\cdot \qquad (3\text{-}156)$$

The decomposition into methyl radicals instead of the original
radicals is due to the lower bond dissociation energy of the C–C
bond (83 kcal/mole versus 96 kcal/mole). Another example involves
the reaction of the I atom with ethyl radical [93]

$$I\cdot + C_2H_5\cdot \rightarrow C_2H_4 + HI \qquad (3\text{-}157)$$

in a reaction similar to disproportionation.

3. Disproportionation

Disproportionation is a reaction in which two radicals combine
to give two stable molecules, one of which is more unsaturated than
the other. One example is the reaction of radicals from ethyl
alcohol,

$$2 \text{ CH}_3\dot{\text{C}}\text{HOH} \rightarrow \text{CH}_3\text{CH}_2\text{OH} + \text{CH}_3\text{CHO} \qquad (3\text{-}158)$$

another is methyl plus ethyl,

$$\text{CH}_3\cdot + \text{C}_2\text{H}_5\cdot \rightarrow \text{CH}_4 + \text{C}_2\text{H}_4 \qquad (3\text{-}159)$$

Disproportionation may compete with simple recombination,

$$2 \text{ C}_2\text{H}_5\cdot \rightarrow \text{C}_2\text{H}_6 + \text{C}_2\text{H}_4 \qquad (3\text{-}160)$$

$$2 \text{ C}_2\text{H}_5\cdot \rightarrow \text{C}_4\text{H}_{10} \qquad (3\text{-}161)$$

In such cases the rate of disproportionation k_d is usually less than the rate of combination k_c, because the energy of activation for disproportionation appears to be slightly higher than that of simple combination.

4. Addition

The addition of radicals to unsaturated molecules to form new radicals is a characteristic free radical reaction. Examples include

$$\text{CH}_3\cdot + \text{C}_2\text{H}_4 \rightarrow \text{C}_3\text{H}_7\cdot \qquad (3\text{-}162)$$

$$\text{H}\cdot + \text{C}_2\text{H}_4 \rightarrow \text{C}_2\text{H}_5\cdot \qquad (3\text{-}163)$$

Reaction (3-163) is quite significant in the radiolysis of hydrocarbons for hydrogen atoms at thermal energies. If this reaction is efficient, the neutral molecule is known as a "radical scavenger." Another example of a scavenger reaction is

$$\text{CH}_3\cdot + \text{NO} \rightarrow \text{CH}_3\text{NO} \qquad (3\text{-}164)$$

Radical-induced polymerization reactions are examples of radical addition, involving repeated monomer addition to the growing chain,

$$-CH_2-\overset{\bullet}{C}HR + CH_2=CHR \rightarrow -CH_2-CHR-CH_2-\overset{\bullet}{C}HR \qquad (3-165)$$

5. Decomposition

Alkyl radicals will break down if enough activation energy is provided. Examples of radical decomposition include

$$C_2H_5 \cdot \rightarrow C_2H_4 + H\cdot \qquad (3-166)$$

$$C_3H_7 \cdot \rightarrow CH_3 \cdot + CH_2CH_2 \qquad (3-167)$$

$$C_4H_9 \cdot \rightarrow C_2H_5 \cdot + C_2H_4 \qquad (3-168)$$

$$(CH_3)_3CO \cdot \rightarrow CH_3 \cdot + CH_3COCH_3 \qquad (3-169)$$

$$CH_3CO_2 \cdot \rightarrow CH_3 \cdot + CO_2 \qquad (3-170)$$

$$CH_3CO \cdot \rightarrow CH_3 \cdot + CO \qquad (3-171)$$

Through loss of CO_2 by acetoxy radical, Eq. (3-170) proceeds easily, helped by the high heat of formation of CO_2. Activation energies for alkyl breakdown are about 40 to 50 kcal/mole for decomposition to hydrogen [Eq. (3-166)] and about 20 to 30 kcal/mole for decomposition into other radicals [Eqs. (3-167) and (3-168)]. For acetoxy radicals, the activation energy of decomposition is only about 5 kcal/mole. For resonance-stabilized radicals, such as $C_6H_5CH_2\cdot$, the energy of activation for decomposition is quite high.

6. Rearrangement

It is possible for reactive radicals to undergo rearrangement and attain more stable structures. Groups that may migrate include phenyl, halogen, hydrogen, and methyl. The following is an example [94] of phenyl migration.

$$Ph_3C-CH_2\cdot \rightarrow Ph_2\overset{\cdot}{C}-CH_2Ph \qquad\qquad (3\text{-}172)$$

Chlorine migrations of the following type have been reported [95], supposedly resulting in a more stable radical

$$Cl_3C-\overset{\cdot}{C}= \rightarrow Cl_2\overset{\cdot}{C}-CCl= \qquad\qquad (3\text{-}173)$$

REFERENCES

1. Platzman, R. L., *Intern. J. Appl. Radiation Isotopes,* 10, 116 (1961).
2. Tickner, A. W. and Knewstubb, P. F., *J. Chem. Phys.,* 38, 464 (1963).
3. Tickner, A. W., Dawson, P. H., and Knewstubb, P. F., *J. Chem. Phys.,* 38, 1031 (1963).
4. Dawson, P. H. and Tickner, A. W., *J. Chem. Phys.,* 40, 3745 (1964).
5. Kebarle, P. and Hogg, A. M., *J. Chem. Phys.,* 42, 798 (1965).
6. Hogg, A. M. and Kebarle, P., *J. Chem. Phys.,* 43, 449 (1965).
7. Benyon, J. H., *Mass Spectrometry and Its Applications to Organic Chemistry,* Elsevier, Amsterdam, 1960, p. 263.
8. Tal'roze, V. L. and Lyubrinova, A. K., *Dokl. Akad. Nauk, SSSR,* 86, 909 (1952).
9. Abramson, F. P. and Futrell, J. H., *J. Chem. Phys.,* 45, 1925 (1966).
10. Giardini-Guidoni, A. and Friedman, L., *J. Chem. Phys.,* 45, 937 (1966).
11. Abramson, F. P. and Futrell, J. H., *J. Chem. Phys.,* 46, 3264 (1967).

12. Wagner, C. D., Wadsworth, P. A., and Stevenson, D. P., *J. Chem. Phys.* <u>28</u>, 517 (1958).

13. Bone, L. I. and Futrell, J. H., *J. Chem. Phys.*, <u>46</u>, 4084 (1967).

14. Munson, M. S. B. and Field, F. H., *J. Amer. Chem. Soc.*, <u>87</u>, 3294 (1965).

15. Menendez, G. B., Thomas, B. S., and Bailey, T. L., *J. Chem. Phys.*, <u>42</u>, 802 (1965).

16. Munson, M. S. B., Franklin, J. L., and Field, F. H., *J. Phys. Chem.*, <u>68</u>, 3098 (1964).

17. Gaydon, A. G. *Dissociation Energies and Spectra of Diatomic Molecules*, Chapman and Hall, London, 1953.

18. Henglein, A. and Muccini, G. A., *Z. Naturforsch.*, <u>17a</u>, 454 (1962).

19. Champion, R. L., Doverspike, L. D., and Bailey, T. L., *J. Chem. Phys.*, <u>45</u>, 4377 (1966).

20. Henglein, A., Lacmann, K., and Knoll, B., *J. Chem. Phys.*, <u>43</u> 1048 (1965).

21. Field, F. H. and Lampe, F. W., *J. Amer. Chem. Soc.*, 5589 (1958).

22. Lias, S. G. and Ausloos, P., *J. Chem. Phys.*, <u>37</u>, 877 (1962).

23. Ausloos, P., Lias, S. G., and Scala, A. A., *Advan. Chem. Ser.*, <u>58</u>, 264 (1966).

24. Borkowski, R. O. and Ausloos, P., *J. Chem. Phys.*, <u>40</u>, 1128 (1964).

25. Draper, L. M. and Green, J. H., *J. Phys. Chem.*, <u>68</u>, 1439 (1964).

26. Borkowski, R. P. and Ausloos, P., *J. Chem. Phys.*, <u>38</u>, 36 (1963).

27. Ausloos, P. and Lias, S. G., *J. Chem. Phys.*, <u>45</u>, 524 (1966).

28. Doepher, K. D. and Ausloos, P., *J. Chem. Phys.*, <u>44</u>, 1591 (1966).

29. Sieck, L. W. and Futrell, J. H., *J. Chem. Phys.*, <u>45</u>, 560 (1966).

30. Derwish, G. A. W., Galli, A., Giardini-Guidoni, A., and Volpi, G. G., *J. Chem. Phys.*, <u>41</u>, 2998 (1964).

31. Pettersson, E. and Lindholm, E., *Arkiv Fysik*, 24, 49 (1963).

32. Chupka, W. A. and Lindholm, E., *Arkiv Fysik* 25, 349 (1963).

33. Ausloos, P., Lias, S. G., and Gorden, R., *J. Chem. Phys.*, 39, 818 (1962).

34. Ausloos, P., Lias, S. G., and Gorden, R., *J. Chem. Phys.* 40. 1854 (1964).

35. Ausloos, P. and Lias, S. G., *J. Chem. Phys.*, 40, 3599 (1964).

36. Ausloos, P. and Lias, S. G., *Discussions Faraday Soc.*, 39, 36 (1965).

37. Aquilanti, V. and Volpi, G. G., *J. Chem. Phys.*, 44, 2307 (1966).

38. Sandoval, T. B. and Ausloos, P., *J. Chem. Phys.*, 38, 2454 (1963).

39. Doepher, K. D. and Ausloos, P., *J. Chem. Phys.*, 42, 3746 (1965).

40. Ausloos, P. and Lias, S. G., *J. Chem. Phys.*, 43, 127 (1965).

41. Ausloos, P., Scala, A. A., and Lias, S. G., *J. Amer. Chem. Soc.*, 88, 1583 (1966).

42. Johnson, G. R. A. and Warman, J. M., *Discussions Faraday Soc.*, 37, 87 (1964).

43. Dondes, A., Harteck, P., and von Weyssenhoff, H., *Z. Natur-forsch* , 19, 13 (1964).

44. Anderson, A. R., Best, J. V. F., and Willett, M. J., *Trans. Faraday Soc.*, 62, 395 (1966).

45. Anderson, A. R. and Best, J. V. F., *Trans. Faraday Soc.*, 62 610 (1966).

46. Dawes, D. H. and Back, R. A., *J. Phys. Chem.*, 69, 2385 (1965).

47. Milton, C. E., Rapp, G. A., and Rudolph, P. S., *J. Chem. Phys.*, 29, 968 (1958).

48. Henglein, A. and Muccini, G. A., *J. Chem. Phys.*, 31, 1426 (1959).

49. Magee, J. L. and Burton, M., *J. Amer. Chem. Soc.*, 73, 523 (1951).

50. Magee, J. L., *Discussions Faraday Soc.*, 12, 33 (1952).

51. Hamill, W. H. and Forrestal, L. H., *J. Am. Chem. Soc.*, 84, 3648 (1962).

52. Van Dusen, W. and Hamill, W. H., *J. Amer. Chem. Soc.*, 84, 3648 (1962).

53. Platzman, R. L., *Radiation Res.*, 2, 1 (1955).

54. Platzman, R. L., *Vortex*, 23, 372 (1962).

55. Lipsky, S. and Burton, M., *J. Chem. Phys.*, 31, 1221 (1959).

56. De Mayo, P. and Reid, S. T., *Quart. Rev.*, 15, 393 (1961).

57. Franck, J. and Rabinowitch, E., *Trans. Faraday Soc.*, 30, 120 (1934).

58. Wilson, J. E. and Noyes, W. A., Jr., *J. Amer. Chem. Soc.*, 65, 1547 (1943).

59. Calvert, J. G. and Pitts, J. N., Jr., *Photochemistry*, Wiley, New York, 1966.

60. Anderson, A. R., Knight, B., and Winter, J. A., *Nature*, 201, 1026 (1964).

61. Cottin, M., *J. Chem. Phys.*, 56, 1024 (1959).

62. Moruzzi, J. L. and Phelps, A. V., *J. Chem. Phys.*, 45, 4617 (1966).

63. Jones, E. T. and Sevorski, T. J., *Trans. Faraday Soc.*, 63, 2411 (1967).

64. Thynne, J. C. J. and Harrison, A. S., *Trans. Faraday Soc.*, 62, 2468 (1966).

65. Hogg, A. M. and Kebarle, P., *J. Chem. Phys.*, 43, 449 (1965).

66. Milton, C. E., unpublished work reported by T. W. Sworski in *Proceedings of the Fifth Informal Conference on the Radiation Chemistry of Water (Notre Dame)*, AEC Document No. C00-38-519, p. 41, 1966.

67. Hoffmann, A. S., Diehl, N., Rupp, W., and Sizman, R., *Radiochem. Acta*, 1, 203 (1963).

68. Anderson, A. R., Knight, B., and Winter, J. A., *Trans. Faraday Soc.*, 62, 359 (1966).

69. Firestone, R. F., *J. Amer. Chem. Soc.*, 79, 5593 (1957).

70. Anderson, A. R., in *Fundamental Processes in Radiation Chemistry* (P. J. Ausloos, ed.) Interscience, 1968.

71. Johnson, G. R. A. and Simic, M., *J. Phys. Chem.*, 71, 1118 (1967).

72. Yang, J. Y. and Marcus, I., *J. Amer. Chem. Soc*, 88, 1625 (1966).

73. Baxendale, J. H. and Gilbert, B. O., *J. Amer. Chem. Soc.*, 86 516 (1964).

74. McNesby, J. R., Tanaka, I., and Okabe, H., *J. Chem. Phys.*, 36, 605 (1962).

75. Firestone, R. F., *J. Amer. Chem. Soc.*, 79, 5593 (1957).

76. Anderson, A. R. and Winter, J. A., in *Radiation Research* (G. Silini, ed.), North Holland, 1967.

77. Baxendale, J. H. and Gilbert, G. P., *Science*, 147, 1571 (1965).

78. Johnson, G. R. A. and Simic, M., *Nature*, 212, 1570 (1966).

79. Dorfman, L. M. and Noble, P. C., *J. Phys. Chem.*, 63, 980 (1959).

80. Derwish, G. A. W., Galli, A., Giardini-Guidoni, A., and Volpi, G. G., *J. Chem. Phys.*, 39, 1599 (1963).

81. Milton, C. E., *J. Chem. Phys.*, 45, 4414 (1966).

82. Cleaver, D., Collinson, E., and Dainton, F. S., *Trans. Faraday Soc.*, 56, 1640 (1960).

83. Horscroft, R. C., *Trans. Faraday Soc.*, 60, 323 (1964).

84. Smith, C. and Essex, H., *J. Chem. Phys.*, 6, 188 (1938).

85. Burtt, B. P. and Baurer, T., *J. Chem. Phys.*, 23, 466 (1955).

86. Dainton, F. S., Skevarski, T., Smithies, D., and Wezranowski, E., *Trans. Faraday Soc.*, 60, 1068 (1964).

87. Sutherland, J. W. and Kramer, H., Abstracts 153 *Amer. Chem. Soc. Mtg.*, April, 1967.

88. Steacie, E. W. R., *Atomic and Free Radical Reactions,* ACS Monograph No. 125, Reinhold, New York, 1954.

89. Trotman-Dickenson, A. F., *Free Radicals*, Methuen, London, 1959.

90. Franklin, J. L. and Broida, H. P., *Ann. Rev. Phys. Chem.*, 10, 145 (1959).

91. Berlis, M. R. and LeRoy, P. J., *Discussions Faraday Soc.*, 14, 50 (1953).

92. Darwent, R. D. and Steacie, E. W. R., *J. Chem. Phys.*, 16, 381 (1948).

93. Bunbury, D. L., Williams, R. R., and Hamill, W. H., *J. Amer. Chem. Soc.*, 78, 6228 (1956).

94. Curtin, D. Y. and Hurwitz, M. J., *J. Amer. Chem. Soc.*, 74, 5381 (1952).

95. Nesmeyanov, A. N., Freidlina, R. K., and Zakharkin, L. I., *Quart. Rev.*, 10, 330 (1956).

Chapter 4

RADIATION CHEMISTRY OF SMALL MOLECULES

I.   INTRODUCTION

A study of the radiation chemistry of small* molecules is
helpful in understanding the radiation chemistry of polymers. For
example, small molecules contain characteristic groupings such as
carboxyl, carbonyl, ester, amide, and halide. The radiation chemistry of such groupings in small molecules is similar to their
chemistry in large molecules, but the smaller molecules may be more
convenient to investigate because of greater ease of analysis and
handling. Furthermore, the radiation chemistry of the olefins is
obviously pertinent to polymer chemistry, because olefinic monomers
are commonly employed in radiation-induced polymerization reactions.

As another example, the radiation chemistry of saturated
aliphatic hydrocarbons has been studied extensively, and it shows
many features exhibited by hydrocarbon polymers such as polyethylene and polypropylene. Rapid progress in the radiation chemistry
of hydrocarbons has taken place only in recent years. Advancement
has been promoted by the availability of techniques such as mass
spectroscopy and gas chromatography for analysis, and the availability of radioactive iodine for use as a radical detector. The

---

*For the purposes of this chapter, a "small" molecule is any mole-
cule that is less than polymeric in size.

185

interpretation of the data has been facilitated by a knowledge of
the reactions of hydrocarbon free radicals and of the reactions of
hydrocarbon ions in the mass spectrometer.

Early radiolysis studies were mostly concerned with establish-
ing the products of irradiation, but more recent work has involved
the identification of the primary and intermediate species taking
part in the reaction. Luminescence studies showed the presence of
electronically excited molecules in irradiated liquids [1]. The
use of radical scavengers [2, 3] indicated that radicals were
formed in sizeable yields. The irradiation of liquids with electron beams from high intensity accelerators enable the formation of
radicals in sufficient concentration that their identities and
reactions can be determined through spectroscopy or electron-spin-
resonance (ESR) studies.

The significance of ionic intermediates in organic liquids was
demonstrated by a series of investigations in the 1960s. In 1961,
evidence was obtained that hydrogen production in alcohol radiolysis was associated with a reaction of electrons [4, 5]. In 1963,
evidence from pulse radiolysis indicated the presence of the
solvated electron in alcohols [6, 7], following similar observations of the hydrated electron in water. Although a considerable
yield of free ions was detected in polar liquids, conductivity
studies showed a lower yield in hydrocarbons [8, 9]. Studies were
also made on the detection of ions in liquids by the employment of
chemical scavengers [10, 11].

When an ionizing ray or particle passes through an organic
liquid, the primary processes that take place lead to the production of excited molecules, electrons, ions, and free radicals.
These primary processes can be summarized as follows [12]:

$$(excitation) \quad M \rightarrow M^* \tag{4-1}$$

$$(ionization) \quad M \rightarrow M^+ + e^- \tag{4-2}$$

$$M^+ \begin{cases} \nearrow A^+_\bullet + B\bullet & (4\text{-}3) \\ \searrow C^+ + D & (4\text{-}4) \end{cases}$$

(ion dissociation)

(neutralization) $\quad M^+ + e^- \longrightarrow M^{**} \qquad\qquad (4\text{-}5)$

$$A^* \text{ and } A^{**} \begin{cases} \nearrow A\bullet + B\bullet & (4\text{-}6) \\ \searrow C + D & (4\text{-}7) \end{cases}$$

(dissociation)

where A• and B• are free radicals and C and D are molecular products. The products from these primary processes are distributed
along the track of the ionizing particle in spurs (small groups),
which may overlap adjacent spurs as the active species diffuse
apart. Whether the primary species react before the expanding
spurs overlap depends on the closeness of the spacing of the spurs
along the particle track and, therefore, on the LET of the ionizing
particle. For γ-radiolysis, the spur is probably a small, spherical volume which contains several primary species. For the more
densely ionizing radiation of α-particles, the spurs may be close
enough together to form a continuous track of ionized and excited
species.

Electrons ejected in the ionization step (4-2) generally remain within a 20-Å radius of the parent ion and return to neutralize the parent ion by reaction (4-5) within 10^{-13} sec. This mechanism is essentially that of the Samuel-Magee model [13] and is
favored in organic liquids because of the strong attraction between
the ions owing to the low dielectric constant. (Electrostatic
attractive force between the parent ion and the electron is inversely proportional to the dielectric constant.) The alternative
Lea-Gray-Platzman [14] model would be less likely in organic
liquids because it postulates a more distant removal (150 Å) of the
electron from the parent positive ion.

The highly excited molecule (M^{**}) formed by ion recombination will probably quickly dissociate into free radicals, although molecular products can also be formed by this means. Primary excited molecules (M^{*}) may also dissociate into radical or molecular products, but such products will possess less energy than those from the dissociation of M^{**}. A radical pair without enough energy to escape from its solvent cage may quickly recombine, a process known variously as caging, geminate pairing, or the Franck-Rabinowitch effect. Such recombination is quite probable if both radicals are large. Molecular dissociation products will not undergo geminate pairing, since they will not react together should they collide. The intermediate positive ion may also dissociate to radical or molecular products by (4-3) or (4-4) before neutralization takes place. Ion-molecule reactions are omitted from the simple mechanism shown above.

In the radiation chemistry of small molecules, the primary processes are not markedly different for substrates in the gas, liquid, or solid phase. In fact, if the vapor, liquid, and solid states of a particular substance are placed in the same flux of γ-radiation, the energy deposition per unit time is approximately the same for each state. Absorption differences caused by changes in energy levels resulting from stronger intermolecular forces in the condensed phases are insignificant. The γ-ray energy is converted to the kinetic energy of electrons by Compton and photoelectric processes, and is dissipated by ionization and excitation of molecules in about 10^{-15} sec.

However, the chemical reactions that follow the primary processes are often phase dependent. The density of the medium determines the molar concentrations and spatial distribution of the reaction intermediates, thus affecting the relative likelihood of the possible competing reactions. With regard to solid phase reactions, solid substances are often able to transfer energy or charge over many molecular diameters and to influence the fate of reaction intermediates by trapping them at points of imperfection in the solid.

The phenomenon of ion combination shows differences in the
gas, liquid, and solid phases. When a gaseous molecule is ionized
by a high energy electron, the resulting positive ion has a very
low probability of recapturing its lost electron, because the mean
free path of the latter is so long. Hence, the positive ion may
undergo several collisions with neutral molecules before neutralization and therefore have a chance to decompose or take part in an
ion-molecule reaction. The same ionization event in a liquid
phase will probably terminate by the quick return of the electron
to its parent positive ion, because the kinetic energy of the
electron is reduced to thermal energies at a distance such that the
coulombic attraction can still cause the recombination. Thus the
G value for escape of electrons from recombination with parent
cations in liquid hydrocarbons is only 0.1-0.2 [15, 16], as compared to 3-4 in the gas phase. In solid hydrocarbons at 77°K the
electrons must become thermalized at even shorter distances from
the parent ion than in liquids, but more electrons avoid quick
recombination. For example, G(trapped e^-) in γ-irradiated 3-
methylpentane glass observed by infrared absorption is 0.8 at 77°K
[17]. This relatively high G value has been attributed to structural features of the rigid amorphous medium which may serve as
potential wells for trapping electrons or to induced traps made by
solvation of the electron by induced dipoles. Such traps have
enough strength to compete with the coulombic attractive forces,
which return nearly all electrons to their parent positive ions in
irradiated liquids at room temperature. The polar organic solid
2-methyl tetrahydrofuran is even more effective in low temperature
trapping of electrons, giving a G(trapped e^-) value of about 3, as
determined by ESR and infrared absorption studies [18, 19].

An increase in the density of the medium increases the likli-
hood that the products formed by the breakdown of a particular
molecule will remain trapped in the solvent cage and recombine with
each other [20-23] or undergo geminate disproportionation [24, 25].
The steady-state molar concentration of a reaction intermediate,

which is homogeneously distributed and which is undergoing second order recombination, is proportional to the square root of the density of the medium, if it is assumed that dose rate is constant and the rate constant is independent of density. The average life-time of such an intermediate is inversely proportional to the square root of the density. Therefore, the probability that the intermediates will react with the solvent or with scavengers rather than with each other decreases as the density increases. Such effects are important in comparing a gas-phase reaction with the same reaction in a condensed phase [26].

For reactions occurring in organic media, it is often possible to differentiate between "radical" and "molecular" products. Radical products are those formed by way of free radicals and are eliminated by the use of an effective free radical scavenger. This term has also been applied to the scavengeable free radicals themselves. Molecular products are those which are not influenced by free radical scavengers and which are produced by unimolecular or nonradical processes or by radical reactions which are not accessible to a scavenger. It is possible for a product to be formed partly by a radical mechanism and partly by a molecular mechanism. Also, radical-radical reactions can form products which are called "molecular" if the reaction takes place within a spur or solvent cage, and "radical" if the radicals diffuse into the bulk of the medium before reacting. Molecular products may also result from reactions involving ions or excited states rather than free radicals.

Photochemical and mass-spectrometric data are often useful in the understanding of radiolysis mechanisms. Photochemical data are less useful for gases than for liquids, because in the liquid it is more probable that excited states will be quenched before any re-action occurs. Mass spectrometric data can also be employed with more confidence in gaseous systems, which involve conditions more similar to those found in the mass spectrometer. Often in liquid-phase reactions the radical and ion intermediates react so rapidly

(microseconds) that they can only be observed, if at all, by very
fast measurements using pulse radiolysis [27, 28]. The advantage
of solid-state studies at 77°K and below is that the removal of
reaction intermediates is slowed to such a rate that their spectra
and reactions can be studied by common spectrophotometric, ESR,
electrical conductivity, and luminescence measurements.

Aromatic organic compounds are considered separately, because
of the special phenomena associated with the irradiation of such
compounds. Aromatic compounds are much more stable to radiation
than aliphatic compounds, because the absorbed energy is imparted
to electrons in π-orbitals and is usually lost in various ways
without causing a significant amount of chemical reaction. One of
the products that may result is a small amount of polymer of some
sort. Energy transfer phenomena are especially noticeable in mixtures of aromatic compounds with other substances. Some of the
reactions that may follow energy transfer include oxidation, halo-
genation, amination, and hydrogenation. The mechanism of reaction
in aqueous solution involves an attack on the aromatic compounds by
free radicals formed from water. Products from benzene include
diphenyl and phenol. Monosubstituted benzenes in water containing
oxygen are hydroxylated in the ring in the three possible positions,
and the substituent is attacked.

When the addition of an aromatic substance to a second compound is found to retard the radiation-induced decomposition of the
second compound, the phenomenon is called "protection." This type
of protection is essentially "physical" in nature and involves
either charge transfer or transfer of excitation energy from an
energy-rich species to the aromatic protective agent, thus removing
energy from a molecule or species which would otherwise dissociate
or react. A second type of protection is known as "chemical" protection, which usually implies radical scavenging by the additive,
thus preventing or minimizing the usual reactions involving radicals. In a dilute solution the solute may be protected by adding a
"chemical" protective agent, and the solvent may be protected by a

"physical" protective agent. The physical protective agent would
also have some effect in protecting the solute by reducing the rate
of free radical formation in the solution.

As an example of physical protection, the rate of radiation-
induced polymerization of acetylene is reduced by benzene vapor
[29, 30]. This has been explained by a charge transfer between
acetylene ions and benzene.

$$C_2H_2^+ + C_6H_6 \rightarrow C_2H_2 + C_6H_6^+ \qquad\qquad (4-8)$$

The radiolysis of other aliphatic compounds including cyclohexane
is known to be inhibited by benzene. Processes yielding physical
protection can be classified [31] into four groups: charge transfer,
quenching, energy transfer, and negative ion formation. Charge
transfer and energy transfer are most effective when the protecting
substance has a slightly lower lying ionized or excited state than
the activated substrate. In quenching, the excited molecule is
converted to the ground state or to a more stable (triplet) excited
state by loss of energy to the protective additive. Negative ion
formation results from electron capture by the protective additive,
resulting in protection by interfering with the normal ion-
neutralization reaction. (Neutralization of a positive ion by a
negative ion usually requires less energy than neutralization by
an electron.) Physical protection can be intermolecular or intra-
molecular, that is, in mixtures of solvent and protecting agent or
in pure compounds when the protective agent is part of the molecule.
The latter type is found in alkyl benzenes, where the phenyl group
inhibits decomposition in the side-chain alkyl group.

Whereas the radiolysis of organic liquids produces radicals
that can react with the solvent, the radiolysis of water solutions
forms hydrogen atoms and hydroxyl radicals (the *radical products*),
which do not react with water but only between themselves or with
dissolved substances. In aqueous solutions, product yields are
dependent on LET. If the radiation dissipates large amounts of

energy along a short length of track, most of the atoms and radicals react within the track to give mainly molecular hydrogen and
hydrogen peroxide (the *molecular products*). If the radiation dissipates its energy more sparsely, most of the atoms and radicals
react with dissolved substances.

The stability and reactions of ions in aqueous solution will
be affected by solvation. It is ordinarily assumed that any ions
present will become solvated if they have a lifetime greater than
approximately 10^{-11} sec. In this process the surrounding water
molecules become oriented around the ion so that the parts of the
molecules which are oppositely charged to the ion are closest to
it. For example, a positive ion attracts the negative oxygen of
the water molecules. The solvated ion will become surrounded by a
shell of water molecules, held loosely by electrostatic attraction
between the ion and the solvent molecules. Since solvation causes
a reduction in potential energy, it releases energy in aqueous
solutions of the order of 1 eV/ion, which may be enough to supply
the energy required by reactions which are endothermic in the gas
phase, where solvation does not take place.

The hydrogen ion H^+ forms a hydrate H_3O^+ known as the hydronium ion. In water solutions after the ions become solvated, the
symbols H^+ and H_3O^+ both represent the same thing, namely, the
solvated hydrogen ion H_{aq}^+ (or H^+, n H_2O). The solvated electron
is probably an exception to this simple concept of solvation and
may be bettter represented as an electron spread over several
water molecules.

When water is irradiated, the excited molecules formed directly (the primary excited molecules) are generally ignored, the
idea being that they return directly to the ground state without
decomposition or else dissociate into H· and OH· radicals, which
have little extra energy and recombine within their cage of water
molecules without causing any net chemical change. However, it has
been postualted that excited water molecules may react to give part

of the molecular product yield by the reaction [32, 33]

$$H_2O^* + H_2O^* \rightarrow H_2 + H_2O_2 \qquad\qquad (4\text{-}9)$$

Gaseous substances are more simple to investigate by radiation-chemical techniques than are liquids and solids. Their lower density essentially eliminates the effect of LET because the active species are not held in the tracks where they are formed, with the result that α-particles and γ-rays give almost the same product yields. The net result of irradiation is to form positive ions, electrons, and excited molecules homogeneously distributed throughout the gas. This is quite different from the corresponding effect in solids and liquids, where the active species are formed close together in spurs and are less able to move apart because of the confining effect of the adjacent molecules.

Techniques can be used to study gases that are not applicable to liquids and solids. Such techniques include measurement of ionization produced in the gas, mass spectrometric studies of ions formed, and the photochemical study of gaseous decomposition, which affords evidence concerning excited molecules and the reactions of radicals produced by their decomposition.

The number of ion pairs (N) that is produced in a gas by irradiation can be determined by ionization measurements. If the energy absorbed by the gas is known, the quantity W indicating the mean energy absorbed per ion pair formed can be computed. The *ion-pair yield* of a product formed by irradiation is defined as M/N, where M is the number of molecules of product. In the early years of radiation chemistry it was common to use M/N to express product yields in irradiated gases, but this practice was dropped as emphasis shifted to solids and liquids, since neither N nor W can be measured directly in the latter systems. With the passage of time, the G value came to be the common method of expressing

yields in gases as well as in solids and liquids. Some have
argued [34] that gaseous yields are still best expressed as M/N,
and such figures can readily be converted to G values by the equation

$$G = M/N \times 100/W \tag{4-10}$$

When W is not known, it is often taken as 32.5 eV, the value for
air, which yields on substitution in Eq. (4-10)

$$G = M/N \times 100/32.5 \tag{4-11}$$

$$G \cong 3M/N \tag{4-11A}$$

Early studies of the radiation chemistry of gases made use of
radiolysis mechanisms involving the formation and neutralization
of ions, as developed by S. C. Lind. In 1936, attention was called
to the formation of free radicals in radiation chemistry by Eyring,
Hirschfelder, and Taylor, and for several years following the
postulated theories of mechanism were dominated by free radicals.
More recently, the pendulum has swung back toward ion reactions in
radiation chemistry, largely as a result of extensive evidence
obtained in mass spectrometric studies. Most investigators now
believe that both ion and free radical reactions are of significance in radiation chemistry.

Electric discharges in gases can result in reactions analogous
to those caused by ionizing radiation [35]. Reactions which occur
in the upper atmosphere involve oxygen, nitrogen, nitrogen oxides,
water vapor, and carbon dioxide and are similar to the reactions
of these substances as revealed by radiation chemistry studies.

II.   FREE RADICAL DETECTION AND IDENTIFICATION

In view of the importance of free radicals in radiation chemistry, it is necessary at this point to summarize the methods of
free radical detection and identification. So far as their role
in reaction mechanisms is concerned, a distinction must be made
between the reactions of radicals in spurs and the reactions of
radicals homogeneously distributed throughout the substance undergoing irradiation. The homogeneously distributed radicals can be
detected by the methods discussed here, but the radicals of very
short lifetime which react within the spur cannot ordinarily be
detected by these methods. In general, radicals can be detected
by physical methods such as electron-spin-resonance (ESR) spectroscopy or by chemical methods involving titration or radical sampling
[36].

One of the most common physical methods is ESR spectroscopy.
The unpaired electron of a free radical possesses a magnetic moment
which can become reoriented at a higher energy level in a magnetic
field through the absorption of microwave radiation of the proper
frequency. There is no background absorption in the sense that
molecules without an unpaired electron do not absorb such radiation.
In the case of a free radical, hyperfine structure is observed due
to the effect on the electrons of neighboring nuclei, thus enabling
the radical to be identified by means of the hyperfine structure.
Use of ESR was originally restricted to the detection of radicals
in solids, but was extended by Fessenden and Schuler [36] in 1960
to the detection of radicals in liquids during irradiation. The ESR
absorption lines in solids are broad, but are narrow and well resolved in liquids, enabling the frequent identification of free
radicals that may be present.

Use of ESR to detect radicals in liquids necessitates a
relatively high rate of radical formation. For example, assume
that radicals disappear by a second-order recombination having a

rate constant k,

$$R\cdot + R\cdot \xrightarrow{k} R\text{--}R \tag{4-12}$$

Then assuming a value of k of about 10^9 M^{-1} sec^{-1} and a detectable radical concentration of $10^{-7}M$ the rate of radical disappearance would be

$$(2)(10^9)(10^{-7})^2 = 2 \times 10^{-5} \text{ M sec}^{-1} \tag{4-13}$$

Hence, the rate of radical production taking into account G_R and source invtensity must be about 2×10^{-5} M sec^{-1} in order for the radicals to be detected by the ESR technique. Radical formation rate can be expressed as

$$10G_R \times (\text{dose rate})/N = 2k[R\cdot]^2 \tag{4-14}$$

where dose rate is in electron volts per milliliter per second.

A number of hydrocarbon radicals have been definitely identified by ESR in radiolysis experiments [37]. The method has not been used to detect radicals in polar liquids because of the high loss of microwave power in such liquids which physically fill the ESR cavity.

Equation (4-14) has been employed to compute either the rate of radical recombination or the yield of free radicals. For example, the rate of recombination of ethyl radicals was computed [36] from the ethyl radical concentration determined by ESR and the yield of $G(C_2H_5) = 4.4$ [38]. On the other hand, radical yields can be calculated by substituting the lifetime and concentration of radicals and the dose rate into Eq. (4-14). The yield of ethyl free radicals was determined by this method to be 3.8 radicals/100 eV at -175°C [39].

There are two techniques for estimating radical lifetime from

ESR work [39]. One is similar to the rotating sector technique of photochemistry and involves the use of an intermittently pulsed beam of electrons from a Van de Graaf accelerator. Radical lifetime is estimated from the manner in which radical concentration changes as the pulsing frequency is varied. The second method is to estimate radical lifetimes by a "sampling technique," so-called because the ESR signal is observed only at a given delay time after each pulse of irradiation. The entire radical decay curve is obtained by employing several delay times. By this technique the lifetime of the ethyl radical in liquid ethane at -177°C was estimated to be 7.3 msec [39].

Free radicals can also be detected in organic liquids by optical absorption spectroscopy in pulse radiolysis experiments [27]. The pulses of microsecond duration must be of high intensity in order to result in an adequately high concentration of radicals to allow their detection by optical absorption. The association of an optical absorption spectrum with a particular radical is not as straightforward as the identification of a radical by ESR. Often additional facts such as the lifetime of the radical are needed in order to assign a spectrum. The absorption spectra of benzyl, ketyl, and phenyl radicals have been determined previously in flash photolysis investigations [40]. Aliphatic radicals are not as easily identified as aromatic radicals, because less is known about their absorption spectra.

Another method of detecting radicals is through their emission spectrum in cases where they are formed in excited states and then lose energy by emission of light. The method can only be used for radicals such as the benzyl radical [41] for which the emission spectrum is known. If the detection equipment is sufficiently sensitive, detection of radicals by emission may be accomplished during steady-state irradiation as well as in pulse radiolysis.

Chemical methods of free radical detection include titration and sampling techniques. In the titration method, a selected

substance is added which scavenges (titrates) the free radicals
present. The sampling method consists in the use of a radioactive
radical such as $^{14}CH_3$ to combine with a small fraction of the free
radicals present in the reaction mixture. Either technique may be
employed to identify and measure individual radical yields or to
measure the total radical yields.

Iodine has been widely employed to scavenge (titrate) radicals
in the radiolysis of hydrocarbons. Iodine is useful for this
purpose because it reacts very quickly with radicals,

$$R\cdot + I_2 \rightarrow RI + I \qquad\qquad (4\text{-}15)$$

and the yields of the resulting alkyl iodides can be measured to
provide a determination of the individual radical yields. The
equation employed is that total radical (TR) yield equals one-half
the yield of iodine molecules removed:

$$G(TR) = G(-\tfrac{1}{2}I_2) \qquad\qquad (4\text{-}16)$$

The concentration of iodine should not exceed 10^{-3} M, since at
higher concentration there are complicating reactions such as
electron capture [42]. One point to note is that the alkyl iodide
formed in Eq. (4-15) may be unstable. For example, tert-alkyl
iodides have not been satisfactorily detected by this method.
Hence, the technique is limited to those hydrocarbons in which such
labile iodides are not formed. Pulse radiolysis investigations
have provided extensive information on the interaction of iodine
with hydrocarbon radicals [43].

Hydrogen iodide has also been employed as a radical scavenger,
as discussed by Forrestal and Hamill [44] and Schuler [45]. Its
reaction with many radicals

$$R\cdot + HI \rightarrow RH + I\cdot \qquad\qquad (4\text{-}17)$$

is almost as fast as reaction (4-15). Total yield of radicals is
indicated by the amount of iodine formed, and the technique is complicated by the fact that the iodine formed also acts as a scavenger for free radicals. Corrections can easily be applied to take
account of the iodine scavenging action [46, 47].

If either deuterium or tritium iodide is employed, then labeled hydrocarbons are produced in reaction (4-17). Such labeled
hydrocarbons can be used to identify the radicals as well as measure their individual yields.

A variety of other radical scavengers have been employed such
as DPPH, anthracene, nitric oxide, etc. Such substances react less
readily with radicals than iodine or HI. For example, anthracene
is about 100 times less reactive to radicals than iodine [48].

Another method for measuring radical yields is embodied in the
radical sampling technique [49, 50]. In this technique, a radio-
actively labeled radical is generated from solutes such as $^{14}C_2H_4$
or $^{14}CH_3I$, which are present at millimolar concentrations, by the
following reactions:

$$e^- + {}^{14}CH_3I \rightarrow {}^{14}CH_3 + I^- \qquad (4\text{-}18)$$

$$H + {}^{14}C_2H_4 \rightarrow {}^{14}C_2H_5 \qquad (4\text{-}19)$$

The resulting labeled radical is present at $<10^{-7}$ M during radiolysis and combines with a small fraction of the free radicals
present to form labeled hydrocarbons, $R_i{}^{14}CH_3$:

$$^{14}CH_3 + R_i \rightarrow R_i{}^{14}CH_3 \qquad (4\text{-}20)$$

In order for this technique to succeed, conditions must be used
which favor radical-radical reactions. This implies high dose
rates over 10^{18} eV g^{-1} sec^{-1} for saturated hydrocarbons, and even
higher dose rates for olefins, ketones, and other reactive substrates (unless low temperature is used).

The relative yields of the labeled hydrocarbons are proportional to the relative radical yields, so that the ratio of yields of two radical types R_1 and R_2 are given by

$$\frac{G(R_1)}{G(R_2)} = \frac{G(R_1{}^{14}CH_3)}{G(R_2{}^{14}CH_3)} \times \frac{1 + D_1/C_1}{1 + D_2/C_2} \tag{4-21}$$

where the D/C ratios are the disproportionation to combination ratios for the reaction of methyl radicals with each radical, that is, k_{22}/k_{20}:

$$^{14}CH_3 + R_i \rightarrow {}^{14}CH_4 + R_i(-H) \tag{4-22}$$

Although these disproportionation ratios are not known accurately, it can generally be assumed that they are quite small, making the second factor on the right-hand side of Eq. (4-21) approximately equal to 1.

Absolute yields of radicals can be computed from Eq. (4-21) if the absolute yield of any one radical is known. One yield that can be measured absolutely is $G(^{14}CH_3)$, since this is the total yield of labeled hydrocarbons. This method of measuring radical *yields* should not be confused with the ESR technique discussed above for measuring radical *lifetimes*.

III.   RADIATION CHEMISTRY OF ALIPHATIC COMPOUNDS

A.   Saturated Hydrocarbons

The radiation chemistry of the simplest hydrocarbon, methane, has been extensively studied. Among the major products are hydrogen, ethane, ethylene, propane, and butane. There is very little

change in the total number of gas molecules during irradiation,
which explains why very small pressure changes are noted. Another
product is a polymeric liquid-containing unsaturated double bonds,
and this explains the poor material balance of the gaseous products. No effect of LET has been observed in the work on methane
or on other gaseous hydrocarbons. Mass spectrometric and photochemical data have been helpful to an understanding of the intermediates and steps involved in the radiolyses of methane.

In the low pressure mass spectrum of methane, the relative
abundances of ions formed include CH_4^+ (46%), CH_3^+ (40%), and CH_2^+
(7.5%), which together make up about 93% of the total ionization
at low pressures and electron energies in the range 50-70 eV [51].
These ions react with methane, the principal reaction of CH_4^+ being

$$CH_4^+ + CH_4 \rightarrow CH_5^+ + CH_3 \tag{4-23}$$

Reaction (4-23) has been investigated by several methods including
pulsing techniques to determine rate constants for the reactions
of thermal CH_4^+ ions. The evidence indicates a value of k_{23} =
1.2×10^{-9} or 7.4×10^{14} cm^3 $mole^{-1}$ sec^{-1} [52, 53]. For CH_4^+ ions
of higher energy, other reactions with CH_4 may take place resulting
in the production of CH_3^+ [54, 55].

The reaction of methane with methyl ion produces mainly ethyl
ion,

$$CH_3^+ + CH_4 \rightarrow C_2H_5^+ + H_2 \tag{4-24}$$

a reaction that was determined very early in the sequence of investigations [56]. The CH_2^+ ion can take part in several reactions
[57], the relative extents being strongly dependent on ion energy,
with lower energies tending to result in a more complex distribution. At a nominal energy of 0.3 eV for CH_2^+, the reaction of CH_2^+
with methane is reported for form 11% $C_2H_2^+$, 25% $C_2H_3^+$, 37% $C_2H_4^+$,
and 27% $C_2H_5^+$ [58].

Investigations of the mass spectrum of methane at higher
pressures up to a few Torr indicate that CH_5^+ and $C_2H_5^+$ do not react with methane to form new products [59, 60] Their contribution
to total ionization is approximately that expected from the relative
intensities of the precursor ion CH_4^+ and CH_3^+, as noted in the low
pressure mass spectrum. Other observations in the high pressure
mass spectrum are the disappearance of $C_2H_3^+$ and the appearance of
$C_3H_5^+$. The variation of ion yields with increasing pressure shows
that ion-molecule reactions are favored by increasing pressure, and
that as the pressure is increased the secondary ions react further
by successive ion-molecule reactions giving C_3 and higher ions. At
pressures of the order of 200 Torr, the $C_3H_7^+$ ion becomes important.

The net effect of the photochemical and mass spectrometric
evidence is to indicate that about a dozen principal intermediates
and a great number of minor ones play some role in product formation during methane radiolysis. Neglecting for a moment the higher
molecular weight products, the principal ionic reactions that occur
during methane radiolysis can be summarized as follows:

$$CH_4 \rightarrow CH_4^+ + e^- \tag{4-25}$$

$$CH_4 \rightarrow CH_3^+ + H\cdot + e^- \tag{4-26}$$

$$CH_4 \rightarrow CH_2^+ + H_2 \text{ (or 2 H·)} + e^- \tag{4-27}$$

$$CH_4^+ + CH_4 \rightarrow CH_5^+ + CH_3\cdot \tag{4-28}$$

$$CH_3^+ + CH_4 \rightarrow C_2H_5^+ + H_2 \tag{4-29}$$

$$CH_2^+ + e^- \rightarrow CH_2 \tag{4-30}$$

$$CH_5^+ + e^- \rightarrow CH_3\cdot + H_2 \tag{4-31}$$

$$C_2H_5^+ + e^- \rightarrow C_2H_5\cdot \tag{4-32}$$

$$C_2H_5^+ + e^- \rightarrow C_2H_4 + H\cdot \qquad\qquad\qquad (4\text{-}33)$$

The free radicals formed then take part in further reactions such
as

$$H\cdot + CH_4 \rightarrow CH_3\cdot + H_2 \qquad\qquad\qquad (4\text{-}34)$$

$$2\ CH_3\cdot \rightarrow C_2H_6 \qquad\qquad\qquad (4\text{-}35)$$

$$2\ C_2H_5\cdot \rightarrow C_4H_{10} \qquad\qquad\qquad (4\text{-}36)$$

or

$$2\ C_2H_5\cdot \rightarrow C_2H_6 + C_2H_4 \qquad\qquad\qquad (4\text{-}37)$$

$$CH_3\cdot + C_2H_5\cdot \rightarrow C_3H_8 \qquad\qquad\qquad (4\text{-}38)$$

$$CH_3\cdot + C_2H_5\cdot \rightarrow CH_4 + C_2H_4 \qquad\qquad\qquad (4\text{-}39)$$

$$2\ CH_2 \rightarrow C_2H_4 \qquad\qquad\qquad (4\text{-}40)$$

$$CH_2\cdot + CH_4 \rightarrow C_2H_6 \qquad\qquad\qquad (4\text{-}41)$$

$$R\cdot + C_2H_4 \rightarrow R\text{--}CH_2\text{--}CH_2\cdot \xrightarrow{C_2H_4} \text{polymer} \qquad (4\text{-}42)$$

where R· represents a radical or a hydrogen atom. In some of these
reactions, third bodies (not shown) may be needed to carry off excess energy. Higher hydrocarbons may be formed by radical-radical
combinations as indicated, or by additional ion-molecule reactions
such as

$$C_2H_5^+ + CH_4 \rightarrow C_3H_7^+ + H_2 \qquad\qquad\qquad (4\text{-}43)$$

Much use has been made of free radical scavengers to indicate

the presence of free radicals in methane during radiolysis. Gevantman and Williams [61] and Meisels et al. [62] employed iodine
to detect and measure radicals produced in pure methane or in
methane/noble gas mixtures. The main radical products indicated
by the appearance of the corresponding iodide were H$\cdot$, CH$_3\cdot$, C$_2$H$_5\cdot$,
and CH$_2$. An increased yield of ethylene in the presence of scavengers was observed, possibly due to the removal of radicals, which
would otherwise initiate the polymerization of ethylene as in Eq.
(4-42). Some radicals such as hydrogen may be formed with excess
energy (that is, they are "hot") and are more likely to react with
methane than with iodine. In such cases, the yields of HI and
CH$_3$I will not give an accurate indication of the number of hydrogen
atoms and methyl radicals produced.

In spite of such questions, there is small doubt that methyl
radicals play an important role in methane decomposition by radiation. They may be produced in several ways, including dissociation of excited CH$_4$ into hydrogen atoms and CH$_3$ and dissociation of
excited ethane. Reactions (4-28) and (4-34) are also possible
sources of methyl radicals. One problem in determining the total
yield of methyl radicals is that every additive used in determining
its total yield will also interfere with some of the steps leading
to its formation. Meisels listed a yield of 3.3 ± 0.2 CH$_3$ radicals
per 100 eV for the propylene-scavenged radiolysis [63].

Based on further computations, Meisels concluded that direct
dissociation of excited methane produces methyl radicals with a
yield of 1.4 dissociations per 100 eV. Other investigators insist
that all products of methane radiolysis can be explained without
postulating the dissociation of excited methane. The existence of
such widely different viewpoints is one indication of the complexity of the methane radiolysis mechanism.

Liquid products which result when methane is subjected to very
large doses of radiation have been studied by Hummel [64], and it
has been observed that the liquid contains saturated hydrocarbons

with chain lengths ranging from C_6 to C_{27}. It has been postulated
that such hydrocarbons are formed by a series of ion-molecule
reactions,

$$CH_3^+ + CH_4 \rightarrow C_2H_5^+ + H_2 \qquad\qquad (4\text{-}44)$$

$$C_2H_5^+ + CH_4 \rightarrow C_3H_7^+ + H_2 \text{ , etc.} \qquad\qquad (4\text{-}45)$$

Saturated hydrocarbons are also produced by polymerization as in
Eq. (4-42).

The effect of radiation on higher hydrocarbons is similar to
the effect on methane to a considerable extent. Again it is observed that hydrogen is an important product, and perhaps the
principal product. Dimer is formed, analogous to formation of
ethane from methane. As for methane, irradiation of lower molecular weight hydrocarbons in the gas phase results in little change
in total pressure.

To approximate terms, all C—C bonds are about equally likely
to break during irradiation. Similarly, all C—H bonds have about
the same probability of breaking. Evidence for these statements
is the fact that methane becomes a relatively less abundant product as the proportion of terminal methyl groups decreases with
increasing chain length, whereas the hydrogen yield shows much
less change corresponding to the rather constant proportion of
C—H bands.

Irradiation of hydrocarbons also forms saturated hydrocarbons
of molecular weight having the same number or fewer carbon atoms
than the parent molecule, higher molecular weight hydrocarbons with
the number of carbons intermediate between the parent and its
dimer, dimeric products, and unsaturated hydrocarbons exhibiting a
range of molecular weights. The G(-alkane) value usually is in the
6-10 range. As the molecular weight of irradiated straight-chain
alkanes increases, low molecular weight products become less

important, and irradiation causes much cross-linking, giving products of higher molecular weights than the parent, with a corresponding increase in melting point and changes in physical properties. For liquid or solid hydrocarbons, extensive radiation links
the molecules, forming an infusible and insoluble gel which extends throughout the hydrocarbon. This explains the property
changes observed when polyethylene is irradiated and accounts for
the early discovery that paraffin becomes infusible when irradiated
and accounts for the early discovery that paraffin becomes infusible when irradiated for a sufficient length of time.

When dodecane is irradiated, the melting point drops below the
original -9°C until a total dose of about 5.5×10^{22} eV/g has been
reached, at which dose the melting point is about -26°C, and then
rises rapidly to a melting point of over 200°C after a total dosage
of approximately 7.0×10^{22} eV/g. Neglecting processes other than
dimerization (cross-linking), the gel point is reached when there
is an average of half a cross-link per molecule originally present.
For straight-chain hydrocarbons, the absorbed dose required to
produce a gel is inversely proportional to molecular weight in the
molecular weight range from 100 to 500. Approximately the same
relation is reportedly true for polyethylenes of molecular weight
of the order of 10^5. This suggests that the G value for the formation of cross-links is approximately independent of molecular
weight. It is generally more convenient to study the cross-linking
reaction in polymers than in low molecular weight materials.

Differences exist in the rate of irradiation breakage of different types of C—H bonds. Studies of product distributions and
the fact that more secondary than primary parent iodides are formed
in the presence of iodine, indicate that secondary C—H bonds are
more easily broken than primary. This is in line with data on
relative bond strengths and with pertinent mass spectroscopic
evidence. Tertiary C—C bonds are more likely to break than primary
or secondary C—C bonds, as shown by the unusually high yields of
methane observed in the radiolysis of neopentane and 2,2-dimethyl-

butane. Examination of the products from n-hexane shows a tendency for C—C bonds away from the chain ends to break, and this tendency has also been observed in the mass spectrometer. In general, the preference for radiation C—C bond breakage appears to be what one would expect from the well-known relative strengths of the three types of bonds, namely primary > secondary > tertiary. The tendency for tertiary C—C bonds to break is in line with the results from mass spectroscopy, where the proportion of parent ions decreases in the order n-alkane > iso-alkane >> neo-alkane. For example, n-octane yields 1.8% parent ions, 2-methylheptane yields 1.19%, and 2,2-dimethylhexane yields 0.012%.

This picture is supported by evidence from ESR spectroscopy [65]. For frozen alkanes irradiated at low temperature, straight-chain alkanes give ESR spectra corresponding to the loss of a hydrogen atom, with small indication of C—C bonds breaking. Branched chain alkanes under the same conditions give spectra indicative of the breaking of C—C bonds.

Bond breakage and methyl yield during irradiation were studied for 18 branched hydrocarbons by Schuler and Kuntz [66]. Fragment alkyl radicals made up an increasing percentage of the total radical yield as the number of side chains increased. Schuler developed an empirical relationship for computing methyl radical yields for any alkane. A similar effect of structure was found to apply to fragment radicals other than methyl in various branched alkanes. The effect can be summarized by saying that the C—C bond most likely to break is the one with the most C—C bonds adjacent to it.

No satisfactory theory is available to explain such structural effects on radical yields. However, it was noted in the photolysis of liquid 2,2,4-trimethylpentane at 1470 Å that the ratio of quantum yields of fragment radicals is quite similar to the ratio of fragment radical yields found in the radiolysis of the same liquid [67]. Such similarity indicates that neutral excited molecules are possibly the precursors of fragment radicals in alkanes.

The type of radiation does not appear to be a significant variable in determining the distribution of products from irradiated hydrocarbons. In comparing liquid-phase experiments, irradiation with 2 MeV electrons, 14 MeV deuterons, and 35 MeV α-particles all resulted in the same yield of hydrogen from cyclohexane, G = 5.25. These results over a range of LET values indicate that combination of hydrogen atoms in tracks is a relatively insignificant source of hydrogen. Also, product yields are often independent of dose rate over wide ranges. In this connection, the application of an electric field increases the decomposition of ethane by α-rays, possibly because ions are neutralized at the electrodes yielding reaction products, instead of recombining to form ethane.

Irradiation of a hydrocarbon mixed with a small amount of some other substance often provides information of interest. When the hydrocarbon is the major component, direct radiation action on the minor component is usually of little significance. In fact, if the minor constituent is involved in the reaction, it is probably because it is being attacked by free radicals or other active species formed from the hydrocarbon or because energy is transferred to it from the hydrocarbon. An important example involves attack of the minor constituent by radicals formed from the hydrocarbon, as in the use of a minor amount of iodine or a radical scavenger. Another important case is the irradiation of hydrocarbon solutions of sulfur dioxide, where the SO_2 functions as a bivalent radical scavenger.

Another important example of a mixture is hydrocarbon/oxygen. When such a mixture is irradiated, free radical reactions play an important role in the oxidation mechanism. The products include carbonyl compounds, alcohols, peroxides, and acids. n-Heptane yields methylbutyl ketone and other carbonyl compounds, alcohols, peroxides, and acids. The reaction is not a chain reaction, and the products differ from those of the autoxidation. The possible explanation is that at high dose rates the RO_2 radicals do not abstract hydrogen from other molecules to yield hydroperoxides and chain-propagating radicals, but rather react with each other. Above 100°C the reaction becomes a free radical chain reaction similar to that normally observed in the autoxidation of hydrocarbons.

Early studies on the α-ray-induced oxidation of methane showed carbon dioxide and water are produced. The addition of dimethylselenium accelerates the oxidation of irradiated methane. When oxidation of methane is induced by fast electrons, the products include carbon monoxide, hydrogen, and formic acid.

Free radicals formed in the irradiation of hydrocarbons often result in free radical chain reactions, with G values reaching 10^5 to 10^6. Radiation-induced reactions of hydrocarbons include chlorination, sulfochlorination, bromination, and isomerization. Although such reactions differ in mode of initiation, the propagation steps responsible for product distribution are the same for all types of initiation. Large G values are important in commercial reactions because they imply large product yields for small radiation energy input.

A useful reaction of the hydrocarbon/additive type is represented by the irradiation of hydrocarbon/tritium mixtures to form tritium-labeled compounds. This reaction has not been explained in terms of any reasonable free radical mechanism, because of the high bond strength of the H—H bond. It has been postulated that the tritium is attacked by ions produced from the hydrocarbon.

Hexane is one of the hydrocarbons that has undergone an extensive investigation of its radiolysis. The major products are H and C_{12} compounds, plus minor amounts of various hydrocarbons ranging from methane to C_{11} in chain length. The product yields are little affected by the LET of the radiation, with similar relative yields being given by electrons, deuterons, and helium ions [68].

Free radical scavengers reduce the hydrogen yield from hexane vapor by 60% and the hydrogen yield from liquid hexane by 40%, the residue of the hydrogen presumably being produced by "molecular" processes [69]. Iodine was used to scavenge and identify the radicals formed in irradiated hexane [69]. The radicals identified and G values of the corresponding iodide were as follows: methyl (0.10); ethyl (0.40); propyl (0.70); butyl (0.30); hexyl (0.70); and 1-methyl pentyl (2.60).

Futrell [70] has developed a mechanism for hexane radiolysis involving several ion-molecule reactions of the hydride-ion-transfer type. Thirteen primary ions, identified by mass spectrometer from CH_3^+ to $C_5H_{11}^+$, are postulated to be able to take part in the reaction

$$R^+ + C_6H_{14} \rightarrow C_6H_{13}^+ + RH \tag{4-46}$$

In some cases the neutral product keeps enough energy to dissociate as follows:

$$C_2H_3^+ + C_6H_{14} \rightarrow C_2H_4^* + C_6H_{13}^+ \tag{4-47}$$

$$C_2H_4^* \rightarrow C_2H_2 + H_2 \tag{4-48}$$

It is further postulated that essentially all the primary ions are converted by such reactions to $C_6H_{13}^+$, which then undergoes the sequence

$$C_6H_{13}^+ + e^- \rightarrow C_6H_{13}^{**} \rightarrow C_6H_{12} + H\cdot \tag{4-49}$$

Ethyl, propyl, and hexyl radicals are formed in appreciable amounts, and by further reactions such as combination and disproportionation they can form many of the observed products. By employing the mass spectral pattern to indicate the initial relative abundance of ions and by postulating reasonable ion-formation, ion-molecule, and interradical reactions, G values were computed for the radiolysis products formed from hexane vapor. The calculated G values closely approximated the observed G values.

Hardwick [71] adopted an entirely different view of the radiolysis of hexane, emphasizing free radical reactions and postulating that ion-molecule reactions are relatively insignificant in the
radiolysis of liquid hexane. Hardwick suggested the formation of
radicals by the dissociation of primary and secondary excited
molecules as follows.

$$C_6H_{14} \rightarrow \cdot C_6H_{13} + H \cdot \qquad\qquad (4\text{-}50)$$

$$C_6H_{14} \rightarrow \cdot C_5H_{11} + CH_3 \cdot \qquad\qquad (4\text{-}51)$$

$$C_6H_{14} \rightarrow \cdot C_4H_9 + C_2H_5 \cdot \qquad\qquad (4\text{-}52)$$

$$C_6H_{14} \rightarrow 2C_3H_7 \cdot \qquad\qquad (4\text{-}53)$$

A "molecular" dissociation was also postulated.

$$C_6H_{14} \rightarrow H_2 + \text{hexene or other product} \qquad (4\text{-}54)$$

Hydrogen atoms formed in reaction (4-50) will possess a range of
energies. Those formed as hot atoms will have very high energy
and be capable of reacting at once with an adjacent molecule. In
the latter case the reactions,

$$C_6H_{14} \rightarrow \cdot C_6H_{13} + \underline{H} \cdot \qquad\qquad (4\text{-}55)$$

and

$$\underline{H} \cdot + C_6H_{14} \rightarrow \cdot C_6H_{13} + H_2 \qquad\qquad (4\text{-}56)$$

where $\underline{H} \cdot$ indicates a hot hydrogen atom, will take place among adjacent molecules enclosed in the solvent cage. The organic radi-
cals produced will react together under these conditions by
disproportionation or combination:

$$2 \; \cdot C_6 H_{13} \rightarrow C_6 H_{14} + C_6 H_{12} \tag{4-57}$$

$$2 \; \cdot C_6 H_{13} \rightarrow C_{12} H_{26} \tag{4-58}$$

Isomeric hexenes and dodecanes can result from the various isomeric structures of the hexyl radical. Such reactions occurring in a solvent cage will not be modified by moderate concentrations of a free radical scavenger, and the products would therefore be considered "molecular" products. This conclusion holds true for the reaction sequence (4-50) plus (4-56), and also for reactions (4-51) through (4-53) providing the radicals formed react with each other before escaping from the cage. After the radicals have left the solvent cage they may react with each other or with the solvent, or react with an added radical scavenger if one is present. Hydrogen atoms of thermal energy may abstract hydrogen from the solvent or add to an unsaturated product.

The radiolysis of hexane in the presence of osyben produces carbon monoxide, carbon dioxide, perocides, and water. There is some indication hexanone-2 and hexanone-3 may also be formed. Oxygen functions as a radical scavenger and essentially eliminates C_7 products and above. Above 235°C, a thermal chain decomposition becomes prominent, masking the radiation-induced reaction.

Radiolysis studies of cyclohexane are of special interest because all the hydrogen atoms are equivalent and loss of any one of them yields the same product, thus possibly simplifying the reaction mechanism. The principal products from the radiolysis of liquid cyclohexane include hydrogen, cyclohexene, and bicyclohexyl. Various mechanisms for the radiolysis have been advanced, but the following rather simple one accounts for the main products [72]:

$$c\text{-}C_6 H_{12} \rightarrow c\text{-}C_6 H_{11} \cdot + H \cdot \text{ (about 85\%)} \tag{4-59}$$

$$c\text{-}C_6 H_{12} \rightarrow c\text{-}C_6 H_{10} + H_2 \text{ (about 15\%)} \tag{4-60}$$

$$H\cdot + c\text{-}C_6H_{12} \rightarrow c\text{-}C_6H_{11}\cdot + H_2 \tag{4-61}$$

$$2\ c\text{-}C_6H_{11} \rightarrow C_{12}H_{22} \tag{4-62}$$

$$2\ c\text{-}C_6H_{11} \rightarrow c\text{-}C_6H_{12} + c\text{-}C_6H_{10} \tag{4-63}$$

High dose rates result in polymeric products (C_{18}, C_{24}, and higher), possibly by reaction of cyclohexyl radicals with cyclohexene. The cyclohexyl radicals have been identified and studied by ESR [36], absorption spectroscopy, and chemical techniques. The yield of cyclohexyl radicals as determined by iodine scavenger is about 4 at 10^{-4} M iodine [73]. Such a large yield of cyclohexyl in the radiolysis contrasts with a very small yield in the photolysis of cyclohexane, thus casting doubt on the formation of cyclohexyl by the decomposition of neutral excited cyclohexane [Eq. (4-59)].

The hydrogen yield does not depend on the LET of the radiation [74] nor on the temperature [75], The yield of hydrogen and cyclo-hexene is decreased by scavengers but not reduced to zero, indicat-ing that these products are formed by "molecular" as well as "radical" processes. Iodine at less than 10^{-3} M acts as a scav-enger, but at higher concentrations can have other effects. More organic iodide is formed than can be explained by its action as a radical scavenger, and three times as much hydrogen iodide is formed as corresponds to the reduction in hydrogen yield. The transfer of excitation energy may be involved in producing these effects.

Benzene as an additive in cyclohexane causes a pronounced decrease in the hydrogen yield, possibly because of the transfer of the transfer of excitation energy from cyclohexane to benzene.

The irradiation of cyclohexane-containing oxygen forms cyclo-hexanol, cyclohexanone, cyclohexene, cyclohexyl hydroperoxide [76], a peroxide, and water. The reaction is not a chain reaction and products differ from those formed in autoxidation, possibly because at high dose rates the RO_2 radicals do not propagate chains but instead react with each other. At very high temperatures, oxidation takes place at such a high rate that radiation produces only a slight effect on the reaction.

B. Olefins

When olefins are irradiated, the yield of hydrogen is less than from the corresponding saturated hydrocarbons. Other major differences include a greater overall yield and a higher yield of polymeric material. Some of these results can be explained by assuming that hydrogen atoms and other radicals formed will add to C—C double bonds, in line with the fact that olefins are known to be fairly efficient radical scavengers. The effect of such an addition is to initiate a chain polymerization reaction, although generally one of very short kinetic length.

The effect of α-rays and β-rays on ethylene is qualitatively similar, both giving polymeric products plus gas. An analysis of the polymer from 1-hexene with doses of about 10^{21} eV/g revealed the presence of dimer (G = 0.98), trimer (G = 0.76), tetramer (G = 0.22), and pentamer (G = 0.35). The gas formed contains hydrogen, methane, ethane, and acetylene.

The γ-irradiation of ethylene was studied very extensively from 1950-1956, with the idea of making polyethylene commercially. At room temperature and atmospheric pressure a solid polymer was formed with G(-ethylene) of about 45. Increasing the pressure to 21 atm increased G for ethylene loss to about 2500 and gave a solid, waxy polymer. At 21 atm and 237°C, a more liquid polymer was formed with G(-ethylene) of about 12,000. The reaction appears to be a free radical chain reaction.

The free radicals in irradiated liquid ethylene were identified by ESR and sampling techniques [36, 77] as ethyl, vinyl, and

3-butenyl. Interestingly, vinyl radicals were found below -160°C
but not at -130°C. Rather, 3-butenyl radicals were found at
approximately -130°C. Apparently, this effect results from a fast
addition of vinyl radicals to ethylene,

$$C_2H_3 + C_2H_4 \rightarrow 3\text{-}C_4H_7 \tag{4-64}$$

Based on its rate change with temperature, reaction (4-64) had an
energy of activation of only 3.4 kcal/mole, an amount well below
that for adding alkyl radicals to ethylene. The great reactivity
of the vinyl radical may be due to its unusual sp^2 structure [36].

Addition of other substances to ethylene sometimes affects the
rate of polymerization, although the reason for such effects is
often not known. The rate of ethylene polymerization is increased
when it is irradiated in methanol, n-heptane, cyclohexane, or
acetone solution. In such solutions, the yield ranges up to
G = 2000 molecules polymerized per 100 eV. In CCl_4 the principal
products are low molecular weight telomers such as $CCl_3(C_2H_4)_2Cl$.
Metal oxides also enhance the polymerization reaction, probably
because of a higher rate of initiation on the metal oxide surface.

The irradiation of ethylene in aqueous solution forms aldehydes as well as polymer. Acetaldehyde, glycolaldehyde, and hydrogen peroxide are each produced with a G value of about 2.5.
Minor products include formaldehyde and a hydroperoxide. At high
pressures of ethylene and oxygen the G value for acetaldehyde is
reported to reach 200.

The irradiation of other alkenes has been studied, and the
results are similar to those for ethylene. For example, irradiation of isobutylene above 0°C causes polymerization in a reactopm
apparently involving free radicals. However, liquid isobutylene
polymerizes in better yield at -80°C, at which temperature free
radical propagation is relatively unimportant.

The polymerization at -80°C is proportional to dose rate,

inhibited by oxygen and benzoquinone, and exhibits a G value for monomer conversion of about 820. Since the ionic polymerization of polybutylene is known to take place rapidly at low temperature, the radiation-induced polymerization mechanism is probably ionic at -80°C.

The allylic radical is an important free radical in irradiated alkenes other than ethylene, as shown by the radical sampling technique [78] and by ESR studies of liquids [37] and solids [79]. Investigations have shown that in propylene about 45% of the radicals are allyl, whereas in the 2-butenes and in cyclohexene about two-thirds of the radicals (based on G values for radical yields) are allylic [78, 80]. The 1-methallyl radicals formed in the radiolysis of *cis*- and *trans*-2-butene have essentially the same configuration as the parent olefin, as illustrated in the structure of the following radical from trans-2-butene:

$$\begin{array}{ccc} CH_3 & CH & \\ \diagdown & \diagup \diagdown & \\ & CH & CH_2 \end{array}$$

Another radical of considerable significance that is formed in alkenes is the H atom adduct. For example, cyclohexyl radicals are produced in cyclohexene, butyl radicals in butene, and propyl radicals in propylene [78]. Such radical formation cannot result entirely from the addition of thermal hydrogen atoms to alkenes, as shown by the large yields of normal (straight chain) radicals that are produced in the radiation-induced decomposition of 1-alkenes [78]. As an illustration, the $G(iso\text{-}C_3H_7)/G(n\text{-}C_3H_7)$ is only 2.7 in irradiated propylene, whereas the thermal hydrogen addition to liquid propylene is known to yield an isopropyl/n-propyl radical ratio of 10:1 [49]. An additional piece of evidence is the generally low yield of thermal hydrogens in radiolyzed alkenes: In cis-2-butene $G(H) \lesssim 0.4$ atoms/100 eV [78].

This line of reasoning indicates that H atom addition is not adequate to explain the production of alkyl radicals in irradiated

alkenes.   The following reactions have been suggested for isobutene
[81]

$$C_4H_8^+ + C_4H_8 \rightarrow C_4H_9^+ + C_4H_7 \qquad\qquad (4\text{-}65)$$

$$C_4H_9^+ + e^- \rightarrow C_4H_9^* \qquad\qquad (4\text{-}66)$$

and similar sequences have been proposed for other alkenes [82, 83].
The excited butyl radicals produced in (4-66) may be deactivated
by collision, but some would be expected to undergo decomposition,

$$C_4H_9^* \rightarrow C_2H_4 + C_2H_5\cdot \qquad\qquad (4\text{-}67)$$

In line with this hypothesis, the yield of alkyl radicals is less
than that of allyl radicals for most alkenes.

Radiation can be used to induce a number of the usual free
radical type reactions of alkenes, such as *cis-trans*-isomerization.
cis-trans-Isomerization of polybutadiene in benzene solution takes
place on irradiation, and the same sensitizers are effective in
both the photo-induced and radiation-induced reactions.   The yield
may exceed G = 1000 when diphenyl disulfide is used as the
sensitizer.

Free radical additions can be induced by x- or γ-rays, such
as the addition of *n*-butyl mercaptan to 1-pentene:

$$C_4H_9SH + C_3H_7CH{=}CH_2 \rightarrow C_3H_7CH_2CH_2SC_4H_9 \qquad\qquad (4\text{-}68)$$

The addition of silicon hydrides to alkenes may also be radiation
induced.   Addition of sulfur dioxide to alkenes (co-polymerization)
and the bromination of cyclohexene with N-bromosuccinimide may be
radiation induced.   Radiation-induced addition reactions generally
resemble those caused by other modes of initiation, although exceptions have been noted.   For example, γ-rays will induce the
addition of bromotrichloromethane to styrene, but ultraviolet light
reportedly will not.

Yield of acetylene and its derivatives which disappear during irradiation is even higher than the yields for alkenes. While methylacetylene and other derivatives have been investigated, most of the research in this field has been done on acetylene. The principal product from acetylene irradiation is a yellow polymer known as cuprene (80-85%) and a lesser amount of benzene (15-20%). The polymer is insoluble in all solvents tested, and readily absorbs up to about 25% of its weight of oxygen on standing in air. Any gaseous product apparently results from the decomposition of the primary products. The yield for loss of acetylene is about $G = 75$ for α-rays and a little higher for electrons. Use of deuteroacetylene in place of acetylene gives about the same polymerization rate.

An early mechanism for acetylene radiolysis assumed that each acetylene ion formed was quickly surrounded by a cluster of about 20 acetylene molecules, and was known as the "cluster theory." The cluster theory attempted to rationalize the fact that the G value for acetylene disappearance is independent of dose rate, pressure, and partial pressure of added inert gases. The polymerization is also unusual in being very reproducible and not easily inhibited and very little affected by a change in temperature. By contrast, the photochemical or mercury-sensitized photochemical polymerization of acetylene is temperature dependent. An attempt to explain the facts by postulating a free radical or chain mechanism has been only partially successful.

When inert gases such as nitrogen or oxygen are added to α-ray-irradiated acetylene, the rate of polymerization is always proportional to the total number of ions formed from the inert gas and acetylene together. Such inert gases have higher ionization potentials than acetylene, so that transfer of positive charge to acetylene is probable. Benzene, with an ionization potential less than that of acetylene, is found to retard the polymerization of irradiated acetylene. Carbon dioxide and hydrogen function in a fashion similar to that of the inert gases. The addition of oxygen, however, results in the formation of a different polymer, $(C_2H_3)_n$, plus some carbon monoxide and carbon dioxide.

C. Halides

With the exception of the fluorides, the radiolysis of the aliphatic halides is unusual in that the carbon-halide bond tends to break on irradiation, since it is weaker than either the C—C or C—H bond. The reactivity of the halide atoms, and corresponding tendency to abstract hydrogen, decreases in the order F· > Cl· > Br· > I . The chloride atoms readily abstract hydrogen from organic molecules, yielding hydrogen chloride. The similar reaction for iodine would be endothermic, so iodine atoms react with each other or with other radicals. Bromine atoms are intermediate between chlorine and fluorine inactivity, so that radiolysis of bromo compounds may give both bromine and hydrogen bromide at times. The radiation chemistry of the aliphatic halides is of evident interest in connection with radiation effects on polyvinyl chloride and polytetrafluoroethylene.

The radiolysis of chloroform forms hydrogen chloride, hexachloroethane, and other products. The following mechanism suggested by Ottolenghi and Stein [84] accounts for the observed products in their observed proportions:

$$CHCl_3 \rightarrow \cdot CHCl_2 + Cl\cdot \tag{4-69}$$

$$\cdot CHCl_2 + CHCl_3 \rightarrow \cdot CCl_3 + CH_2Cl_2 \tag{4-70}$$

$$\rightarrow CHCl_2 - CHCl_2 + Cl\cdot \tag{4-71}$$

$$Cl\cdot + CHCl_3 \rightarrow \cdot CCl_3 + HCl \tag{4-72}$$

$$\cdot CHCl_2 + \cdot CCl_3 \rightarrow CHCl_2 - CCl_3 \tag{4-73}$$

or

$$\rightarrow Cl_2C=CCl_2 + HCl \tag{4-74}$$

$$2 \cdot CHCl_2 \rightarrow CHCl_2-CHCl_2 \tag{4-75}$$

$$2 \cdot CCl_3 \rightarrow CCl_3-CCl_3 \tag{4-76}$$

This is not a chain reaction and does not indicate formation of Cl_2 or CCl_4, which were not found among the products.

Isomerization is one of the more important reactions of alkyl chlorides that is induced by radiolysis. Thus, n-propyl chloride yields the more stable isopropyl chloride with a G value of about 60. As another example, 1,3-dichloropropane forms 1,2-dichloro-propane on radiolysis. Such isomerizations can be explained by chain reactions based on radical rearrangements,

$$CH_3CH_2CH_2Cl \rightarrow CH_3CH_2CH_2 \cdot + Cl \cdot \tag{4-77}$$

$$CH_3CH_2CH_2 \cdot \rightarrow CH_3\overset{\cdot}{C}HCH_3 \tag{4-78}$$

$$CH_3\overset{\cdot}{C}HCH_3 + CH_3CH_2CH_2Cl \rightarrow CH_3CHClCH_3 + CH_3CH_2CH_2 \cdot \tag{4-79}$$

Carbon tetrachloride lacks hydrogen atoms, so that any chlorine atoms formed in radiolysis must yield chlorine molecules. The only other product is hexachloroethane, and the following mechanism [85] accounts for these facts:

$$CCl_4 \rightarrow \cdot CCl_3 + Cl \cdot \tag{4-80}$$

$$\cdot CCl_3 + Cl \cdot \rightarrow CCl_4 \tag{4-81}$$

$$\cdot CCl_3 + Cl_2 \rightarrow CCl_4 + Cl \cdot \tag{4-82}$$

$$2 \cdot CCl_3 \rightarrow C_2Cl_6 \tag{4-83}$$

$$2 \ Cl\cdot \rightarrow Cl_2 \qquad\qquad\qquad\qquad (4\text{-}84)$$

After some chlorine has built up, it will scavenge the $\cdot CCl_3$ radicals, which escape from the spur, and the observed yields will be those of the "molecular" products (C_2Cl_6 and Cl_2). The overall yield of $G(-CCl_4)$ in the liquid phase is 0.8 molecules per 100 eV absorbed.

The radiolysis of alkyl bromides forms the corresponding alkanes and dibromides. For example, n-butyl bromide yields n-butane (G = 3.4) and 1,2-dibromobutane (G = 1.0). The alkane may be formed from hydrogen abstraction by an alkyl radical, whereas the dibromide comes from combination of a bromine atom with a radical formed by loss of a hydrogen from an alkyl bromide. Most investigators find little or no hydrogen and HBr among the products, although some report appreciable HBr. In contrast to chlorides, the bromides appear to undergo little isomerization. Although much of the data can be explained by the ease of carbon-halogen bond scission, the subsequent reactions of the radicals formed are not well understood.

In the radiation chemistry of alkyl iodides, the halogen atoms formed can only react with other atoms or radicals, whereas organic radicals react with iodine, resulting in exchange as the main reaction. Hamill has suggested the mechanism [86]

$$C_2H_5I \rightarrow C_2H_5\cdot + I\cdot \qquad\qquad\qquad (4\text{-}85)$$

$$C_2H_5\cdot + I\cdot \rightarrow C_2H_4 + HI \qquad\qquad\qquad (4\text{-}86)$$

$$C_2H_5\cdot + I\cdot \rightarrow C_2H_5I \qquad\qquad\qquad (4\text{-}87)$$

$$C_2H_5\cdot + I_2 \rightarrow C_2H_5I + I\cdot \qquad\qquad\qquad (4\text{-}88)$$

$$C_2H_5\cdot + HI \rightarrow C_2H_6 + I\cdot \qquad\qquad\qquad (4\text{-}89)$$

$$I\cdot + I\cdot \rightarrow I_2 \qquad\qquad\qquad (4\text{-}90)$$

The exchange yield is a function of iodide type and iodine concentration, and G ranges from 4 to 6 iodine atoms converted to organic forms per 100 eV absorbed. Where isomeric iodides are possible, the main product is the parent iodide, showing that most of the organic radicals do not rearrange before taking part in reaction (4-88).

If reactions (4-88) and (4-89) in combination are responsible for the production of iodine, then the addition of excess iodine would be expected to decrease the iodine yield and increase the exchange yield, and addition of hydrogen iodide should increase the iodine yield and decrease the exchange yield. Such effects have been observed, but the addition of much iodine never decreases iodine formation to less than half the original value.

In alkyl fluoride radiolysis, the C—F bond is strong so the C—C bond usually undergoes scission. For example, radical yields in irradiated C_2F_6 have been studied by ESR, and the radical spectrum observed was that of CF_3 instead of C_2F_5 [87]. Hence, C—C bond cleavage took place rather than C—F bond cleavage in fluoroethane.

The ESR spectrum of CF_3 indicated that its structure was nonplanar, with the electron in an orbital of considerable s-character approaching sp^3. With this structure, CF_3 may react without rearrangement to form CF_3H by hydrogen abstraction, perhaps explaining the lower energy of activation than for the corresponding reaction of CH_3 radicals [88].

The extensive work on the radiolysis of Teflon, perfluoropolyethylene-$(CF_2)_n$, is of particular interest to polymer chemists. Teflon is rather resistant to radiation in the absence of oxygen, but breaks down quickly when oxygen is present. The degradation products [89] include CF_4, C_2F_6, and C_3F_8 in the absence of oxygen and COF_2 when oxygen is present. The suggested mechanism involves the reactions

$$-CF_2-CF_2-CF_2- \rightarrow -CF_2-CF_2\cdot + \cdot CF_2- \qquad (4-91)$$

or

$$-CF_2-CF_2-CF_2- \rightarrow -CF_2-\overset{\cdot}{C}F-CF_2- + F\cdot \qquad (4-92)$$

When oxygen is present,

$$-CF_2-CF_2\cdot + O_2 \rightarrow -CF_2-CF_2OO\cdot \qquad (4-93)$$

$$2 -CF_2-CF_2OO\cdot \rightarrow 2 -CF_2-CF_2O\cdot + O_2 \qquad (4-94)$$

followed by formation of carbonyl fluoride,

$$-CF_2-CF_2O\cdot \rightarrow -CF_2\cdot + COF_2 \qquad (4-95)$$

plus reactions of secondary radicals,

$$-CF_2-\overset{\cdot}{C}F-CF_2- \xrightarrow{\ O_2\ } -CF_2-C(O\cdot)F-CF_2- \qquad (4-96)$$

$$-CF_2-C(O\cdot)F-CF_2- \rightarrow -CF_2\cdot + OFC-CF_2- \qquad (4-97)$$

leading to the rapid breakdown of the entire polymer molecule, after the formation of one free radical center.

D. Alcohols

The usual result of the radiolysis of primary or secondary alcohols is the loss of a hydrogen atom from the α-carbon atom. Another result which may also occur is the loss of the hydroxyl hydroxyl hydrogen atom. The products from these two principal reactions are α-glycols and carbonyl compounds, respectively.

The more important ions in the radiolysis of methanol and their modes of formation have been postulated [90] to be

$$CH_3OH + e^- \rightarrow CH_3OH^+ + 2\ e^- \tag{4-98}$$

$$\rightarrow CH_2OH^+ + H\cdot + 2\ e^- \tag{4-99}$$

$$\rightarrow CHO^+ + H_2 + H\cdot + 2\ e^- \tag{4-100}$$

Suggested ion-molecule reactions are

$$CH_3OH^+ + CH_3OH \rightarrow CH_3OH_2^+ + \cdot CH_2OH\ (or\ CH_3O\cdot) \tag{4-101}$$

$$CH_2OH^+ + CH_3OH \rightarrow CH_3OH_2^+ + HCHO \tag{4-102}$$

$$CHO^+ + CH_3OH \rightarrow CH_3OH_2^+ + CO \tag{4-103}$$

or
$$\rightarrow CH_2OH^+ + HCHO \tag{4-104}$$

or
$$\rightarrow CH_3^+ + CO + H_2O \tag{4-105}$$

$$CH_3^+ + CH_3OH \rightarrow CH_2OH^+ + CH_4 \tag{4-106}$$

Reactions (4-104) and (4-106) are examples of hydride-ion transfer, while the other ion-molecule reactions are proton transfer reactions. Positive ions may be neutralized by the return of the ejected electron,

$$CH_3OH_2^+ + e^- \rightarrow CH_3OH + H\cdot \tag{4-107}$$

or
$$\rightarrow CH_3O\cdot + H_2 \tag{4-108}$$

or
$$\rightarrow \cdot CH_3 + H_2O \tag{4-109}$$

The above scheme accounts for the main products of methanol radiolysis, which are hydrogen, carbon monoxide, methane, formaldehyde,
and glycols.

Scavengers that have been used to study the hydrogen atom
yield in methanol radiolysis include benzoquinone, pentadiene [91],
acetone, and chloroacetic acid [92], and all yield $G(H)$ values of
about 2.4 or 2.5. The reduction of methane yield with added benzoquinone indicates $G(GH_3) = 1.0$ for methanol and $G(CH_3) = 0.44$ for
ethanol. The presence of CH_3O and CH_2OH radicals has been shown by
use of benzene as a scavenger. The H, OH, CH_3, and CH_3O radicals
will abstract H from methanol to form CH_2OH radical, which is the
major radical produced in irradiated methanol and has been detected
in pulse radiolysis studies. In ethanol radiolysis, the 1-hydroxy-
ethyl radical was found in pulse radiolysis studies. The yield of
this radical was estimated to be 9 radicals/100 eV by assuming its
molar extinction coefficient at 297 nm to be the same in ethanol
as in water [93].

E. Ethers

The radiation chemistry of ethers is similar to that of the
alcohols. The main bonds broken are the α–C–H bond and the C–O
bond. While some scission of C–O bond occurs in alcohols, the
more extensive scission in ethers is in line with its lower bond
strength in these compounds.

The principal products from the irradiation of diethyl ether
with helium ions [94] are hydrogen (G = 3.62), ethylene (G = 1.07),
and ethane (G = 0.62), plus lesser amounts of acetylene, carbon
monoxide, methane, alkanes, and alkenes. Other ethers generally
also yield alcohols, carbonyl compounds, and polymeric materials on
irradiation. Both free radical and molecular rearrangement reactions seem to be involved in the mechanism. The distribution of
products can be changed by altering the temperature, adding a free

radical scavenger such as iodine, or using a different type of
radiation.

F. Carboxylic Acid

   The bond most easily broken on irradiation of carboxylic acids
is the one joining the COOH group to the main chain. The main
products resulting from this scission are carbon dioxide, the
saturated hydrocarbon with one carbon less than the original acid,
and the saturated hydrocarbon of twice this size. Water is also
produced in considerable yield, with a G value of the order of 1.5
to 2.0. Acids of high molecular weight resemble hydrocarbons in
radiolysis, with the hydrogen yield increasing with molecular
weight as the CO_2 yield decreases. Above about C_{16} acids, some
reversals in the trend of H_2 and CO_2 yields take place which are
difficult to explain. A minor product from acids is the corresponding ketone, with acetic acid giving acetone in a yield of
about 0.5 to 1.0 molecules per 100 eV absorbed.

   To determine the origin of the methane product, Burr [131]
irradiated deuterated and tritiated acetic acids and from the isotopic composition of the methane and hydrogen formed, and taking
into account the rate of H_2 production, showed that methane is
produced mainly by

$$\cdot CH_3 + CH_3COOH \rightarrow CH_4 + \cdot CH_2COOH \qquad\qquad (-110)$$

and that hydrogen arises by an analogous reaction

$$H\cdot + CH_3COOH \rightarrow H_2 + \cdot CH_2COOH \qquad\qquad (4-111)$$

Burr further concluded that no more than 15% of the methane could
be formed by direct dissociation of an excited molecule,

$$CH_3COOH^* \rightarrow CH_4 + CO_2 \qquad\qquad (4\text{-}112)$$

Unsaturated carboxylic acids undergo the typical decarboxyla-
tion reaction, with oleic acid yielding 8-heptadecene. They also
give the common reactions of unsaturated compounds. For example,
oleic acid undergoes polymerization and also exhibits isomerization
to the trans form. Hydrogenation of the double bond takes place,
with oleic acid yielding some stearic acid. When oxygen is present,
radiation induces oleic acid to oxidize by a chain reaction.

G. Esters

Radiolysis of liquid methyl acetate, a typical ester, produces
fair yields [95] of hydrogen, methane, ethane, carbon monoxide,
carbon dioxide, dimethylether, formaldehyde, methanol, and acetic
acid. Based on known free radical reactions and photochemical
evidence, the primary reactions probably are

$$CH_3COOCH_3 \rightarrow CH_3CO\cdot + \cdot OCH_3 \rightarrow \cdot CH_3 + CO + \cdot OCH_3 \qquad (4\text{-}113)$$

$$CH_3COOCH_3 \rightarrow \cdot CH_3 + CH_3COO\cdot \rightarrow 2 \cdot CH_3 + CO_2 \qquad (4\text{-}114)$$

$$CH_3COOCH_3 \rightarrow H\cdot + \cdot CH_2COOCH_3 \text{ (or } CH_3COOCH_2\cdot) \qquad (4\text{-}115)$$

These may be followed by a number of free radical reactions, probably including

$$\cdot CH_3 + CH_3COOCH_3 \rightarrow CH_4 + \cdot CH_2COOCH_3 \qquad (4\text{-}116)$$

$$H\cdot + CH_3COOCH_3 \rightarrow H_2 + \cdot CH_2COOCH_3 \qquad (4\text{-}117)$$

$$\cdot CH_3 + \cdot OCH_3 \rightarrow CH_3OCH_3 \qquad (4\text{-}118)$$

or

$$\rightarrow CH_4 + HCHO \tag{4-119}$$

$$H\cdot + \cdot OCH_3 \rightarrow H_2 + HCHO \tag{4-120}$$

$$2 \cdot OCH_3 \rightarrow CH_3OH + HCHO \tag{4-121}$$

$$2 \cdot CH_3 \rightarrow C_2H_6 \tag{4-122}$$

with the above being only a partial list of all possible reactions.

Scavengers have been used to show that methane is formed almost solely by free radical reactions. The results indicate that hydrogen is formed in part by unimolecular dissociation, or possibly through hydrogen abstraction by hot hydrogen atoms (which are not scavenged by iodine). Dimethyl ether may be formed by the caging and recombination of radicals produced in reaction (4-113).

H. Carbonyl Compounds

The bond next to the carbonyl group is very susceptible to scission by radiation. Irradiation breaks down propionaldehyde to give carbon monoxide, hydrogen, and ethane; acetone to give carbon monoxide and methane; and diethyl ketone to give carbon monoxide and ethane. Other important products from ketones include hydrogen and polymeric materials.

The methyl radical is a major intermediate in the radiolysis of acetone. The effect of scavengers on the methane yield [96] indicate that most of the methane is produced by reactions of thermal methyl radicals. Early studies gave high values for the yield of methane, apparently due to traces of moisture. Later work with dry aceton gave a $G(CH_4)$ value of 1.7. Iodine scavenger reduced the $G(CH_4)$ to 0.4, indicating a G value of 1.3 for methyl radicals [97]. The presence of methyl indicates that acetyl

radical is also formed. Further evidence for acetyl formation is
the production of acetaldehyde (G = 0.09), biacetyl (G = 0.56) [97],
and acetylacetone [98].

It is not likely that thermal hydrogen atoms are important in
acetone radiolysis. In the event of their formation, they would
be expected to abstract H from acetone forming H_2. However, since
$G(H_2)$ is only 0.44 [97], it is estimated that $G(H) \lesssim 0.6$.

The fact that acetonylacetone is an important liquid product
[97, 98] shows that acetonyl radical is probably present. Acetonylacetone and methylethyl ketone are possibly formed by the combination of acetonyl with other radicals. Acetonyl radical is probably
formed by the abstraction of H from acetone by CH_3 radicals [99].
It is also possible that acetonyl may be formed by abstraction of
hydrogen from acetone by excited molecules [100]

$$(CH_3COCH_3)^* + CH_3COCH_3 \rightarrow CH_3\overset{\bullet}{C}OHCH_3 + CH_3COCH_2\cdot \qquad (4\text{-}123)$$

This reaction forms the radical $CH_3COCH_2\cdot$, which must be present in
view of the formation of the dimer (pinacol) as a product [98].
The total radical yield is about 4.2 per 100 eV, based on iodine
scavenger experiments.

I. Amines and Other Nitrogen Compounds

The radiolysis of methyl- , dimethyl- , and trimethylamine
produces hydrogen and methane [101, 102]. As methyl substitution
increases, the yield of hydrogen decreases while that of methane
increases. About 81% of the hydrogen formation from CH_3NH_2 and 41%
of that from $(CH_3)_3N$ can be scavenged by adding ethylene. Another
product from all three amines is ammonia.

Steif and Ausloos [103] have studied the radiolysis of azomethane. The main products are nitrogen, ethane, methane, and

hydrogen. Oxygen as an inhibitor prevents the formation of ethane
and greatly reduces the formation of methane. The sum of the yields
of ethane and methane drops a great deal (relative to nitrogen) as
the azomethane pressure increases from 2 to 200 Torr. Most of the
results can be explained by a primary dissociation of excited
azomethane,

$$(CH_3N_2CH_3)^* \rightarrow 2 \ CH_3 + N_2 \qquad\qquad (4\text{-}124)$$

followed by abstraction and combination of methyl radicals to produce methane and ethane. Some sort of molecular processes must be
postulated to explain the unscavengeable yield of hydrogen.

J. Aromatic Compounds

In studying the radiolysis of benzene, it is instructive to
compare the products from benzene with those from cyclohexane.
Liquid benzene gives an appreciably smaller yield of products than
cyclohexane. The main product from benzene is a viscous yellow
liquid polymer (G = 0.75), which has been shown to contain biphenyl,
phenylcyclohexadiene, phenylcyclohexene, hydrogenated terphenyls,
and higher molecular weight material.

The following reactions have been postulated [104, 105] in the
radiolysis of benzene:

$$C_6H_6 \rightarrow C_6H_6^+ + e^- \rightarrow C_6H_6^* \qquad\qquad (4\text{-}125)$$

$$C_6H_6 \rightarrow C_6H_6^* \qquad\qquad (4\text{-}126)$$

$$C_6H_6^* + C_6H_6 \rightarrow 2 \ C_6H_6 \qquad\qquad (4\text{-}127)$$

$$C_6H_6^* \rightarrow \cdot C_6H_5 + H \qquad\qquad (4\text{-}128)$$

or

$$\rightarrow C_2H_2 + ? \qquad\qquad (4\text{-}129)$$

$$H\cdot + C_6H_6 \rightarrow C_6H_7 \qquad\qquad (4\text{-}130)$$

$$2\ H\cdot \rightarrow H_2 \qquad\qquad (4\text{-}130a)$$

$$\cdot C_6H_5 \text{ and } \cdot C_6H_7 \xrightarrow{+C_6H_6} \text{ polymer} \qquad\qquad (4\text{-}131)$$

$$C_6H_6{}^* + C_6H_6 \rightarrow \quad \text{polymer} \qquad\qquad (4\text{-}132)$$

Polymer is produced either by radical reactions or from excited molecules. Hydrogen atoms react more by addition than by
hydrogen abstraction, which accounts for the low yield of hydrogen
gas $[G(H_2) = 0.036]$. Low hydrogen yield is a general property of
aromatic and unsaturated compounds, in contrast to the high hydrogen yield from saturated aliphatic compounds. The greater stability of liquid benzene compared to gaseous benzene is perhaps explained by the quenching of excited benzene molecules [Eq. (4-127)]
in the liquid state. Quenching of excited benzene molecules in the
photolysis of benzene may contribute to the very low quantum yields
of products [106].

The hydrogen yield shows a marked increase as the LET of the
radiation increases. To explain this effect, Burns and Reed
[107, 108] have suggested the following reactions:

$$C_6H_6{}^* + C_6H_6{}^* \rightarrow H_2 + C_{12}H_{10} \text{ (biphenyl)} \qquad\qquad (4\text{-}133)$$

$$\rightarrow H_2 + \text{radicals} \qquad\qquad (4\text{-}134)$$

Second-order reactions (4-133) and (4-134) will be favored when the
concentration of excited benzene molecules in the track is increased by using radiation of higher LET. On the other hand, reaction (4-127) (quenching) being first-order in excited benzene
molecules would not be favored by an increase in LET.

By employing a similar argument, the increase in acetylene yield with increasing LET can be accounted for by the following reaction:

$$C_6H_6^* + C_6H_6^* \rightarrow nC_2H_2 + \text{radicals or stable products} \qquad (4\text{-}135)$$

Mention was made earlier of the "protective effect" of aromatic compounds, by which they sometimes share their stability toward radiation with other substances to which they may be added. Extensive studies have been made of this effect for cyclohexane/benzene solutions, starting with the work of Schoepfle and Fellows [109]. For example, the H_2 yield from a 50/50 solution of benzene and cyclohexane is only a small fraction of the average H_2 yield from pure benzene and pure cyclohexane, and similar results hold true for the yields of cyclohexene, bicyclohexyl, and cyclohexylhexene. Many explanations have been advanced for the effect, but one that was proposed early [110] and widely accepted is that energy is transferred from excited cyclohexane molecules to benzene molecules,

$$c\text{-}C_6H_{12}^* + C_6H_6 \rightarrow c\text{-}C_6H_{12} + C_6H_6^* \qquad (4\text{-}136)$$

This sort of energy transfer requires that the accepting molecule have an excited state of lower energy than the excited states of the donor. In this connection, benzene is known to have lower-energy excited states than the saturated hydrocarbons. Addition of benzene also protects acetone [111], ethanol vapor [112], and methyl acetate [111] to some extent against radiation damage.

Low product yields are also obtained from the radiolysis of alkyl benzenes. Scission of bonds in the side chains increase as these become larger, leading to increased product yields. However, it appears that a considerable amount of energy absorbed by the side chain is transferred to the ring, in order to account for the unexpectedly low yields from the side chains. The methane yield is found to increase as the number of terminal methyls in the side

chain increases, but does not increase so much with increased
methyl substitution in the benzene ring.

IV. AQUEOUS SOLUTIONS

A. General Theory

The LET of liquid water is about 10^3 times as great as that
of water vapor, so that ions and excited molecules formed in the
liquid will be much closer together than those formed in the vapor.
Factors such as LET which affect the distribution of primary
product species in the tracks will also affect the tendency of
primary species to react among themselves before diffusing from the
track into the body of the liquid. These facts plus the complicating effect of solvation indicate that the radiolysis of liquid
water involves a very complex mechanism.

The ions formed in the irradiation of liquid water are
probably the same as those produced in water vapor, namely, H_2O^+,
H_3O^+, OH^+, and H^+, as shown by mass spectrometric studies [113,
114]. Although the fragment ions H^+ and OH^+ may exist momentarily,
they probably undergo geminate recombination with $\cdot OH$ and $H\cdot$,
respectively, to give H_2O^+ [32].

$$H^+ + \cdot OH \rightarrow H_2O^+ \qquad\qquad (4-137)$$

$$OH^+ + H\cdot \rightarrow H_2O^+ \qquad\qquad (4-138)$$

Hence, little attention is given to ions other than H_2O^+, and the
ionization of liquid water during irradiation is considered to be

$$H_2O \rightarrow H_2O^+ + e^- \qquad\qquad (4-139)$$

There are two well-known "models" of what takes place after irradiation, namely, the Samuel-Magee model and the Lea-Gray-Platzman model. In the Samuel-Magee model [13] it is estimated that a 10 eV electron [in Eq. (4-139)] will move about 20 Å from the H_2O^+ ion before being reduced to thermal energy and will be drawn back by coulombic attraction to neutralize the parent ion and form a highly excited water molecule which dissociates,

$$H_2O^+ + e^- \rightarrow H_2O^{**} \rightarrow H\cdot + OH\cdot \qquad (4\text{-}140)$$

the radicals formed having enough energy to avoid geminate recombination.

The other model, which was proposed by Lea [115], Gray [14], and Platzman [116] estimates that during thermalization the electron in Eq. (4-139) moves about 150 Å from the parent ion on the average. The ion and electron are then assumed to react independently, the ion giving a hydroxyl radical,

$$H_2O^+ + H_2O \rightarrow H_3O^+ + \cdot OH \qquad (4\text{-}141)$$

and the electron producing a hydrogen atom at a distance from the track,

$$e^- + H_2O \rightarrow H\cdot + OH^- \qquad (4\text{-}142)$$

Reaction (4-142) is endothermic and can only take place owing to the availability of the solvation energy of the hydroxyl ion. Whether one assumes the Samuel-Magee model or the Lea-Gray-Platzman model, the resulting products are hydrogen atoms and hydroxyl radicals.

With either model, the radicals will originally be distributed in or near spurs along the track of the ionizing particle. When water is ionized by fast electrons, the spurs will be spaced at intervals of about 5000 Å along the electron track [13]. For high

LET radiation such as α-particles, the distance between spurs will be only about 10 Å, so that the spurs will overlap from the instant of formation forming a column with the particle track as its axis.

In the spurs or columns along the track, the original radical concentration will be high, leading to radical-radical reactions,

$$H\cdot + \cdot OH \rightarrow H_2O \tag{4-143}$$

$$2 \cdot OH \rightarrow H_2O_2 \tag{4-144}$$

$$2 H\cdot \rightarrow H_2 \tag{4-145}$$

or

$$2 e^-_{aq} \rightarrow H_2 + 2 OH^- \tag{4-146}$$

where each electron is hydrated in the last reaction. The products indicated, H_2 and H_2O_2, are known as "molecular products" with yields of G_{H_2} and $G_{H_2O_2}$. The H· and ·OH radicals which escape from the track zone are known as "radical products," with yields indicated by G_H and G_{OH}. High LET radiation favors radical combination reactions giving high yields of molecular products, while low LET radiation favors radical products. However, at increasing intensities of low LET radiation the spurs begin to overlap, leading to an increase in molecular product yields.

The low radical concentration outside the particle tracks makes radical-radical reactions unlikely if any other substance is present with which they can react.

In irradiation studies on liquid water, the radical and molecular yields usually measured include G_{-H_2O}, G_{H_2}, $G_{H_2O_2}$, G_H, and G_{OH}. For radiation of a certain LET at a fixed pH, the molecular and radical yields remain approximately constant. The quantity G_{-H_2O} indicates the water molecules converted to radical and molecular products. It is found that G_{-H_2O} is approximately independent of the type (LET) of radiation used.

The relations between the G values for the various products can be explained by summarizing the total decomposition in terms of two reactions,

$$2 H_2O \rightarrow H_2 + H_2O_2 \tag{4-147}$$

$$H_2O \rightarrow H\cdot + \cdot OH \tag{4-148}$$

Therefore,

$$(2 \, a+c \text{ or } 2 \, b+d)H_2O \rightarrow aH_2 + bH_2O_2 + cH\cdot + dOH\cdot \tag{4-149}$$

indicating

$$G_{-H_2O} = 2G_{H_2} + G_H \tag{4-150}$$

$$G_{-H_2O} = 2G_{H_2O_2} + G_{OH} \tag{4-151}$$

For example, in the radiolysis of liquid H_2O by γ-rays or fast electrons [117]

$$G_{-H_2O} = 3.64 \tag{4-152}$$

$$G_{H_2} = 0.42 \tag{4-153}$$

$$G_{H_2O_2} = 0.71 \tag{4-154}$$

$$G_{OH} = 2.22 \tag{4-155}$$

$$G_H = 2.80 \tag{5-156}$$

These G values can be shown to satisfy Eqs. (4-150) and (4-151) as follows:

$$3.64 = 2(0.42) + 2.80 \tag{4-150A}$$
$$3.64 = 2(0.71) + 2.22 \tag{4-151A}$$

In a closed system with no solute to react with any of the products, the $\cdot$H and $\cdot$OH radicals will react with the other products to form water,

$$H\cdot + H_2O \rightarrow \cdot OH + H_2O \qquad\qquad (4\text{-}157)$$

$$\cdot OH + H_2 \rightarrow H\cdot + H_2O \qquad\qquad (4\text{-}158)$$

Hence, the result of irradiating water in a closed system is to reach a steady state involving an equilibrium concentration of hydrogen and hydrogen peroxide. The use of low LET radiation favoring high radical yields results in a small equilibrium concentration of molecular products (H_2 and H_2O_2). High LET radiation favors low radical yields and produces a relatively high equilibrium concentration of H_2 and H_2O_2.

Irradiation of an open system where hydrogen escapes causes first an increase in hydrogen peroxide concentration, followed by a constant H_2O_2 concentration where its rate of decomposition balances its rate of formation.

$$H\cdot + H_2O_2 \rightarrow \cdot OH + H_2O \qquad\qquad (4\text{-}159)$$

$$\cdot OH + H_2O_2 \rightarrow HO_2\cdot + H_2O \qquad\qquad (4\text{-}160)$$

$$2\ HO_2\cdot \rightarrow H_2O_2 + O_2 \qquad\qquad (4\text{-}161)$$

Hence, the net result is decomposition into hydrogen and oxygen, as observed in nuclear reactors moderated with heavy water.

Recent evidence shows that the solvated electron (rather than H$\cdot$) is the major reducing species in aqueous solution of pH > 3. Although the symbol e_{aq}^- is often employed to indicate the solvated electron, it may be more convenient to use H_2O^- of $(H_2O)^-$ in balancing chemical equations. When solvated electrons are present, hydrogen atoms may be formed slowly by the reaction

$$e_{aq}^- + H_2O \rightarrow H\cdot + OH^- \tag{4-162}$$

when there is no solute present which would otherwise react with e_{aq}^-. The solvated electrons may react with hydrogen ions in acid solution to form hydrogen atoms,

$$e_{aq}^- + H^+ \rightarrow H\cdot \tag{4-163}$$

a reaction much faster than (4-162). Reaction (4-163) may also be prevented by solutes, which react faster with e_{aq}^- than does H^+. It often happens that both solvated electrons and hydrogen atoms react so as to yield the same final product, but the rate of reaction of solvated electrons and hydrogen atoms with a particular solute may be different.

Interesting evidence that the reducing species in neutral solutions is the solvated electron was obtained from relative reaction rate measurements in salt solutions of different ionic strengths [118, 119]. The method is based on the fact that rate constants for reactions between ions of like charge increase with increasing ionic strength, whereas rate constants for reactions between ions of unlike charge decrease, and rate constants for reactions between ions and neutral molecules are unchanged. Czapski and Schwarz [118] investigated the ratio of rate constants for the reaction of the reducing radical with H^+, O_2, and NO_2^- to the rate constant for the reaction with hydrogen peroxide over a range of ionic strength values. Assuming that the reducing species is the solvated electron, the respective reactions would be

$$e_{aq}^- + H^+ \xrightarrow{\;k_1\;} H\cdot \tag{4-164}$$

$$e_{aq}^- + O_2 \xrightarrow{\;k_2\;} O_2^- \tag{4-165}$$

$$e_{aq}^- + NO_2^- \xrightarrow{\;k_3\;} NO_2^{-2} \tag{4-166}$$

$$e_{aq}^{-} + H_2O_2 \xrightarrow{\ k_4\ } \cdot OH + OH^{-} \qquad\qquad (4\text{-}167)$$

The results showed that k_2/k_4 is unchanged by increasing ionic
strength, but k_1/k_4 decreases and k_3/k_4 increases. This shows
clearly that the reducing radical has a negative charge in slightly
acid solutions, and from the slope of the log rate versus square
root of ionic strength curves it was shown that the value of the
vharge is -1. Using a comparable approach, Collinson et al. [119]
investigated the effect of ionic strength on the relative rate of
the reactions between the reducing species and Ag^{+} and acrylamide
and deduced that at pH 4 and 2 the reducing species has a charge
of -1 and 0, respectively.

The relative numbers of hydrogen atoms and solvated electrons
in irradiated solutions can be estimated by employing solutes which
scavenge one or the other selectively. Robani and Stein [120]
studied a number of scavenger pairs such as acetone/isopropanol,
ferricyanide/formate, and bicarbonate/methanol, where the first
component scavenges solvated electrons and the second scavenges
hydrogen atoms. Baxendale et al. [121] used oxygen and/or hydrogen
peroxide to estimate the total yield of reducing species, and used
acetic acid to measure hydrogen atoms.

B. Organic Compounds in Aqueous Solution

As a general rule, when aqueous solutions are irradiated,
almost all the energy absorbed is taken up by water molecules, and
the resulting changes in the solute are brought about indirectly
through reaction with the molecular products and especially the
radical products from the water. Any direct effect because of
absorption of radiation by the solute is usually unimportant in
dilute solution (less than 0.1 M). At higher concentrations,
direct action becomes more important, and possibly excited water

molecules may transfer energy directly to the solute.

The following sections discuss in more detail the effect of
radiation on aqueous solutions of various organic compounds.

1. Formic Acid

The following reactions have been postulated when aqueous
formic acid solution is irradiated in the presence of oxygen:

$$\cdot OH + HCOOH \rightarrow \cdot COOH + H_2O \tag{4-168}$$

$$\cdot COOH + O_2 \rightarrow HO_2\cdot + CO_2 \tag{4-169}$$

$$H\cdot + O_2 \rightarrow HO_2\cdot \tag{4-170}$$

$$2\ HO_2\cdot \rightarrow H_2O_2 + O_2 \tag{4-171}$$

The number of molecules of CO_2 formed and measurable equals the
number of $\cdot COOH$ radicals originally formed by reaction (4-168). The
number $\cdot COOH$ radicals formed also equals the number of $\cdot OH$ radicals
originally formed [reaction (4-168)], so that the measured yield of
carbon dioxide, $G(CO_2)$, equals the radical yield, G_{OH}, or

$$G_{OH} = G(CO_2) \tag{4-172}$$

By similar lines of reasoning, the other molecular and radical
yields can be related to the measured values of $G(H_2)$, $G(H_2O_2)$,
$G(CO_2)$, and $G(-O_2)$ as follows [122, 123]:

$$G_{H_2} = G(H_2) \tag{4-173}$$

$$G_{H_2O_2} = G(H_2O_2) - G(-O_2) \tag{4-174}$$

$$G_H = 2\ G(-O_2) - G(CO_2) \tag{4-175}$$

The reaction is more complex in the absence of oxygen. The following steps have been proposed for the condition of no oxygen, pH below 3, and concentration less than 5 mM [124, 125]:

$$H\cdot + HCOOH \rightarrow \cdot COOH + H_2 \qquad\qquad (4\text{-}176)$$

$$\cdot OH + HCOOH \rightarrow \cdot COOH + H_2O \qquad\qquad (4\text{-}177)$$

$$H_2O_2 + \cdot COOH \rightarrow \cdot OH + CO_2 + H_2O \qquad\qquad (4\text{-}178)$$

$$2 \cdot COOH \rightarrow HCOOH + CO_2 \qquad\qquad (4\text{-}179)$$

At higher values of pH, the yields of hydrogen and carbon dioxide decrease. Only hydrogen is evolved above pH 7, and oxalic acid being the other major product, the total reaction becomes

$$2 \; HCOOH \rightarrow H_2 + (COOH)_2 \qquad\qquad (4\text{-}180)$$

The products of the irradiation depend strongly on the conditions employed. Irradiation of dilute oxygen-free solutions of formic acid with cyclotron-produced helium ions forms glyoxal, glyoxylic acid, and formaldehyde [126]. Formaldehyde is also produced at high formic acid concentrations where the irradiation is by γ-rays. In the range from molar to pure formic acid, the yield of carbon dioxide rises and $G(CO_2)$ values up to 12 can result [127].

2. Acetic Acid

Irradiation of dilute solutions of acetic acid containing no oxygen yields hydrogen, hydrogen peroxide, and succinic acid [128]. These products could be accounted for by the reactions

$$H\cdot + CH_3COOH \rightarrow \cdot CH_2COOH + H_2 \qquad\qquad (4\text{-}181)$$

$$\cdot OH + CH_3COOH \rightarrow \cdot CH_2COOH + H_2O \qquad\qquad (4\text{-}182)$$

$$2 \; \cdot CH_2COOH \rightarrow (CH_2COOH)_2 \qquad\qquad (4\text{-}183)$$

When acetic acid solutions of concentration over 1 M are irradiated with heavy particle radiation or x-rays, considerable amounts of carbon dioxide and methane are formed, plus smaller quantities of carbon monoxide, ethane, acetaldehyde, and biacetyl. Hayon and Weiss [129] explained these products through the action of "positive polarons," $(H_2O)^+$, and "negative polarons," $(H_2O)^-$. The negative polarons would presumably react as follows:

$$(H_2O)^- + CH_3COOH \rightarrow (CH_3COOH)^- + H_2O \qquad\qquad (4\text{-}184)$$

$$CH_3CO^- + \cdot OH \qquad\qquad (4\text{-}185)$$

$$(CH_3COOH)^- \longrightarrow CH_3COO^- + H \cdot \qquad\qquad (4\text{-}186)$$

$$CH_3CO \cdot + OH^- \qquad\qquad (4\text{-}187)$$

The reactions of the acetyl radical would form carbonyl compounds and carbon monoxide. The positive polaron would probably react as follows:

$$(H_2O)^+ + CH_3COOH \rightarrow CH_3COO \cdot + H^+ + H_2O \qquad\qquad (4\text{-}188)$$

followed by

$$CH_3COO \cdot \rightarrow \cdot CH_3 + CO_2 \qquad\qquad (4\text{-}189)$$

$$\cdot CH_3 + CH_3COOH \rightarrow \cdot CH_2COOH + CH_4 \qquad\qquad (4\text{-}190)$$

$$2 \; \cdot CH_3 \rightarrow C_2H_6 \qquad\qquad (4\text{-}191)$$

3. Ethanol

The x-ray or γ-ray irradiation of oxygen-free aqueous ethanol

solutions yields hydrogen, acetaldehyde, butane-2,3,-diol, and
hydrogen peroxide, probably by the following reactions [130]:

$$H\cdot + CH_3CH_2OH \rightarrow CH_3\overset{\cdot}{C}HOH + H_2 \qquad\qquad (4\text{-}192)$$

$$\cdot OH + CH_3CH_2OH \rightarrow CH_3\overset{\cdot}{C}HOH + H_2O \qquad\qquad (4\text{-}193)$$

$$H\cdot + CH_3\overset{\cdot}{C}HOH \rightarrow CH_3CH_2OH \qquad\qquad (4\text{-}194)$$

$$\cdot OH + CH_3\overset{\cdot}{C}HOH \rightarrow CH_3CHO + H_2O \qquad\qquad (4\text{-}195)$$

$$2\ CH_3\overset{\cdot}{C}HOH \rightarrow CH_3\text{--}CHOH\text{--}CHOH\text{--}CH_3 \qquad\qquad (4\text{-}196)$$

or

$$\rightarrow CH_3CH_2OH + CH_3CHO \qquad\qquad (4\text{-}197)$$

The hydrogen yield is larger (G = 4.2) in the pH range from
1.2 to 4 than above pH 4.5. Also, the acetaldehyde yield drops
from G = 2 in the lower pH range to G = 0.2 at pH 7. These changes
may be related to the formation of H_2^+ at low pH. Hence, at low
pH, reaction (4-194) may become

$$H_2^+ + CH_3\overset{\cdot}{C}HOH \rightarrow CH_3CHO + H_2 + H^+ \qquad\qquad (4\text{-}198)$$

and reaction (4-192) may be replaced by

$$H_2^+ + CH_3CH_2OH \rightarrow CH_3\overset{\cdot}{C}HOH + H_2 + H^+ \qquad\qquad (4\text{-}199)$$

Butanediol is not formed in the presence of oxygen, but more
acetaldehyde is formed, possibly through the reaction sequence

$$H\cdot + O_2 \rightarrow HO_2\cdot \qquad\qquad (4\text{-}200)$$

$$CH_3\overset{\cdot}{C}HOH + O_2 \rightarrow CH_3C(O_2\cdot)HOH \qquad\qquad (4\text{-}201)$$

$$HO_2\cdot + CH_3C(O_2\cdot)HOH \rightarrow CH_3CHo + H_2O_2 + O_2 \qquad\qquad (4\text{-}202)$$

REFERENCES

1. Burton, M., Berry, P. J., and Lipsky, S., *J. Chim. Phys.*, <u>52</u>, 657 (1955).

2. Weber, E. N., Forsyth, P. F., and Schuler, R. H., *Radiation Res.*, <u>3</u>, 68 (1955).

3. Prevost-Bernas, A., Chapiro, A., Cousin, C., Landler, Y., and Magat, M., *Discussions Faraday Soc.*, <u>12</u>, 98 (1952).

4. Baxendale, J. H. and Mellows, F. W., *J. Amer. Chem. Soc.*, <u>83</u>, 4720 (1961).

5. Hayon, E. and Weiss, J. J., *J. Chem. Soc.*, <u>1961</u>, 3962.

6. Taub, I. A., Cauer, M. C., and Dorfman, L. M., *Discussions Faraday Soc.*, <u>36</u>, 206 (1963)

7. Taub, I. A., Harter, D. A., Sauer, M. C., and Dorfman, L. M., *J. Chem. Phys.*, <u>41</u>, 979 (1964).

8. Allen, A. O. and Hummel, A., *Discussions Faraday Soc.*, <u>36</u>, 95 (1963).

9. Freeman, G. R., *J. Chem. Phys.*, <u>39</u>, 988 (1963).

10. Scholes, G. and Simic, M., *Nature*, <u>202</u>, 895 (1964).

11. Williams, F., *J. Amer. Chem. Soc.*, <u>86</u>, 3954 (1964).

12. Burton, M., Magee, J. C., and Samuel, A. H., *J. Chem Phys.*, <u>20</u>, 760 (1952).

13. Samuel, A. H. and Magee, J. L., *J. Chem. Phys.*, <u>21</u>, 1080 (1953).

14. Gray, L. H., *J. Chim. Phys.*, <u>48</u>, 172 (1951).

15. Hummel, A. and Allen, A. O., *J. Chem. Phys.*, <u>44</u>, 3426 (1966); <u>46</u>, 1602 (1967).

16. Freeman, E. R. and Fayadh, J. M., *J. Chem. Phys.*, <u>43</u>, 86 (1965).

17. Skelly, D. W. and Hamill, W. H., *J. Chem. Phys.*, <u>44</u>, 2891 (1966).

18. Smith, D. R. and Pierone, J. J., *Can. J. Chem.*, <u>43</u>, 2141 (1965).

19. Ronayne, M. R., Guarino, J. P., and Hamill, W. H., *J. Amer. Chem. Soc.*, <u>84</u>, 4230 (1962).

20. Lampe, F. W. and Noyes, R. M., *J. Amer. Chem. Soc.*, <u>76</u>, 2140 (1954).

21. Luebbe, R. H., Jr. and Willard, J. E., *J. Amer. Chem. Soc.*, <u>81</u>, 761 (1959).

22. Strong, R. L. and Willard, J. E., *J. Amer. Chem. Soc.*, <u>79</u>, 2098 (1957).

23. Claridge, R. F. C. and Willard, J. E., *J. Amer. Chem. Soc.*, <u>87</u>, 4992 (1965).

24. Scala, A. A. and Ausloos, P., *J. Chem. Phys.*, <u>47</u>, 5129 (1967).

25. Rebbert, R. E. and Ausloos, P., *J. Chem. Phys.*, <u>46</u>, 4333 (1967).

26. Takehisa, M., Levey, G., and Willard, J. E., *J. Amer. Chem. Soc.*, <u>88</u>, 5694 (1966).

27. Matheson, M. S. and Dorfman, L. M., in *Progress in Reaction Kinetics* (G. Porter, ed.), Vol. III, Pergamon Press, New York, 1965.

28. Ebert, M., Keene, J. C., Swallow, A. J., and Baxendale, J. H., Eds., *Pulse Radiolysis*, Proc. Intern. Symp., Manchester, 1965, Academic Press, New York, 1965.

29. Lind, S. C. and Rudolph, P. S., *J. Chem. Phys.*, <u>26</u>, 1768 (1957).

30. Rudolph, P. S. and Melton, C. E., *J. Chem. Phys.*, <u>32</u>, 586 (1960).

31. Burton, M. and Lipsky, S., *J. Phys. Chem.*, <u>61</u>, 1461 (1957).

32. Dewhurst, H. A., Samuel, A. H., and Magee, J. L., *Radiation Res.*, <u>1</u>, 62 (1954).

33. Kelly, P., Rigg, T., and Weiss, J., *Nature*, <u>173</u>, 1130 (1954).

34. Back, R. A., Woodward, T. W., and McLauchlan, K. A., *Can. J. Chem.*, <u>40</u>, 1380 (1962).

35. Glockler, G. and Lind, S. C., *The Electrochemistry of Gases and Other Dielectrics*, John Wiley and Sons, New York, 1939.

36. Fessenden, R. W. and Schuler, R. H., *J. Chem. Phys.*, <u>39</u>, 2147 (1963).

37. Schuler, R. H. and Fessenden, R. W., *Proceedings of the Third International Conference on Radiation Research*, Cortina, Italy, 1966, from RRL-2310-180.

38. Gillis, H. A., *J. Phys. Chem.*, <u>67</u>, 1399 (1963).

39. Fessenden, R. H., *J. Phys. Chem.*, <u>68</u>, 1508 (1964).

40. Land, E. J., *Progress in Reaction Kinetics*, Vol. III, Pergamon Press, Oxford, 1965, p. 386.

41. Watts, A. T. and Walker, S., *J. Chem. Soc.*, <u>1962</u>, 4323.

42. Schuler, R. H., *J. Phys. Chem.*, <u>62</u>, 37 (1958).

43. Ebert, M., Keene, J. P., Land, E. J., and Swallow, A. J., *Proc. Roy. Soc. (London)*, <u>A287</u>, 1 (1965).

44. Forrestal, L. J. and Hamill, W. H., *J. Amer. Chem. Soc.*, <u>83</u>, 1535 (1961).

45. Schuler, R. H., *J. Phys. Chem.*, <u>61</u>, 1472 (1957).

46. Mani, I. and Hamahan, R. J., *J. Phys. Chem.*, 70, 2233 (1966).

47. Perner, D. and Schuler, R. H., *J. Phys. Chem.*, <u>70</u>, 2224 (1966).

48. Charlesby, A. and Lloyd, P. G., *Proc. Royl Soc. (London)*, <u>A249</u>, 51 (1958).

49. Holroyd, R. A. and Klein, G. W., *Intern. J. Appl. Radiation Isotopes*, <u>13</u>, 493 (1962).

50. Holroyd, R. A. and Klein, G. W., *Intern. J. Appl. Radiation Isotopes*, <u>15</u>, 633 (1964).

51. *Catalog of Mass Spectral Data*, American Petroleum Institute Project 44, Texas A and M University, College Station, Texas, 1947-1961.

52. Franklin, J. L., Wada, Y., Natalis, P., and Hierl, P. M., *J. Phys. Chem.*, 70, 2353 (1966).

53. Gupta, S. K., Jones, E. G., Harrison, A. G., and Myher, J. J., *Can. J. Chem.*, <u>45</u>, (1967).

54. Giardini-Guidoni, A., and Friedman, L., *J. Chem. Phys.*, <u>45</u>, 937 (1966).

55. Abramson, F. P. and Futrell, J. H., *J. Chem. Phys.*, <u>46</u>. 3624 (1967).

56. Lampe, F. W., *J. Amer. Chem. Soc.*, <u>79</u>, 1055 (1957).

57. Field, F. H., Franklin, J. L., and Mureson, M. S. B., *J. Amer.
     Chem. Soc.*, 85, 3575 (1963).

58. Abramson, F. P. and Futrell, J. H., *J. Chem. Phys.*, 45, 1925
     (1966).

59. Wexler, S. and Jesse, N., *J. Amer. Chem. Soc.*, 84, 3425 (1962).

60. Field, F. H. and Munson, M. S. B., *J. Amer. Chem. Soc.*, 87,
     3829 (1965).

61. Gevantman, C. H. and Williams, R. R., *J. Phys. Chem.*, 56, 569
     (1952).

62. Meisels, G. G., Hamill, W. H., and Williams, R. R., *J. Phys.
     Chem.*, 61, 1456 (1957).

63. Meisels, G. G., *Abstracts 150th Meeting of the American Chemi-
     cal Society,* Atlantic City, New Jersey, 1965, p. 17V.

64. Hummel, R. W., *Nature*, 192, 1178 (1961).

65. Matheson, M. S., *Nucleonics*, 19, 57 (1961).

66. Schuler, R. H. and Kuntz, R. R., *J. Phys. Chem.*, 67, 1004
     (1963).

67. Holroyd, R. A., *Abstracts 153rd Meeting of the American Chemi-
     cal Society*, Miami, 1967, R55.

68. Dewhurst, H. A. and Schuler, R. H., *J. Amer. Chem. Soc.*, 81,
     3210 (1959).

69. Dewhurst, H. A., *J. Phys. Chem.*, 62, 15 (1958).

70. Futrell, J. H., *J. Amer. Chem. Soc.*, 81, 5921 (1959).

71. Hardwick, T. J., *J. Phys. Chem.*, 64, 1623 (1960).

72. Dewhurst, H. A., *J. Phys. Chem.*, 63, 813 (1959).

73. Holroyd, R. A., in *Fundamental Processes in Radiation Chem-
     istry* (P. J. Ausloos, ed.), Interscience, 1968.

74. Schuler, R. H. and Allen, A. O., *J. Amer. Chem. Soc.*, 77, 507
     (1955).

75. Hamashima, M., Reddy, M. P., and Burton, M., *J. Phys. Chem.*,
     62, 246 (1958).

76. Bach, N., *Proc. Intern. Conf. Peaceful Uses Atom. Energy,
     United Nations, New York*, 1956, Vol. 7, p. 538.

77. Holroyd, R. A. and Klein, G. W., *Intern. J. Appl. Radiation
     Isotopes*, 15, 633 (1964).

78. Holroyd, R. A. and Klein, G. W., *J. Phys. Chem.*, 69, 194 (1965).

79. Smith, D. R. and Pieroni, J. J., *J. Phys. Chem.*, 70, 2379 (1966).

80. Burns, W. G., Holroyd, R. A., and Klein, G. W., *J. Phys. Chem.*, 70, 910 (1966).

81. Lampe, F. W., *J. Phys. Chem.*, 63, 1986 (1959).

82. Chang, P. C., Yang, N. C., and Wagner, C. D., *J. Amer. Chem. Soc.*, 81, 2060 (1959).

83. Wagner, C. D., *Tetrahedron*, 14, 164 (1961).

84. Ottolinghi, M. and Stein, G., *Radiation Res.*, 14, 281 (1961).

85. Chen, T. H., Wong, K. U., and Johnston, F. J., *J. Phys. Chem.*, 64, 1023 (1960).

86. Gillis, H. A., Williams, R. R., and Hamill, W. H., *J. Amer. Chem. Soc.*, 83, 17 (1961).

87. Fessenden, R. W. and Schuler, R. H., *J. Chem. Phys.*, 43, 2704 (1965).

88. Majer, J. R. and Simons, J. P., *Advances in Photochemistry*, Vol. II, Interscience, New York, 1964, p. 165.

89. Golden, J. H., *J. Polymer Sci.*, 45, 534 (1960).

90. Wilmenius, P. and Lindholm, E., *Arkiv Egs.*, 21, 97 (1962).

91. Myron, J. J. and Freeman, G. R., *Can. J. Chem.*, 43, 381 (1965).

92. Adams, G. E. and Sedgewick, R. D., *Trans. Faraday Soc.*, 60, 865 (1964).

93. Taub, I. A. and Dorfman, L. M., *J. Amer. Chem. Soc.*, 84, 4053 (1962).

94. Newton, A. S., *J. Chem. Phys.*, 61, 1485 (1957).

95. Hummel, R. W., *Trans. Faraday Soc.*, 56, 236 (1960).

96. Ausloos, P. and Paulson, J., *J. Amer. Chem. Soc.*, 80, 5117 (1958).

97. Barker, R., *Trans. Faraday Soc.*, 59, 375 (1963).

98. Kucera, J., *Collrvyions Czech. Chem. Common.*, 30, 3080 (1965).

99. Barker, R., *Nature*, 192, 62 (1961).

100. Noyes, W. A., Jr., *Radiation Res.*, Suppl. No. 1, 164 (1959).

101. Collin, J., *Bull. Soc. Chim. Belg.*, 67, 549 (1958).

102. Takumuka, S. and Sakurai, H., *Bull. Chem. Soc. Japan*, 38, 791 (1965).

103. Steif, L. J. and Ausloos, P., *J. Phys. Chem.*, 65, 877 (1961).

104. Patrick, W. N. and Burton, M., *J. Amer. Chem. Soc.*, 76, 2626 (1954).

105. Gorden, S., Van Dyken, A. R., and Doumani, T. F., *J. Phys. Chem.*, 62, 20 (1958).

106. Wilson, J. E. and Noyes, W. A., Jr., *J. Amer. Chem. Soc.*, 63, 3025 (1941).

107. Burns, W. G., *Trans. Faraday Soc.*, 58, 961 (1962).

108. Burns, W. G. and Reed, C. R. V., *Trans. Faraday Soc.*, 59, 101 (1963).

109. Schoepfle, C. S. and Fellows, C. H., *Ind. Eng. Chem.*, 23, 1386 (1931).

110. Burton, M. and Patrick, W. N., *J. Phys. Chem.*, 58, 421 (1954).

111. Ausloos, P. J., *J. Amer. Chem. Soc.*, 83, 1056 (1961).

112. Ramardhya, J. M. and Freeman, G. R., *Can. J. Chem.*, 39, 1843 (1961).

113. Smyth, H. D. and Mueller, D. W., *Phys. Rev.*, 43, 116 (1933).

114. Mann, M. M., Hustrulid, A., and Tate, J. L., *Phys. Rev.*, 58 340 (1940).

115. Lea, D. E., *Actions of Radiations on Living Cells*, Cambridge U. P., Cambridge, 1946.

116. Platzman, R. L., *Basic Mechanisms in Radiobiology*, National Research Council Publication 305, Washington, D. C., 1953.

117. Allen, A. O. and Schwarz, H. A., *Proc. Second Intern. Cong. Peaceful Uses Atom. Energy, United Nations, Geneva*, 1958, Vol. 29, p. 30.

118. Czapski, G. and Schwarz, H. A., *J. Phys. Chem.*, 66, 471 (1962).

119. Collinson, E., Dainton, F. S., Smith, D. R., and Tazaké, S., *Proc. Chem Soc.*, 140 (1962).

120. Robani, J. and Stein, G., *J. Chem. Phys.*, 37, 1865 (1962).

121. Baxendale, J. H., Thomas, J. K., and Woodward, T., as reported by M. S. Matheson, *Amer. Rev. Phys. Chem.*, 13, 77 (1962).

122. Hart, E. J., *J. Amer. Chem. Soc.*, 76, 4198 (1954).

123. Hart, E. J., *J. Amer. Chem. Soc.*, 76, 4312 (1954).

124. Hart, E. J., *J. Amer. Chem. Soc.*, 73, 68 (1951).

125. Fricke, H., Hart, E. J., and Smith, H. P., *J. Chem. Phys.*, 6, 229 (1938).

126. Garrison, W. M., Bennett, W., and Jayko, M., *J. Chem Phys.*, 24, 631 (1956).

127. Smithies, D. and Hart, E. J., *J. Amer. Chem. Soc.*, 82, 4775 (1960).

128. Garrison, W. M., Bennett, W., Cole, S., Haymond, H. R., and Weeks, B. M., *J. Amer. Chem. Soc.*, 77, 2720 (1955).

129. Hayon, E. and Weiss, J., *J. Chem. Soc.*, 5091 (1960).

130. Jayson, G. G., Scholer, G., and Weiss, J., *J. Chem. Soc.*, 1358 (1957).

131. Burr, J. B., *J. Phys. Chem.*, 61, 1481 (1957).

Chapter 5

RADIOLYTIC POLYMERIZATION IN HOMOGENEOUS SYSTEMS

I. INTRODUCTION AND BACKGROUND THEORY

Radiolytic polymerization of vinyl monomers has been known since 1938, when Hopwood and Phillips produced polymers by γ-ray and neutron irradiation of liquid methyl methacrylate, styrene, and vinyl acetate [9].

Following World War II, there was a considerable increase in interest in radiolytic activation because of the increased availability of radioisotope sources. Early postwar investigations were carried out by Dainton [10], who studied the effects of γ-rays on aqueous solutions of acrylonitrile and methacrylonitrile [10]. In 1948, additional work was done with γ-rays and reactor radiation in the polymerization of styrene, methyl methacrylate, and acrylonitrile [11]. This line of research has expanded greatly through the years, and at present extensive investigations of radiolytic polymerization are being carried out in the United States, Japan, France, Roumania, Italy, Czechoslovakia, Russia, and other countries.

The principal experimental techniques used in this field of investigation include electron-spin-resonance (ESR) spectroscopy, optical spectroscopy, chemical scavenging, luminescence determinations, mass spectrometry, and measurement of radiation-induced conductivities.

In order to discuss radiation-induced polymerization, it is
necessary to present some background information on polymerization
mechanisms. The following section covers three types of polymerization mechanisms that are promoted by the presence of a suitable initiator or catalyst, namely, free radical, cationic, and
anionic mechanisms. Although radiation-induced polymerication
does not generally involve the use of a chemical catalyst, these
catalytic mechanisms represent a background of knowledge that is
useful as a starting point in the discussion of radiation-induced
polymerization.

A. Free Radical Mechanism

Free radicals may be generated by thermal or photochemical decomposition of an initiator molecule or by radiolytic decomposition
of almost any kind of molecule. Assume some molecule A is broken
down to form free radicals,

$$A \rightarrow 2 \; R\cdot \qquad\qquad\qquad\qquad (5\text{-}1)$$

and such radicals add to a vinyl monomer to produce another radical

$$R\cdot \; + \; CH_2{=}CHX \; \rightarrow \; RCH_2\underset{X}{\overset{H}{C}}\cdot \qquad\qquad (5\text{-}2)$$

Such radical regeneration is a characteristic of chain reactions.

The radical formed in the initiation step may then add additional monomer molecules, resulting in chain growth or propagation.

$$R{-}(CH_2CHX{-})_n CH_2\underset{X}{\overset{H}{C}}\cdot \; + \; CH_2{=}CHX \; \rightarrow \; R{-}(CH_2CHX{-})_{n+1}CH_2\underset{X}{\overset{H}{C}}\cdot \qquad (5\text{-}3)$$

The chain would continue to grow if it were not terminated in some
way, which often takes place by a combination or coupling of the
growing radical chains

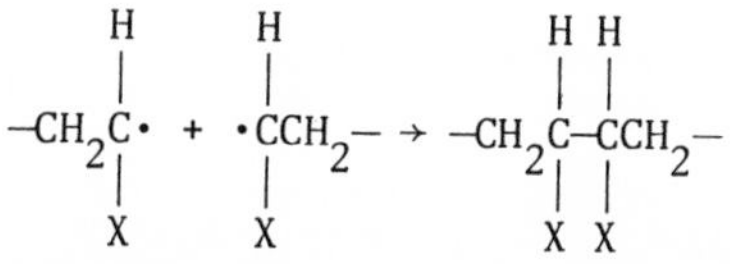

$$-CH_2\overset{\overset{\displaystyle H}{|}}{\underset{\underset{\displaystyle X}{|}}{C}}\cdot \;+\; \cdot\overset{\overset{\displaystyle H}{|}}{\underset{\underset{\displaystyle X}{|}}{C}}CH_2- \;\rightarrow\; -CH_2\overset{\overset{\displaystyle H}{|}}{\underset{\underset{\displaystyle X}{|}}{C}}-\overset{\overset{\displaystyle H}{|}}{\underset{\underset{\displaystyle X}{|}}{C}}CH_2- \qquad\qquad (5\text{-}4)$$

The above reactions can be symbolized for mathematical consideration by a series of steps

Step	Rate

Initiation:

$$A \rightarrow 2\ R\cdot \qquad\qquad\qquad\qquad R_i$$

Propagation:

$$R\cdot\ +\ M \rightarrow RM_1\cdot$$

$$RM_1\cdot\ +\ M \xrightarrow{\ k_p\ } RM_2\cdot \qquad\qquad k_p[M\cdot][M]$$

$$M_n\cdot\ +\ M \xrightarrow{\ k_p\ } M_{n+1}\cdot$$

Termination:

$$M_n\ +\ M_m \xrightarrow{\ k_t\ } \text{dead polymer} \qquad\qquad k_t[M\cdot]^2$$

where k_t is the rate constant for either combination or disproportionation. In most cases a steady state is reached in the
concentration of free radicals $[M\cdot]$, in which the rate of radical
formation equals the rate of radical disappearance,

$$R_i = k_t\ [M\cdot]^2 \qquad\qquad\qquad\qquad (5\text{-}5)$$

In this steady state the concentration of free radicals is therefore

$$[M\cdot] = (R_i/k_t)^{\frac{1}{2}} \tag{5-6}$$

The rate of propagation is virtually the same as the rate of monomer consumption,

$$R_p = -\frac{d[M]}{dt} = k_p[M\cdot][M] = k_p k_t^{-\frac{1}{2}} R_i^{\frac{1}{2}}[M] \tag{5-7}$$

This is the classic or "normal" kinetic equation for free radical polymerization, which shows that the rate is proportional to the square root of the rate of initiation and to the first power of monomer concentration.

The kinetic chain length ν is defined as the average number of monomer molecules polymerized per initiating radical. This is equal to the ratio of polymerization rate to the initiation rate or to the termination rate since the latter two are equal,

$$\nu = \frac{R_p}{R_i} = \frac{R_p}{R_t} \tag{5-8}$$

Employment of Eqs. (5-5), (5-7), and (5-8) gives

$$\nu = \frac{k_p[M]}{k_t[M\cdot]} \tag{5-9}$$

which, in view of Eq. (5-7), can be expressed

$$\nu = \frac{k_p^{2}[M]^2}{k_t R_p} \tag{5-10}$$

Equation (5-10) is very significant because it shows that the kinetic chain length is directly proportional to monomer concentration squared and inversely proportional to the radical concentration or polymerization rate. An increase in polymerization rate or radical concentration will produce smaller polymer molecules. On the other hand, Eqs. (5-9) and (5-10) show that the kinetic chain length is a characteristic of a particular monomer and does not depend on the method of initiation.

B. Cationic Mechanism

Most monomers undergo free radical polymerizations, whereas
ionic polymerizations are more selective. Monomers with electron-
releasing substituents such as alkoxy, phenyl, or vinyl are more
liable to take part in cationic polymerization. Anionic polymerization is more probable with monomers having electron-withdraw-
ing groups such as nitrile, carboxyl, phenyl, and vinyl.

Experimental difficulties are greater in the study of ionic
polymerizations. It is difficult in many cases to obtain reproducible rate data, since ionic polymerizations proceed at very fast
rates which are sensitive to small concentrations of cocatalysts,
impurities, and other materials. In the case of a monomer which
can polymerize either by free radical or ionic mechanism, the rate
of the ionic polymerization is usually faster.

A typical *catalyst* for a cationic polymerization is a Lewis
acid such as BF_3. A *cocatalyst* such as water must also be present,
which reacts with the catalyst to form a *catalyst-cocatalyst*
complex:

$$BF_3 + H_2O \rightleftarrows H^+(BF_3OH)^- \qquad (5-11)$$

The catalyst-cocatalyst complex then protonates the monomer, which
can be written as follows using isobutylene as an example:

$$H^+(BF_3OH)^- + (CH_3)_2C{=}CH_2 \rightarrow (CH_3)_3C^+(BF_3OH)^- \qquad (5-12)$$

In general terms, the initiation process can be written as follows:

$$C + RH \overset{K}{\rightleftarrows} H^+(CR)^- \qquad (5-13)$$

$$H^+(CR)^- + M \overset{k_i}{\rightarrow} HM^+(CR)^- \qquad (5-14)$$

where C, RH, and M represent catalyst, cocatalyst, and monomer,
respectively, and K is the equilibrium constant for reaction (5-13).
The theoretical concept of the cocatalyst is widely accepted, but
its presence has been shown experimentally in only a few cases.

The ion pair consisting of the *carbonium ion*, HM^+, and its *gegen ion*, $(CR)^-$, remain near each other throughout the polymerization.

In the propagation step, the ion pair formed in Eq. (5-12) or (5-14) grows into a long chain by the addition of monomer molecules. In the case of isobutylene this could be indicated by

$$H\{CH_2C(CH_3)_2\}_n^+ (BF_3OH)^- + (CH_3)_2C=CH_2 \rightarrow$$

$$H\{CH_2C(CH_3)_2\}_n CH_2C^+(CH_3)_2 (BF_3OH)^- \qquad (5\text{-}15)$$

In general this could be expressed as follows:

$$HM_n^+ (CR)^- + M \xrightarrow{k_p} HM_nM^+ (CR)^- \qquad (5\text{-}16)$$

The growth in the chain takes place by inserting a monomer molecule between the carbonium ion and its gegen ion.

Termination may take place by a sort of "spontaneous termination" or "chain transfer to counterion."

$$HM_nM^+ (CR)^- \xrightarrow{k_t} M_{n+1} + H^+(CR)^- \qquad (5\text{-}17)$$

It should be noted that although a particular monomer chain has been brought to an end, the catalyst-cocatalyst species remains free to initiate another chain, and actually many polymer molecules are thought to be produced for each catalyst-cocatalyst species.

Based on Eqs. (5-13), (5-14), (5-16), and (5-17), the rates of initiation, propagation, and termination can be written as

$$R_i = Kk_i[C][RH][M] \qquad (5\text{-}18)$$

$$R_p = k_p[HM^+(CR)^-][M] \qquad (5\text{-}19)$$

$$R_t = k_t[HM^+(CR)^-] \qquad (5\text{-}20)$$

where $[\text{HM}^+(\text{CR})^-]$ is the total concentration of all propagating species of ion pairs.

Assumption of the existence of a steady state in the concentration of the propagating species enables the equating of rates of initiation and termination, giving

$$[\text{HM}^+(\text{CR}^-)] = \frac{Kk_i[\text{C}][\text{RH}][\text{M}]}{k_t} \tag{5-21}$$

Combining Eqs. (5-19) and (5-21) gives the rate of polymerization, R_p:

$$R_p = \frac{Kk_i k_p[\text{C}][\text{RH}][\text{M}]^2}{k_t} \tag{5-22}$$

and the number-average degree of polymerization becomes

$$\bar{X}_n = \frac{R_p}{R_t} = \frac{k_p[\text{M}]}{k_t} \tag{5-23}$$

On the other hand, termination may occur by chain transfer to the monomer,

$$\text{HM}_n\text{M}^+(\text{CR})^- + \text{M} \xrightarrow{k_{tr,M}} \text{M}_{n+1} + \text{HM}^+(\text{CR})^- \tag{5-24}$$

in which case the concentration of $[\text{HM}^+(\text{CR})^-]$ becomes,

$$[\text{HM}^+(\text{CR})^-] = \frac{Kk_i[\text{C}][\text{RH}]}{k_{tr,M}} \tag{5-25}$$

and polymerization rate and degree of polymerization become

$$R_p = \frac{Kk_i k_p[\text{C}][\text{RH}][\text{M}]}{k_{tr,M}} \tag{5-26}$$

$$\bar{X}_n = \frac{k_p}{k_{tr,M}} = \frac{1}{C_M} \tag{5-27}$$

where C_M is the chain transfer constant for monomer, by definition.

The polymerization rate in the cationic mechanism depends on the ration k_p/k_t (or k_p/k_{tr}). Experimental determinations indicate that in cationic polymerizations the ratio of the propagation rate constant to the sum of the rate constants for all modes of termination is generally of the order of 10^2. This is the main factor resulting in generally very fast rates for cationic polymerization.

Cationic polymerizations are usually very exothermic. Examination of Eqs. (5-22) and (5-23) shows that

$$E_R = E_i + E_p - E_t \qquad (5-28)$$

and

$$E_{\overline{X}_n} = E_p - E_t \qquad (5-29)$$

where E_i, E_p, and E_t are activation energies for the initiation, propagation, and termination steps.

Experimentally the values of E_i and E_t are usually found to be greater than E_p. Ordinarily E_R lies in the range from -5 to +10 kcal/mole. With negative E_R the rate of polymerization increases with decreasing temperature. However, the value of E_R varies from one monomer to another, and even using only one monomer its value is influenced considerably by the nature of the catalyst, cocatalyst, and solvent. Values of E_R are generally less than in free radical polymerization, and rates do not change as much with temperature.

The activation energy $E_{\overline{X}_n}$ is always negative because E_t is greater than E_p, and hence the degree of polymerization decreases as temperature increases. The value of $E_{\overline{X}_n}$ is greater in the negative direction when termination is by transfer rather than by spontaneous termination.

C. Anionic Mechanism

There are a variety of types of anionic polymerization. One which has attracted a good deal of attention is the polymerization of vinyl monomers initiated by metal alkyl such as butyl lithium [12]: The catalyst first adds to the monomer,

$$C_4H_9Li + CH_2{=}CHY \rightarrow C_4H_9{-}CH_2{-}\overset{Y}{\underset{H}{C}}{:}^- (Li^+) \qquad (5\text{-}30)$$

followed by propagation,

$$C_4H_9{-}CH_2{-}\overset{Y}{\underset{H}{C}}{:} (Li^+) + {}_nCH_2{=}CHY \rightarrow$$

$$C_4H_9{\{}CH_2CHY{\}}_n CH_2{-}\overset{Y}{\underset{H}{C}}{:}^- (Li^+) \qquad (5\text{-}31)$$

Such anionic polymerizations often have no termination reaction. Propagation takes place with complete consumption of the monomer. The anionic centers remain intact because transfer of a proton or similar positive species from the solvent simply does not take place. Such nonterminated chain anions have been called living polymers [13].

In other cases, a free propagating negative ion is known to be formed. For example, in the polymerization of styrene by potassium amide in liquid ammonia the potassium amide first dissociates,

$$KNH_2 \overset{K}{\rightleftharpoons} K^+ + H_2N{:}^- \qquad (5\text{-}32)$$

The amide ion then adds to the monomer unit.

$$H_2N{:}^- + C_6H_5CH = CH_2 \overset{k_i}{\longrightarrow} H_2N{-}CH_2{-}\overset{H}{\underset{C_6H_5}{C}}{:}^- \qquad (5\text{-}33)$$

The rate of polymerization in anionic systems can be expressed simply as the rate of propagation,

$$R_p = k_p [M^-][M] \qquad\qquad\qquad (5\text{-}34)$$

where $[M^-]$ is the total concentration of living anionic propagating chain ends. If chain transfer agents are absent, $[M^-]$ will simply equal the concentration of catalyst.

Anionic polymerization is much faster than radical polymerization, mainly because of the higher concentrations of the propagating anions and radicals. The concentration of free radicals in a radical polymerization is about 10^{-9} to 10^{-7} M, whereas that of propagating anions in an anionic system is about 10^{-3} to 10^{-2}. Therefore, the rate of anionic polymerization is about 10^4 to 10^7 times as great as for free radical polymerization.

In discussing anionic polymerization kinetics, Battacharyya et al. [14] have expressed the rate as the sum of polymerization rates for the ion pair $P^-(C^+)$ and the free propagation ion P^-. This method gives

$$R_p = k_{P^-(C^+)}[P^-(C^+)][M] + k_{P^-}[P^-][M] \qquad (5\text{-}35)$$

where $k_{P^-(C^+)}$ and k_{P^-} are the rate constants for the propagation of the ion pair $[P^+(C^-)]$ and the free propagating ion P^-, respectively. The symbol C^+ indicates the positive gegen ion.

The propagating species take part in the following equilibrium

$$P^-(C^+) \overset{K}{\rightleftharpoons} P^- + C^+ \qquad\qquad (5\text{-}36)$$

where the dissociation constant K is given by

$$K = \frac{[P^-][C^+]}{[P^-(C^+)]} \qquad\qquad\qquad (5\text{-}37)$$

or

$$K = \frac{[P^-]^2}{[P^-(C^+)]} \qquad\qquad\qquad (5\text{-}38)$$

for the case where $[P^-] = [C^+]$.

Combination of Eqs. (5-35) and (5-38) yields

$$\frac{R_p}{[M][P^-(C^+)]} = k_{p^-(C^+)} + \frac{K^{\frac{1}{2}}k_{p^-}}{[P^-(C^+)]^{\frac{1}{2}}} \tag{5-39}$$

The right side of Eq. (5-39) equals k_p in Eq. (5-34), assuming
that the concentration of P^- is small. Such is generally the case,
and $[P^-(C^+)]$ is approximately equal to the total concentration of
living ends (the catalyst concentration).

Examination of Eqs. (5-34) and (5-39) indicates that a plot of
experimental k_p values versus $[P^-(C^+)]^{-\frac{1}{2}}$ should be linear. This
has been found to be the case, with $K^{\frac{1}{2}}k_{p^-}$ equal to the slope and
$K_{p^-(C^+)}$ equal to the intercept. Studies to date indicate that k_{p^-}
is much larger than $k_{p^-(C^+)}$ in many cases.

The number-average degree of polymerization for an anionic
polymer equals the ratio of the concentrations of monomer and
living ends,

$$\bar{X}_n = \frac{[M]}{[M^-]} \tag{5-40}$$

Generally speaking, all the catalyst is converted into living ends
of one kind or another, so that Eq. (5-40) becomes

$$\bar{X}_n = \frac{[M]}{[C]} \tag{5-41}$$

for cases involving one polymer molecule per catalyst molecule
(using butyllithium, for example).

Experimental results on temperature dependence show the E_R
values for anionic polymerization to be usually low and positive.
This E_R value is merely the energy of activation for propagation.
The activation energy depends to some extent on the solvent employed, just as in the case of cationic polymerization. For
example, the propagation energy of activation is about 9 kcal/mole
for the styrene-sodium naphthalene system in dioxane, but only

about 1 kcal/mole for the same system in tetrahydrofuran [15]. The
molecular weight of the polymer formed in a nontermination polymerization is unaffected by temperature, providing transfer agents
are absent.

D. Deviations from Conventional Models of
Polymerization Kinetics

The above discussion gives the conventional analysis of the
reaction kinetics involved in free radical, cationic, and anionic
polymerization mechanisms. As recognized some time ago by Chapiro
[42], the assumptions employed in these generalized derivations may
not hold true for certain monomers and for certain experimental
conditions, with the result that the observed polymerization behavior for individual cases may deviate considerably from that
outlined above.

Consider first the parallel expressions for ions,

$$R_i = k_t [M^+][M^-] \qquad (5\text{-}42)$$

and for radicals,

$$R_i = k_t [R\cdot][R\cdot] \qquad (5\text{-}43)$$

which are based on the equality of rates of initiation and termination in the steady state. Allowing [Z] to represent the steady
state concentration of either ions or free radicals, the situation
for either case can be summarized by the same general equation,

$$[Z] = (R_i/k_t)^{\frac{1}{2}} \qquad (5\text{-}44)$$

so that the rate of polymerization for long kinetic chains is given
by

$$R_p = k_p [Z][M] \qquad\qquad (5\text{-}45)$$

where $[M]$ equals monomer concentration and k_p is the propagation
rate constant. Combining (5-44) and (5-45) yields

$$R_p = k_p \left(\frac{R_i}{k_t} \right)^{\frac{1}{2}} [M] \qquad\qquad (5\text{-}46)$$

which is the well-known conventional expression for polymerization
rate. If it is further assumed that $R_i k_t = \tau_{ss}^{-2}$ (see page 284),
it is evident that

$$R_p = k_i k_p [M] \tau_{ss} \qquad\qquad (5\text{-}47)$$

where τ_{ss} is the mean lifetime for free ions or free radicals in
the stationary state.

Equation (5-46) can be applied to a wide variety of thermal,
photochemical, and radiolytic polymerizations having free radical
mechanisms. It also applies to ionic polymerizations, providing
free ions are mainly responsible for the propagation of polymerization, which would not be the case for catalyzed polymerizations
(see discussion above). Equation (5-46) indicates a second-order
recombination of the appropriate intermediates, and it has recently
been postulated that such a recombination can apply to free ions as
well as free radicals [43].

The assumptions underlying Eq. (5-46) are worth examining.
They are, first: that k_p does not depend on the chain length of the
active, propagating species; second: that k_t does not depend on the
value of R_i; and third: that all intermediate active species terminate through bimolecular recombination.

The first assumption would be called into question if the reactivity of the radical initially formed was appreciably different
from that of the resulting propagating intermediate. This would
imply that the rate constant of the first initiation reaction k_i

would be different from k_p for succeeding addions of monomer. If k_p should be appreciably greater than k_i, the kinetic chain length would be reduced, and it can be shown [33] that the equation for rate of polymerization would take the modified form

$$R_p' = \frac{k_p}{k_t^{\frac{1}{2}}} [M]R_i^{\frac{1}{2}} - \left(\frac{k_p - k_i}{k_i}\right)R_i \qquad (5\text{-}48)$$

Noting that R_p' is of the form $aR_i^{\frac{1}{2}} - bR_i$, it is evident that the size of the second term increases relative to the first as the dose rate increases. In other words, deviations on account of $k_p > k_i$ will become more noticeable at high dose rates. Experimentally, this will be reflected in a decrease of the exponent of R_i in the equation

$$R_p' \propto R_i^X \qquad (5\text{-}49)$$

below a value of 0.5 as the dose increases and the second term in Eq. (5-48) becomes significant.

The second assumption indicating k_t independent of R_i may also fail at high dose rate. Benson and North [44] have shown that k_t is essentially independent of radical size above a chain length of 100 monomer units. At very high dose rates the kinetic chain becomes shorter, radicals are smaller, and k_t increases as R_i increases. The latter circumstance would cause the exponent X in Eq. (5-49) to drop below 0.5, even though the first assumption (k_p independent of chain length) continued to hold true. When the first and second assumptions both break down, the tendency for X to drop below 0.5 will be correspondingly enhanced.

Assumption three postulates bimolecular termination, and there are at least two cases where this assumption would not apply. One case would involve unimolecular termination of the propagating species. This has been referred to as "degradative chain transfer" in free radical mechanisms and results in the production of a

stable radical that will not propagate. If the unimolecular termination predominates, R_p will depend on the first power of R_i as
shown in the equation

$$R_p = R_i \frac{k_p[M]}{k_1} \qquad (5\text{-}50)$$

where k_1 is the internal or unimolecular termination rate constant,
and the lifetime of the propagating species equals $1/k_1$.

The other important case would be termination by transfer of
the active species to a scavenger molecule, giving rise to a product that would not propagate the chain. Numerous examples of this
type of termination have been found in radiation-induced ionic
polymerizations. It has been shown that rate of polymerization
could then be expressed as follows [43]:

$$R_p = \frac{R_i k_p[M]}{(R_i k_t)^{\frac{1}{2}} + k_x[X]} \qquad (5\text{-}51)$$

where k_x is the rate constant for transfer to the scavenger, X.
Equation (5-51) would become identical with Eq. (5-46) at high dose
rates providing the second term in the denominator can be neglected.
Equation (5-51) shows that at a large scavenger concentration polymerization rate will depend on the first power of R_i, whereas at
zero scavenger concentration the dependence will revert to the 0.5
power of R_i. The dependence on the 0.5 power of R_i is more difficult to attain in free ion polymerizations, because of their
sensitivity to very small amounts of impurities which function as
scavengers.

In radiolytic polymerization, a question often arises as to
whether the mechanism is free radical, cationic, or anionic. The
following points [33] are helpful in distinguishing between the
mechanisms.

1. Evaluate the effect of variables such as temperature, dose
rate, and solvent. The temperature coefficient and energy of

activation are often smaller for ionic polymerizations than for
radical polymerizations. Ionic polymerizations are more sensitive
to solvating power and polarity of the solvent.

2. Compare the behavior in the presence of free radical or
ionic catalysts.

3. Evaluate the effect of free radical or ionic scavengers
at low concentrations. Care should be taken in evaluating the
effect of the additive. For instance, benzoquinone inhibits both
cationic and free radical polymerization.

It should generally be assumed that all methods of initiation
may take place simultaneously during irradiation, and the problem
is to determine which type of mechanism predominates. This determination depends mainly on kinetic measurements.

The known types of initiators for various monomers give some
idea as to whether a monomer will polymerize by a free radical,
cationic, or anionic mechanism. Table 5-1 from Schildknecht [45]
lists the various monomers that are known to respond to certain
types of initiators. Generally speaking, irradiated monomers are
apt to exhibit the same type(s) of polymerization mechanism(s) as
indicated in Table 5-1.

TABLE 5-1

Homopolymerization Initiator Types That Are

Effective for Various Monomers

Cationic	Free radical	Anionic
Isobutylene	Vinyl chloride	Nitroethylene
Cyclopentadiene	Vinyl acetate	Vinylidene cyanide
Alkyl vinyl ethers	Acrylonitrile	Acrylonitrile
β-Pinene	Methyl methacrylate	Methyl methacrylate
α-Methylstyrene	Ethylene	α-Methylstyrene
Styrene	Styrene	Styrene
Butadiene	Butadiene	Butadiene

It might be thought that a study of reactivity ratios would provide a reliable indication as to the type of mechanism obtaining in copolymerization, but such is not the case. Reactivity ratios are known to vary with changes in solvent or initiator. Furthermore, the mechanism of a radiation-induced ionic polymerization can be changed by very minute amounts of an impurity such as water, and the polymerization of one monomer of a pair may be inhibited by the presence of the other monomer.

The measurement of the number-average degree of polymerization $\overline{DP}_n$ provides additional information which supplements that obtained from kinetic measurements. Under steady state conditions where reaction rates are independent of time,

$$G(-m) = 100 \ R_p N/I \tag{5-52}$$

and

$$G_i = 100 \ R_i N/I \tag{5-53}$$

when N equals Avogadro's number and I equals dose rate in eV g^{-1} sec^{-1}. The kinetic chain length ν is given by

$$\nu = G(-m)/G_i = \frac{R_p}{R_i} \tag{5-54}$$

When chain transfer processes are lacking, the number-average molecular weight $<M_n>$ can be used to relate $G(-m)$ and G_i as follows:

$$G(-m) = G_i \overline{DP}_n = G_i \ \frac{<M_n>}{M_1} \tag{5-55}$$

where M_1 equals the molecular weight of the monomer.

The value of $\overline{DP}_n$ is also influenced by any nondegradative chain processes that may be taking place. The general equations for $\overline{DP}_n$ and ν are the following:

$$\overline{DP}_n = \frac{k_p[Z][M]}{k_f[Z][M] + k_s[Z][S] + k_1[Z] + k_x[Z][X] + k_t[Z]^2} \tag{5-56}$$

$$\nu = \frac{k_p[Z][M]}{k_1[Z] + k_x[Z][X] + k_t[Z^2]} \tag{5-57}$$

All processes leading to chain breaking are indicated in the denominator of Eq. (5-56), while all processes leading to chain termination are indicated in the denominator of Eq. (5-57). The symbols k_f and k_s are rate constants for chain transfer to monomer and solvent, respectively. By inverting and combing Eqs. (5-56) and (5-57) it is shown that

$$\frac{1}{\overline{DP}_n} = \frac{1}{\nu} + \frac{k_f}{k_p} + \frac{k_s}{k_p}\frac{[S]}{[M]} \tag{5-58}$$

If there is essentially no chain transfer, the last two terms drop out of Eq. (5-58) and it becomes simply

$$\frac{1}{\overline{DP}_n} = \frac{1}{\nu} \tag{5-59}$$

as indicated earlier in Eqs. (5-54) and (5-55). In that simple case G_i is equal to the experimentally measurable quantity $G(-m)/\overline{DP}_n$.

Turning first to chain transfer in bulk polymerization, the expression for $\overline{DP}_n$ is obtained from Eq. (5-58) by equating [S] to zero, yielding

$$\frac{1}{\overline{DP}_n} = \frac{G_i}{G(-m)} + \frac{k_f}{k_p} \tag{5-60}$$

The importance of transfer to monomer can be estimated from the sign of the quotient $G(-m)/\overline{DP}_n$, as indicated by Eq. (5-60).

For example, in the radiation-induced cationic bulk polymerization
of α-methylstyrene [46], the magnitude of $G(-m)/\overline{DP}_n$ is greater
than 100, indicating that $\overline{DP}_n$ is determined by the transfer constant k_f/k_p in Eq. (5-60). In this example, the value of $\overline{DP}_n$ is
independent of $G(-m)$ even in the presence of enough water or
ammonia to produce strong retardation of polymerization.

In order for $\overline{DP}_n$ to be appreciably less than the limiting
value $(\overline{DP}_n)_0$ determined by k_p/k_f, it is necessary for

$$G(-m) < G_i (\overline{DP}_n)_0 \qquad\qquad (5\text{-}61)$$

It is known that G_i is approximately 0.1 for ionic polymerizations,
so that for α-methylstyrene having a $(\overline{DP}_n)_0$ no greater than 50, the
value of $G(-m)$ must be less than 5 to produce an effect of this
type. The fact that the value of $\overline{DP}_n$ is dependent on chain transfer to monomer for α-methylstyrene, and isobutylene, provides con-
irmatory evidence that radiation-induced polymerization of these
types is mainly cationic in mechanism.

In considering solution polymerization it should be noted that
ν is a function of $k_p[M]\tau$ as shown by Eq. (5-57), where the lifetime τ of the kinetic chain is given by the equation,

$$\tau^{-1} = k_f + k_x[X] + k_t[Z] \qquad\qquad (5\text{-}62)$$

If chain transfer to solvent is ignored, Eq. (5-58) becomes

$$\frac{1}{\overline{DP}_n} = \frac{k_f}{k_p} + \frac{1}{k_p[M]\tau} \qquad\qquad (5\text{-}63)$$

From this it is clear that $k_p\tau$ may be computed by plotting $1/\overline{DP}_n$
versus $1/[M]$, providing that other variables affecting τ such as
[X] and dose rate remain unchanged. Hence G_i may be computed as
follows:

$$G_i = \frac{1}{k_p\tau} \frac{G(-m)}{[M]} \qquad\qquad (6\text{-}63A)$$

The latter procedure has been used in studying [47] the radiation-induced ionic polymerization of styrene in dichloromethane at -78°C.

The results of changing the temperature and dose rate have often been used in attempting to determine the type of mechanism involved in radiation-induced polymerization. However, dependence of rate on the half-power of R_i does not guarantee a free radical mechanism nor does rate dependence on the first power of R_i prove an ionic mechanism; a very low energy of activation suggests but does not prove the existence of an ionic mechanism; and it is not correct to say that ionic mechanisms are important *only* at low temperature.

With regard to the former belief that ionic polymerization rates depend on the first power of R_i, it has been found that under extremely anhydrous conditions the rate depends on the 0.5 power of dose rate as expected for a bimolecular recombination of free ions. Although a free radical mechanism is usually associated with rate dependence of the half-power of I, a first power dependence would hold true if termination took place mainly by degradative chain transfer [Eq. (5-49)]. Hence, R_p can depend on any power of I between 0.5 and 1.0 as determined by the kinetics of the termination reaction, with no reference to whether the mechanism is ionic or free radical. It can be concluded that studies of dose rate and temperature dependence, taken alone, are not reliable guides as to the nature of the mechanism.

What can be deduced in regard to temperature is that radiation-induced polymerizations which take place at -80° to 0°C and with activation energies of less than 2 kcal/mole are apt to be ionic rather than free radical in mechanism. Styrene in dichloromethane, bulk isobutylene, and bulk cyclopentadiene are examples of this class. Because of their melting points some bulk monomers can only be studied in the liquid phase at high temperature, but it would not be correct to conclude that free radical mechanisms always predominate in such cases.

The use of known scavengers for certain types of intermediates can provide a good deal of information about mechanisms, provided the results are properly interpreted. Free radical scavengers include DPPH, galvinoxyl, nitric oxide, oxygen, ferric chloride, and p-benzoquinone, which are capable at concentrations of about 10^{-3} M of intercepting most of the free radicals in the steady state of polymerization. Scavengers, which cause proton transfer from the "primary" ion to the scavenger, include water, ammonia, alcohols, and amines. Substances which cause positive charge exchange from the primary ion include electron donors such as hydrogen sulfide and nitric oxide. Electron scavengers cover a wide variety of aromatic hydrocarbons and their derivatives, carbon monoxide, sulfur hexafluoride, alkyl halides, and nitrous oxide.

Certain precautions must be observed in using scavengers. Some of the free radical scavengers such as DPPH, p-benzoquinone, and oxygen are also at times capable of interfering with ionic reactions, and hence would not give an unambiguous indication as to the nature of the mechanism with which they are interfering. Molecular iodine would be considered a straightforward radical scavenger, but it will react with electrons to some extent and is a mild polymerization catalyst for monomers such as isobutylene [48].

The action of ion scavengers such as water and ammonia is much less ambiguous. Mass spectrometric data indicate that water and ammonia act as bases in proton transfer reactions from certain hydrocarbon ions [38]. This confirms the hypothesis that water and ammonia retard radiation-induced polymerization by chain termination through proton transfer [46]. Hence, inhibition of a radiation-induced polymerization reaction by water at 10^{-3} M or less would be definite evidence of an ionic mechanism. Since water retards both cationic and anionic propagation, these two possibilities can be distinguished by addition of ammonia or amines which function as specific scavengers for cations.

An additional use for scavengers is in the determination of propagation rate constants. This comes about through the

dependence of R_p and $\overline{DP}_n$ on the concentration $[X]$ of a retarder
(scavenger). Noting that $R_p = R_i \nu$, Eq. (5-57) yields

$$R_p = R_i \ \frac{k_p[M]}{k_1 + k_x[X] + k_t[Z]} \tag{5-64}$$

Through inversion this equation becomes

$$\frac{1}{R_p} = \frac{1}{R_i} \ \frac{k_x[X]}{k_p[M]} + f(R_i) \tag{5-65}$$

where $f(R_i)$ includes the dose-rate dependent term associated with
the bimolecular recombination of the intermediate Z. From Eq.
(5-65) it follows that the variation of R_p with $[X]$ at constant
dose rate can be expressed by

$$\frac{1}{R_p} = \frac{1}{(R_p)_0} + \frac{1}{R_i} \ \frac{k_x[X]}{k_p[M]} \tag{5-66}$$

where $(R_p)_0$ is the uninhibited rate of polymerization corresponding
to $[X] = 0$. Equation (5-66) can also be expressed in terms of G
values,

$$\frac{1}{G(-m)} = \frac{1}{[G(-m)]_0} + \frac{1}{G_i} \ \frac{k_x[X]}{k_p[M]} \tag{5-67}$$

where $[G(-m)]_0$ is the unretarded yield at the same dose rate.

Multiplying Eq. (5-66) by R_i gives $\overline{DP}_n$ as a function of $[X]$ in
the form of the Mayo equation,

$$\frac{1}{\overline{DP}_n} = \frac{1}{(\overline{DP}_n)_0} + \frac{k_x[M]}{k_p[M]} \tag{5-68}$$

where $(\overline{DP}_n)_0$ is the degree of polymerization in the absence of
added scavenger. It can be shown that $\overline{DP}_n$ will not be sensitive to

scavenger concentration when $(\overline{DP}_n)_0$ is low due to extensive transfer to monomer, and under these circumstances the Mayo plot will
not give an accurate value for k_x/k_p. For the case of cyclopentadiene [49], where monomer transfer is relatively unimportant, the
ratios $k_x/G_i k_p$ and k_x/k_p can be computed from the functional
dependence of $G(-m)$ and $\overline{DP}_n$, respectively, on the concentration of
ammonia retarder as indicated in Eqs. (5-67) and (5-68).

For strict accuracy, Eqs. (5-67) and (5-68) would have to be
corrected for scavenger depletion during an experimental polymerization. Such a correction is usually not possible, but reasonable accuracy can be attained by extrapolation to initial $G(-m)$
yields before substitution in Eq. (5-67). The use of initial $\overline{DP}_n$
values in Eq. (5-68) would be impractical experimentally, so a
compromise is made by using the lowest feasible total dose.

II. SPECIAL CHARACTERISTICS OF RADIOLYTIC POLYMERIZATION

A.   Introduction

Radiolytic chemical reactions generally result from secondary
processes involving radicals and ions formed by the interaction of
radiation with matter. Since the process of excitation involves
an interaction of matter with a fast-moving particle or ray, the
primary ions and radicals formed must be located along thepath of
the particle or ray.

It has been supposed that the path of a high energy particle
would resemble the pattern of water droplets along the track of
charged particles made visible in a Wilson cloud chamber. It has
been postulated that an α-particle path would be surrounded by a
densely ionized cylindrical column, while isolated "clusters" of
ions would be distributed along the path of a fast electronlike

beads along a string [16]. One might also visualize "spurs" of excited ions or radicals branching off from the main path of the fast-moving high energy particle.

Each "cluster" or "spur" would presumably contain a high local concentration of ions and radicals initially. This high local concentration would decrease quickly because of diffusion of active species and recombination of some of the free radicals. Diffusion causes the initial cluster of active species to increase in volume, and adjacent clusters would probably soon overlap. Thus the distribution of radicals and ions becomes more random, until it is again disturbed by the passage of another track.

Whether radiolytic polymerizations proceed by free radical or ionic mechanism has been a question of controversy and discussion for many years. In the earliest work, it was postulated that "ionizing radiation" leads to ionic reactions by S. C. Lind, sometimes called the father of radiation chemistry. However, during the later 1940s and through most of the 1950s it was widely believed that radiolytic polymerization of vinyl monomers takes place by a free radical mechanism. This interpretation was based on a square-root dependence of the polymerization rate on dose rate, an inverse square-root dependence of molecular weight on dose rate, a temperature coefficient of rate which was approximately the same as for a photo-reaction, and inhibition by known free radical inhibitors. It was thought by most investigators that ions simply did not exist long enough to play a significant part in the initiation and propagation of polymerization.

However, in 1957, Davison published findings on the radiolytic polymerization of isobutene, which could best be explained as an ionic process [18]. This conclusion was confirmed by other investigators, and interest in the possible existence of other ionic radiolytic polymerization processes was reawakened.

Recent investigations have proceeded along three related lines. First, studies at low temperature should reveal whether a process is ionic, since radiolytic free radical polymerization should have an activation energy of 5 to 8 kcal/mole, while the activation energy of ionic processes should be near zero. Second, the use of solvents having higher dielectric constants than vinyl monomers should favor ionic propagation. Third, the removal of impurities such as water should increase the lifetime of reactive ions that might be present.

Studies along such lines appear to indicate that radiolyses of vinyl monomers can potentially produce free radical, cationic, or anionic propagation, and at times two or three of these simultaneously. The relative importance of the various mechanisms at any given time depends on the chemical nature of the monomer and the solvent, the temperature, and the presence or absence of inhibitors for the various types of mechanism.

Recent research has demonstrated that cationic polymerization is more probable at low temperature, and that conversion is increased if the monomer is dissolved in halogenated solvents. Anionic polymerization takes place when acrylonitrile or methyl methacrylate is irradiated at low temperature in amine or amide solvents. It has been postulated that chlorinated solvents "stabilize" the negative charge by converting electrons into Cl^- ions, while amines "stabilize" the positive charge on quaternary ammonium-type ions.

B.   Primary Processes during Irradiation

A better understanding of radiation-induced polymerization has been obtained in recent years throgh more complete information on the part plyed by specific and identifiable reaction intermediates [33]. The theory must take account of the time scale of the primary and secondary processes which result from the passage of an energetic charged particle. The first generation of reactive species to be formed includes excited molecules, ions, and electrons.

Subsequent reactions of these intermediates produce free radicals
and neutral stable molecules, followed by additional free radical
reactions to form stable products. In solid or liquid phases, the
initial distribution of primary products and reaction intermediates
is quite inhomogeneous. For typical secondary electrons produced
by γ-radiation, a number of multiple ionizations and/or excitations
are formed in well-separated localized groups (spurs) lying along
the track [25,27].

Magee [20] has developed a theory concerning the chronological
order of the processes taking place in a spur. The diffusion of
neutral radicals with the passing of time results in an increase
in spur size until eventually a homogeneous distribution is set up
in the steady state. Hence, the competition between radical-
radical and radical-solute (scavenging) reactions depends on the
local radical concentration during the time-dependent expansion of
the spur. It should be noted that very high values of the product
$k_s[S]$, where k_s is the rate constant and [S] is the scavenger concentration, are needed to scavenge those radicals which otherwise
would undergo combination in the early history of the spur, whereas
the scavenging of radicals undergoing recombination in a medium of
homogeneous distribution is effective at $k_s[S]$ values which are
lower by several orders of magnitude [21]. Similar remarks could
be made about the simpler problem regarding the probability of
geminate recombination ("cage effect") following thermal or photochemical generation of radicals in pairs [22].

In view of the coulombic forces between charged species, the
diffusion model for neutral radicals cannot be used in the ion-
electron recombination problem without certain changes [26]. The
proper theoretical treatment to be employed depends on the nature of
the electron in the condensed phase [23]. Experimental work with ion
scavengers indicates that the lifetimes of positive ions in liquids
may be greater than 10^{-10} sec [24]. This may be explained by
assuming that the motion of the electron trapped in the liquid is
similar to that of a molecular ion.

Although most ions and electrons undergo geminate recombination in hydrocarbons, a small portion of the free ions should escape the forces of coulombic attraction and set up a steady state [33]. The fraction of ions that escape should be smaller than that for neutral free radicals in irradiated hydrocarbons. The experimental evidence for free ions comes from measurements of electrical conductivity produced in hydrocarbons during irradiation [29]. Such results show that the yield of separated ions is about 0.1 ions/100 eV (G = 0.1). Additional evidence for the presence of free ions comes from ionic polymerization experiments, which indicate that kinetic chain lengths can be as high as 10^6 in rigorously purified systems [30]. Such long lifetimes make it evident that some ions can exist for a considerable period of time without undergoing any sort of geminate recombination [28].

The present status of the ionic separation problem can be summarized as follows. In solvents having a large dielectric constant ε there is a greater probability of ionic separation because the expression for coulombic attractive force contains ε in the denominator. In solvents having a low dielectric constant, the major portion of ions and electrons will undergo recombination in the parent spur within about 10^{-7} to 10^{-11} sec, with only a small portion (G = 0.1) of charged species escaping the coulombic field and setting up a steady-state concentration of free ions having an average lifetime greater than 10^{-3} sec at normal dose rates.

Based on separate steady-state assumptions for ions and radicals, it is possible to estimate ion and radical concentrations in the radiolysis of hydrocarbon liquids. For ions in the steady state, the rate of initiation equals the rate of termination, or

$$R_i = k_t [M^+][M^-] \qquad (5\text{-}69)$$

where R_i is the rate of initiation and k_t is the rate constant for recombination. The steady-state ion concentration would accordingly be

$$[M^+] = (R_i/k_t)^{\frac{1}{2}}$$ (5-70)

and the average lifetime τ_{ss} in the stationary state would be equal to

$$\tau_{ss} = [M^+]/R_i = (1/k_t R_i)^{\frac{1}{2}}$$ (5-71)

Similar expressions can be derived for radicals if the rate constant k_t for radical recombination is defined as follows:

$$\frac{-d[R\cdot]}{dt} = 2 \, [\frac{k_t}{2}] [R\cdot]^2 = k_t [R\cdot]^2$$ (5-72)

Smoluchowski [31] gave an equation for computing the reaction rate constant k_t for two particles diffusing toward each other,

$$k_t \text{ (radicals)} = 4\pi\sigma D$$ (5-73)

where σ is the distance at which reaction takes place and D is the total diffusion coefficient $(D_1 + D_2)$ for the reacting species.

Debye [32] has computed the correction factor, which must be employed when the particles diffusing toward each other are singly and oppositely charged ions,

$$k_t \text{ (ions)} = 4\pi r_c D/[1 - \exp(-r_c/\sigma)]$$ (5-74)

where r_c is the distance at which electrostatic and thermal energies are equal. In solvents of low dielectric constant where $r_c \gg \sigma$, the latter equation becomes

$$k_t \text{ (ions)} = 4\pi r_c D$$ (5-75)

Therefore, estimated concentrations of ions or radicals in the steady state can be computed from Eq. (5-71) by assuming reasonable values for R_i, relevant diffusion constants, σ, and r_c. Such computations [33] indicate that at the same dose rate for liquid hydrocarbons the stationary-state concentration of ions is approximately two orders of magnitude less than that of radicals, whereas the average lifetimes of ions and radicals are closely similar. For example, at a dose rate of 100 Mrads/h. the concentrations of ions and radicals were estimated to be 2×10^{-9} and 1.6×10^{-7} M, respectively, while the mean lifetimes were 0.9×10^{-3} and 1.3×10^{-3} sec, respectively. It was shown that this situation resulted from lower G_i values and higher k_t values for ions in relation to radicals.

C. Intermediates in Radiation-Induced Polymerization

Radiation-induced polymerization differs from other types of polymerization, in that other more conventional methods generally exhibit the generation of a specific kind of intermediate such as free radicals or ions. Radiation-induced polymerization is much less specific in its initial phases, and a considerable variety of reactive intermediates is formed. In spite of this situation, it is usually found in radiolytic polymerization that one mechanism predominates under a particular set of reaction conditions.

Whereas the preceding remarks apply to addition polymerization of vinyl monomers, it is found that the polymerization of cyclic epoxides, ethers, and sulfides by ring opening differ markedly from addition polymerization. The ring-opening type of polymerization is generally initiated by a wide variety of cationic and anionic catalysts, but hardly ever by free radical catalysts. Early studies on ring-opening polymerization were important in demonstrating the usefulness of radiation in the initiation of ionic polymerization [34].

Whereas free radicals and ions are generally regarded as the intermediate active species in addition polymerization, there is not much direct evidence concerning the structure of intermediates involved in radiolytic polymerization. Although ESR spectroscopy can be used to identify and study free radicals, little or no work of this type has been published concerning the free radicals present in the irradiation of liquid monomer systems. Electron spin resonance has been used to study radicals present in solid-state radiolytic polymerization, but in this case the hyperfine structure in the ESR spectra of the solids is often too poorly resolved to enable unambiguous interpretation of the free radical structure.

The mass spectrometer has made valuable contributions to the study and understanding of radiolytic ionic polymerizations in the gas phase. For ethylene it is concluded that the primary ion-molecule reaction gives a dimer of the formula $C_4H_8^+ \cdot$ [35]. At pressures of a few millimeters of ethylene it has been demonstrated that $C_4H_8^+ \cdot$ adds molecules of ethylene and forms ions up to C_{14} in length [36].

$$(C_2H_4)_m^+ \cdot \quad C_2H_4 \rightarrow (C_2H_4)_{m+1}^+ \cdot \qquad\qquad (5\text{-}76)$$

At low pressures it was found that $C_4H_8^+ \cdot$ dissociates to yield $C_3H_5^+$ and $CH_3 \cdot$. It was postulated that most of the larger polymeric ions are formed by stepwise addition of ethylene to the $C_3H_5^+$ carbonium ion, which acts as the initiator. However, the presence of large carbonium ions does not prove that the polymerization mechanism is predominantly cationic in nature. In fact, the evidence indicates that the radiolytic gas-phase polymerization of ethylene takes place mainly by a free radical mechanism [37]. This is one example of the general truth that the identification of a certain ion does not prove that it plays an important part in the overall reaction.

Mass spectrometry has also produced evidence concerning the transfer of a proton from a hydrocarbon ion to a molecule such as ammonia or water [38]. Reactions of this type would doubtless explain the sensitivity of radiolytic cationic polymerizations to inhibition by water or ammonia. It has been postulated that scavenging of protons by water and ammonia from hydrocarbon ions can take place when these reactions are exothermic, resulting in the formation of H_3O^+ and NH_4^+ ions [38].

The pulse radiolysis technique has been used to study intermediates in the radiaion-induced polymerization of styrene and α-methylstyrene [39, 40]. Briefly, the technique consists of subjecting the system to very short intense bursts of ionizing radiation from a source such as a Van de Graaf generator. The transient species formed is detected and analyzed over a short time period by a suitably fast method such as optical absorption.

Measurement of conductivity and ion mobility permits the steady-state concentration of ions $n = \Sigma[M^+]$ to be estimated by use of the equation,

$$\kappa = ne(\mu_+ + \mu_-) \tag{5-77}$$

where κ equals conductivity produced by irradiation, μ_+ and μ_- are the mobilities of the ions of the charge indicated, e equals the electronic charge, and the assumption is made that the ions are singly charged.

The rate of formation of ions R_i (equal to $G_i I/100$) can be computed from Eq. (5-70) if k_t is known. The latter can be computed from Eq. (5-75). Equation (5-75) can be expressed in the form

$$k_t = (4\pi e/\varepsilon)(\mu_+ + \mu_-) \tag{5-78}$$

From a knowledge of R_i the value of G_i can then be calculated, which is the yield of free ions forming and recombining under steady-state conditions. A method of measuring ion mobilities during radiolytic ionic polymerization would provide valuable information about the size of propagating ions, but such a method is lacking.

Without measuring ionic mobilities, the conductivity values can be employed to compute the average lifetime τ_{ss} of the ions in the stationary state. By eliminating the sum of ionic mobilities from Eqs. (5-77) and (5-78), the following is obtained:

$$\kappa = (\varepsilon/4\pi)nk_t \qquad\qquad (5\text{-}79)$$

and noting from Eq. (5-70) and (5-71) that

$$\tau_{ss} = 1/nk_t \qquad\qquad (5\text{-}80)$$

it follows that

$$\tau_{ss} = \varepsilon/(3.6 \times 10^{12}\,\pi\kappa) \qquad\qquad (5\text{-}81)$$

where the factor of 9×10^{11} is needed to convert κ in $\text{ohm}^{-1}\,\text{cm}^{-1}$ to units of sec^{-1}. The equation for τ_{ss} as a function of ε and κ is generally called the Maxwell time lag for the decay of charges in an electrolytic conductor [41]. The value of R_i can be written

$$R_i = \tau_{ss}^{-2}k_t^{-1} \qquad\qquad (5\text{-}82)$$

where k_t is thought of as an average rate constant for the recombination of various pairs (of various sizes and mobilities). It should be noted that τ_{ss} can be computed directly from the conductivity, while estimation of R_i also requires a knowledge of k_t, which can be obtained from ionic mobilities [Eq. (5-78)].

The following sections discuss specific examples of radiation-induced homogeneous polymerizations.

III.   HOMOGENEOUS BULK POLYMERIZATION

A.   Introduction

From what has already been said, it is evident that a free-
radical mechanism predominates in the radiation-induced polymerization of certain monomers, while an ionic mechanism predominates in
others. Extensive studies of vinyl acetate and methyl methacrylate
have shown that a free radical mechanism applies to the radiation-
induced polymerization of these monomers at or above room temperature [42]. Generally speaking, these radical polymerizations
follow the kinetics of Eq. (5-46). It should be noted that styrene
undergoes a free radical polymerization when not extremely dry, but
when it is extremely dry it polymerizes by an ionic mechanism
during irradiation.

Such studies have shown that the radiation chemical yield of
free radicals from pure monomers can be computed from the known
rate constants of the individual mechanism steps that were determined earlier. It is of interest to compare such yields (G_R) with
those obtained from the consumption of DPPH in irradiated monomer
systems (see Table 5-2). The values obtained by the two techniques
are in reasonably good agreement.

TABLE 5-2

Radical Yields from Irradiated Monomers [33]

Monomer	G_R	G(-DPPH)
Vinyl acetate	12	6-9
Methyl methacrylate	11.5	5.5-6.7
Acrylonitrile	1.2-5.6	5.0
Styrene	0.69	0.66

The values of k_p and k_t for several monomers are listed in
Table 5-3 for reference purposes. These are generally considered
the most reliable values, as determined several years ago by
Matheson et al. [62]. Since the rate of polymerization is proportional to $k_p/k_t^{\frac{1}{2}}$ and also G_R, the large values of these quantities
for methyl methacrylate and vinyl acetate are consistent with the
fact that these monomers are more easily polymerized by the radical
mechanism than styrene. Other monomers having k_p much larger than
styrene include methyl acrylate and the n-alkyl methacrylates,
while monomers having k_p's less than styrene include 4-vinyl
pyridine and p-methoxystyrene [63].

TABLE 5-3

Propagation and Termination Rate Constants
in Free Radical Polymerizations

Monomer	k_p, M^{-1} sec^{-1}	k_t, M^{-1} sec^{-1}	Temp., °C
Vinyl acetate	1012	5.9×10^7	25
Methyl methacrylate	286	2.4×10^7	30
Acrylonitrile	1960	—	60
Styrene	44	4.75×10^7	25

In short, the results for radiation-induced polymerization indicate reasonable agreement with absolute rate constants in polymerizations activated by free radical initiators. Interpretations
of ionic type radiation-induced polymerizations are more difficult
to handle, partly because of the increased sensitivity to the
nature of the catalyst, solvent, and possible impurities. One recent accomplishment is the determination of k_p for the anionic
polymerization of styrene by free ions and ion pairs [64]. Radiation is an especially suitable method for the activation of ionic

polymerization, because it is the only means of generating free
ions in a medium of low dielectric constant. Furthermore, the
contributions of ionic and free radical mechanisms to radiation-
induced polymerizations can be distinguished and evaluated by the
methods discussed above. Also, it has been found that the rates
of radiation-induced cationic polymerization are very large for a
number of monomers, allowing good progress to be made in the study
and interpretation of such reactions.

B.   Bulk Polymerization of Styrene and α-Methylstyrene

Recent work has emphasized the need for rigorously dry monomers in obtaining ionic type radiation-induced polymerizations.
This was shown early in work on β-pinene [65] and α-methylstyrene
[46]. Rates of polymerization increased with improvement in monomer drying technique. Deliberate addition of water caused a strong
inhibition of polymerization. Polymerization rates for thoroughly
dried samples of β-pinene and α-methylstyrene were larger by a
factor of 1000 than for water-saturated samples. It appeared that
ionic mechanisms might be greatly accelerated for other monomers by
careful drying, and this has been confirmed for isobutyl vinyl ether
[66], styrene [7], and cyclohexene oxide [67].

Some question has arisen as to how very small concentrations
of water can persist in their inhibiting effect throughout the
course of a radiation-induced polymerization. This can be explained
by a reaction of the following type,

$$H_3O^+ + Y^- \rightarrow H_2O + Y \tag{5-83}$$

whereby the original scavenger (H_2O) is regenerated. Hence, a
proton scavenger such as water could transfer protons from the
positive hydrocarbon ions to the negative ions in the system without being used up.

Early work on the radiation-induced polymerization of α-methylstyrene indicated the possibility of a free radical mechanism [68], based partly on inhibition by p-benzoquinone and other additives. Bates et al. [46] argued against the free radical mechanism, noting that the inhibitory effect of water was not consistent with such a hypothesis. He further pointed out that the low $\overline{DP}_n$ and high polymerization rate would be in agreement with a cationic mechanism involving regenerative chain transfer through proton transfer reactions, which would also be consistent with water inhibition.

The original work of Bates et al. [46] gave an average G(-m) value of about 8000. Work by other investigators has gradually increased the value of G(-m) obtainable as the efficiency of monomer drying progressively improved. Hirota and Katayama [69] obtained G(-m) values of 3.3×10^3 to 3.7×10^4 using sodium drying, and values up to 7.2×10^4 using 3.0 wt% zinc oxide in the monomer. Apparently the effect of the zinc oxide was due solely to its efficient dehydrating action. Metz [137] employed silica drying plus a bakeout of all glassware under vacuum and obtained G(-m) values of 7.8×10^4 and 5.1×10^5 at dose rates of 2.4×10^5 and 6.4×10^3 rads/h, respectively. This corresponds to a dependence of k_p on the half-power of R_i, in accordance with a bimolecular termination in the presence of negligible impurities [Eq. (5-51) with [X] = 0].

It has been concluded [33] that all results to date are consistent with an ionic mechanism for the radiation-induced polymerization of rigorously dried α-methylstyrene. As to whether the mechanism is cationic or anionic, the strong retardation caused by ammonia and amines indicates the probability of a cationic mechanism, since such bases should not retard an anionic process. Also, the high efficiency of chain transfer to monomer is uncommon for anionic mechanisms, but is easily explained in terms of proton transfer reactions in cationic polymerization.

For kinetics studies, amines are better scavenging agents than water, because of the specificity of their reaction with hydrocarbon cations. With trimethylamine, linear conversion-dose plots were obtained down to 10^{-6} M amine concentration, from which initial values of G(-m) in the presence of the scavenger were computed. The value of k_p has been computed [70] from the linear plot of $1/G(-m)$ versus amine concentration [Eq. (5-67)]. The slope of this plot equals 3.7×10^3 M^{-1} (molecules/100 eV)$^{-1}$ for α-methylstyrene and corresponds to the quotient $k_r/G_i k_p[M]$, where k_x in Eq. (5-67) has now been replaced by k_r. If k_r is estimated [71] from diffusion theory to be 8.4×10^9 M^{-1} sec^{-1} and an assumed value of $G_i = 0.1$ is employed, then k_p is found to be approximately 3×10^6 M^{-1} sec^{-1}. This result agrees well with another estimate of k_p of 4×10^6 M^{-1} sec^{-1} based [43] on measurements of absolute rates (R_p) and ionic lifetimes.

Styrene is one of the most extensively investigated monomers in the area of radiation-induced polymerization (see Table 5-A). Early work on "moderately dry" styrene indicated a free radical mechanism, but more recent work has emphasized "extremely dry" styrene. By drying the monomer with sodium-potassium alloy, Ueno [4, 7] obtained G(-m) values of 1.4×10^4 over a temperature range of -11 to 37°C. A still higher value of G(-m) = 4.4×10^4 was attained by using monomer irradiated in the presence of zinc oxide [4, 7]. At extremely high rates under dry conditions, the dependence of R_p on dose rate approaches the half-power dependence expected for the free ion mechanism. Hence, the developments in experimental technique and theory for styrene have closely paralleled those for α-methylstyrene.

Arguments that a cationic mechanism predominates for styrene are similar to those for α-methylstyrene [4, 7]. The rate of the "dry" ionic polymerization exceeds that of the "wet" free radical polymerization by a factor of about 100. A precise determination of the energy of activation for propagation is prevented by apparent changes in the effective concentration of residual water

as affected by temperature, even for rate measurements on the
same sample.

Independent determinations of k_p for styrene have been made
from rate and conductivity measurements during the steady state
resulting during continuous irradiation [43]. The analysis starts
with Eq. (5-51), giving the rate of polymerization in the presence
of an impurity at concentration [X]. This equation can be rewritten as follows,

$$R_p = k_i \tau' k_p [M] \qquad (5\text{-}84)$$

where τ' may be defined by

$$\frac{1}{\tau'} = \frac{1}{\tau_{ss}} + \frac{1}{\tau_s} \qquad (5\text{-}85)$$

where

$$\tau_{ss} = (1/R_i k_t)^{\frac{1}{2}} \qquad (5\text{-}86)$$

$$\tau_s = 1/k_x [X] \qquad (5\text{-}87)$$

The value of τ' corresponds to the average lifetime of the propagating ion when it can be terminated by both charge recombination
and scavenging by the impurity. The above equation for R_p reduces
o Eq. (5-47) for the case of zero impurity concentration
$(\tau' = \tau_{ss})$, and to the earlier expression

$$R_p = R_i k_p [M]/k_x [X] \qquad (5\text{-}88)$$

for strong retardation [refer to Eq. (5-66) for the case of
$(R_p)_0 >> R_p$].

When there is no termination by retarder, the determination of
τ_{ss}, R_p, and R_i allows the computation of k_p by means of Eq. (5-47).
Since τ_{ss} is the mean lifetime of ions undergoing charge neutralization, it can be computed from conductivity data according to

Eq. (5-81). In making this computation, a value of G_i = 0.1 is
used, since this is the most reliable value for the yield of free
ions in a hydrocarbon. An experimental plot of log R_p versus log I
indicates a dependence of R_p on the 0.62 power of the dose rate.
Although this indicates a small amount of retardation, at the
highest dose rate used (1.0 x 10^{15} eV cm^{-3} sec^{-1}) only small error
is introduced in τ_{ss}. The computation gives τ_{ss} = 0.0514 sec and
a k_p value of 3.5 x 10^6 M^{-1} sec^{-1}.

It is worth noting that the k_p value for free styrene anions
in tetrahydrofuran has been estimated to be 10^5 M^{-1} sec^{-1}. If this
value holds true in bulk polymerization, the anionic propagation
would be contributing about 3% to the overall polymerization rate.
This possibility is not excluded by the evidence available.

IV.   HOMOGENEOUS SOLUTION POLYMERIZATION

A.   Introduction

The propagation of polymerization in solution is influenced
mainly by the ability of a particular solvent to sustain or inhibit a certain mechanism. In solution polymerization, the intermediates and products formed by irradiation of the solvent must be
considered, and complications can result from a reactivity transfer
between solvent and monomer that markedly affects the production
of initiating species. That is, the initiating processes may not
arise from the separate action of the radiation on the monomer and
on the solvent.

Aqueous systems offer the advantage that a considerable amount
is known about the radiation chemistry of water in terms of radical
yields. Also, water is inert to most organic free radicals, and
chain transfer processes do not have to be taken into account.

Furthermore, ionic polymerization mechanisms are completely inhibited in the presence of appreciable concentrations of water.

Organic solvents are known that favor either anionic or cationic mechanisms. Chlorinated hydrocarbons favor cationic polymerization, whereas amines and ethers may be used for anionic
polymerizations. Most organic solvents will support free radical
polymerization.

B.   Styrene in Dichloromethane at -78°C

The original idea in studying this system was that cationic
polymerization of styrene should predominate over the free radical
mechanism at low temperature in a solvent of high electron affinity
such as dichloromethane.

Most workers agree that this polymerization is cationic, but
some differences have arisen in the details of the interpretation.
Charlesby and Mories [96] emphasized the need for efficient drying
of the monomer, and at 40 krads/h for a 2.2 M solution of styrene
in dichloromethane obtained an R_p value of 6×10^{-3} M h^{-1}. A corresponding value of 5.7×10^{-3} M h^{-1} was obtained by Ueno [97]
under similar conditions. However, at a dose rate of about 40
krads/h, Charlesby and Mories found R_p to depend on the 1.5 power
of monomer concentration, whereas Ueno found it to depend on the
1.0 power of monomer concentration.

Most investigators found R_p dependent on the first power of
dose rate at dose rates less than 0.5 Mrad/h. However, R_p in the
presence of silica gel appears to change from a half-power to a
first-power dependence on dose rate above 0.5 Mrad/h, whereas the
opposite trend holds true in the absence of the silica. Also, the
dependence on monomer concentration appears to change from half-
to first-order with increasing dose rate in the presence of silica.
Results of this type are probably too complicated to explain on the
basis of kinetics alone.

A study was made [96] of molecular weight using Mayo plots based on Eq. (5-63),

$$\frac{1}{\overline{DP}_n} = \frac{k_f}{k_p} + \frac{J}{k_p[M]} \qquad\qquad (5\text{-}89)$$

where J is used to indicate the reciprocal of the kinetic chain lifetime τ. By plotting $1/\overline{DP}_n$ versus $1/[M]$, the intercept which equals k_f/k_p is found to be about the same whether silica is present or not. The intercept corresponds to a limiting value of $\overline{DP}_n \cong$ 500 at high monomer concentration. The slope of the Mayo plot, J/k_p, appears to decrease slightly in the presence of the silica, which may be due to a decrease in J with k_p remaining constant, corresponding to an increase in kinetic chain lifetime in the presence of silica. Confirmatory evidence is furnished by the increased rate of polymerization in the presence of the silica, thus indicating that silica may decrease the termination rate.

Also the fact that R_p dependence on dose rate below 0.5 Mrad/h changes from a first-power to a half-power dependence when silica is present is in agreement with a lengthening of the kinetic chain resulting from a decrease in $k_x[X]$, as implied by Eq. (5-51).

A computation of G_i has been made from Eq. (5-63A) which may be rewritten in the form

$$G_i = \frac{JG(-m)}{k_p[M]} \qquad\qquad (5\text{-}90)$$

Evidently J/k_p is known from the Mayo plot, whereas $G(-m)$ can be calculated as a function of monomer concentration from the kinetic results at each dose rate. The computations show G_i at 40 krads/h has values of 0.1 and 0.2, respectively, with and without silica. Charlesby and Mories [96] postulate that these results are typical of the whole dose range studied, but this has been questioned by other investigators [33].

All results obtained to date appear to be consistent with a simple free ion mechanism in the presence of silica. Silica apparently plays the role of reducing chain termination by removing water or other impurities. It is also probable that the lower raa ates and their dependence on the first power of dose rate in the absence of silica are explained by a predominant termination with impurity under such conditions. The hypothesis has been advanced [96] that an intensity exponent of unity implies termination by the "original gegen ion," but other workers have considered this explanation doubtful [30].

In radiation-induced solution polymerization, the exact mechanism of initiation is often a difficult question. In the case of free radical polymerization, the initiating radicals may include species derived from both monomer and solvent, with the yield of each complicated by the question of reactivity transfer. Even more problems are present in the case of ionic polymerization. Ueno et al. [97] provided some interesting information on this subject in observing that radiation-induced ionic polymerization of styrene takes place in several ionic solvents at -78°C, including carbon disulfide, ethyl bromide, monofluorotrichloromethane, as well as dichloromethane. The G_i values range from 0.1 to 0.2 for styrene concentration up to 2 M in several solvents. These results [97] agree with those of Charlesby and Mories [96] and also with the value of G_i for the free ion polymerization of cyclopentadiene at -78°C [71]. Taking all these data into account, it would appear that chemical nature of the ion is of lesser importance in these particular systems, and that physical separation to give free ions is the main requirement. It remains possible that monomer ions may be formed preferentially by reactivity transfer from solvent to monomer.

There seems to be no simple correlation of R_p for the radiation-induced polymerization of styrene with the dielectric constant of the solvent [97]. The dielectric constant usually does have an effect in catalytic cationic polymerization [98], but the mechanism is quite different in radiation-induced polymerization. However, the value of the dielectric constant must affect the ease of

physically separating the ions. The fact that styrene does not
polymerize at an appreciable rate in nitroethane [97], where ε = 28,
may be due to ineffective initiation or retardation by radiolysis
products.

C. Radiolytic Polymerization of
Ethylene in Solution

   Wiley [1] and associates have investigated the γ-ray-induced
liquid-phase polymerization of ethylene in alkyl chloride solution.
This work is important because the kinetics of the polymerization
were studied in a quantitative way with good reproducibility of
results. The investigation was of particular interest because
rates were measured over a range of temperatures, and the effect
of oxygen as an impurity was evaluated. Rate expressions were derived and the order of the reaction with respect to ethylene was
determined.

   Earlier studies of this reaction were made using gaseous
ethylene, and the resulting rate expressions were not easy to explain by any simple mechanism. In the present work, the authors
took into account initiation by free radicals formed from the solvent, since the interaction of γ-rays with matter is proportional
to the electron density of the medium. The resulting rate expression includes initiation by radicals from both monomer and solvent.

   Although the reaction starts in a homogeneous solution, the
formation of solid polymer particles soon forms a heterogeneous
system. Under these conditions, the growing polymer chains may be
occluded in the solid particles and terminated by primary radicals
which are small enough to diffuse into the particles and reach the
growing chain ends. If the radicals are buried deeply enough,
they may remain unterminated, which would result in pseudo-first-
order termination [2]. It can be shown that either of these
effects could lead to a rate that is second order in monomer

concentration. "Normal" polymerization kinetics leads to a rate
first order in monomer concentration. Hence, depending on the
importance of these effects, the exponent of monomer concentration
in the rate expression should range between 1 and 2.

Generally speaking, the rate of polymerization in alkyl chloride solution was found to be faster than in bulk under the same
conditions. This probably results from a change in initiation
and/or termination rates, the important factors being electron
density of the solution, transfer to solvent, reactivity of solvent
radicals, and homogeneity of the reaction.

Rates of polymerization at various monomer concentrations in
n-propyl chloride are shown in Fig. 5-1. The rates can be analyzed

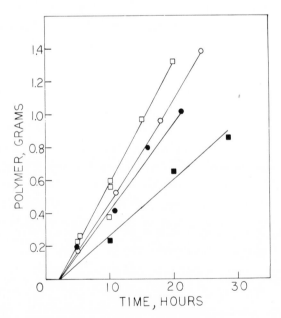

FIG. 5-1. Rate of polymerization of ethylene in n-propyl
chloride: (■) 8.79 moles/l; (●) 8.5 moles/l; (o) 7.99 moles/l; (□)
6.76 moles/l. [From R. H. Wiley, *J. Polym. Sci.*, Part A, **2**, 2503
(1964); John Wiley and Sons, New York.]

by means of the equation

$$\log (\text{rate}) = a + b \log [M] \tag{5-91}$$

The results show an exponent of $[M]$ equal to 1.41, which agrees with the kinetic scheme already discussed.

An induction period of 1.5 h was found for the reaction in the n-propyl chloride solution, and about the same in isopropyl chloride or *tert*-butyl chloride. Reducing the oxygen content of the ethylene from 1000 to 60 ppm caused a decrease in the induction period, as expected from earlier gas-phase polymerization work.

The fact that the solvent takes part in the polymerization is proved by the inclusion of about 0.9 chlorine atoms per polymer chain. The three solvents used were n-propyl chloride, isopropyl chloride, and *tert*-butyl chloride. Changes in rate in the three solvents correspond to changes in electron density of the solution, rate of transfer with the solvent, and reactivity of the solvent radical. The principal factor appears to be the electron density of the solution, a higher electron density indicating more primary radicals formed per volume of solution. Therefore, *tert*-butyl chloride, with the highest electron density of the three solvents, shows the highest polymerization rate.

Studies on oxygen as an additive at 1000 and 60 ppm showed that oxygen not only acts as an inhibitor but also functions as a retarder. (An inhibitor produces an induction period before the reaction starts, whereas a retarder slows down the observed rate after reaction has started.) Other additives known to affect free radicals were tested and found to have predictable effects. The addition of iodine causes an induction period of length proportional to the amount of iodine used. Diphenylamine causes no induction period, but does slow down the polymerization. Both of these effects have been observed earlier in other free radical reactions.

Activation energy computed from the Arrhenius equation was found to be 7.4 kcal/mole in *tert*-butyl chloride and 5.0 kcal/mole in n-propyl chloride. These values are higher than the 4.4 kcal/mole observed for the polymerization of liquid ethylene in bulk [3]. The author postulates that the solvent probably acts to increase the activation energy of the propagation step.

The effect of dose rate, ethylene concentration, and temperature on molecular weight was evaluated. As total dose increased, the molecular weight passed through a maximum and then declined. The initial increase is explained by the consumption of molecular weight-reducing impurities, whereas the eventual decrease may be due to radiation degradation of the polymer.

Over the concentration range used, the molecular weight was approximately proportional to the one-third power of ethylene concentration. The molecular weight increases with temperature, which is explained by postulating termination of the growing chains by primary solvent radicals.

The γ-radiation in this work was provided by a 4.07×10^5 R/h cobalt-60 source. Dosimetry was carried out in the usual way by measuring conversion of ferrous to ferric sulfate. The ethylene polymerization was carried out in a pressure system which had been purged several times with prepurified nitrogen. Each rate reported was taken from the slope of a line obtained by least-squares analysis of four to seven yield determinations over reaction times ranging from 5 to 33 h (see Fig. 1).

D. Effect of Silver Nitrate on the Radiolytic Polymerization of Isoprene in Solution

The effect of an accelerator is illustrated by the addition of silver nitrate to the radiolytic polymerization of isoprene, as described in the work of Fujioka *et al.* [8]. Earlier work had

shown that silver nitrate accelerates the polymerization of
ethylene and acrylonitrile. The investigators had reported that
monomer-metal ion complexes form before polymerization and take
part in a very rapid polymerization.

Fujioka added a mixture of silver nitrate water solution and
isoprene to an L-type ampule. The ampule was cooled with liquid
nitrogen, attached to a vacuum line, and outgassed by the usual
freeze-thaw technique. The ampule was then irradiated with an
800 Ci cobalt-60 source, temperature being held constant within
±1°C. After reaction the ampule was opened, the polymer washed
with water, and dried to constant weight *in vacuo*. Silver was
analyzed and subtracted to determine the net weight of dry polymer.

Polymerization rate was found to be 20 to 50 times as great as
in the radiolytic polymerization of pure isoprene. Increasing
either isoprene concentration or silver nitrate concentration
caused an increase in rate of polymer formation. For all isoprene/
silver nitrate ratios, plots of log rate versus log dose rate were
linear. Arrhenius plots in the range from 2° to 15°C indicated an
apparent activation energy of 8 kcal/mole.

To obtain evidence on the nature of the mechanism, p-benzoquinone and DPPH were used as additives and found to be very
effective inhibitors. On this basis the author concluded that the
reaction proceeds by a free radical mechanism.

Ultraviolet absorption spectra were measured for silver
nitrate solution alone, isoprene solution alone, and combinations
of the two. The composition of the complex, corresponding to two
atoms of silver per molecule of isoprene, was determined by the
continuous variation method using the absorption at 380 nm. The
equilibrium constant K_f and the extinction coefficient E of the
complex were estimated to be 0.18 and 1.4, respectively, where

$$K_f = \frac{[Ag_2B^{2+}]}{[Ag^+]^2[B]} \tag{5-92}$$

and $[Ag^+]$, $[B]$, and $[Ag_2B^{2+}]$ are concentrations of the silver ion, isoprene, and the complex.

The relationship between the polymerization rate and complex concentration was indicated by the fact that a plot of log rate versus log concentration gave a linear curve. The slope indicated a rate dependence on the 1.3 power of complex concentration.

The infrared spectrum of the resulting polymer was determined using the KBr disk method. The data indicated 41.9% of the 1,2-structure, which is much greater than the corresponding percentage in polyisoprene obtained by use of a radical catalyst, hevea, or polyisoprene obtained with sodium catalyst.

Since in all runs $[Ag^+]_0 \gg [B]_0$, the equilibrium expression becomes

$$K_f = \frac{[Ag_2B^{2+}]}{[Ag^+]^2([B]_0 - [Ag_2B^{2+}])} \tag{5-93}$$

which can be rewritten as

$$[Ag_2B^{2+}] = K_f[Ag^+]_0^2[B]_0/(1 + K_f[Ag^+]_0^2) \tag{5-94}$$

In a system with a high concentration of silver nitrate, we have $K_f[Ag^+]_0^2 \gg 1$, so that the equation becomes

$$[Ag_2B^{2+}] = [B]_0 \tag{5-95}$$

This indicates that the concentration of the complex is approximately equal to the original concentration of the isoprene, and that when rate is a function of complex concentration it is also the same function of isoprene concentration. This fact was confirmed by the experimental data.

REFERENCES

1. Wiley, R. H., Lipscomb, N. T., and Parrish, C. F., *J. Pol. Sci.*, Part A, 2, 2503 (1964).

2. Thomas, W. M. and Pellon, J. J., *J. Pol. Sci.*, 13, 329 (1954).

3. Wiley, R. H., Lipscomb, N. T., Johnston, T. J., and Guillet, J. E., *J. Pol. Sci.*, 57, 867 (1962).

4. Ueno, K., Hayashi, K., and Okamura, S., *Polymer Letters*, 3, 363 (1965).

5. Ballantine, D. S., Colombo, P., Glines, A., and Manowitz, B., *Chem. Eng. Prog.*, 50, 267 (1954).

6. Chapiro, A. and Wahl, P., *Compt. Rend.*, 238, 1803 (1954).

7. Ueno, K., Williams, F., Hayashi, K., and Okamura, S., *Trans. Faraday Soc.*, 63, 1478 (1967).

8. Fujioka, S., Ichimura, T., and Shinohara, Y., *J. Pol. Sci., Part A-1*, 6, 1109 (1968).

9. Hopwood, F. L. and Phillips, J. T., *Proc. Phys. Soc. (London)*, 50, 438 (1938).

10. Collinson, E. and Dainton, F. S., *Discussion Faraday Soc.*, 12, 212 (1952).

11. Chapiro, A., Cousin, C., Landler, Y., and Magat, M., *Rec. Trav. Chem.* 68, 1037 (1949)

12. Bywater, S., *Fortschr. Hochpolymer.-Forsch.*, 4, 66 (1965).

13. Szwarc, M., Waak, R., Rimbaum, A., and Coombes, J. D., *J. Amer. Chem. Soc.*, 79, 2076 (1957).

14. Bhattacharyya, D. N., Lee, C. L., Smid, J., and Szwarc, M., *J. Phys. Chem.*, 69, 612 (1965).

15. Allen, G., *Polymer*, 2, 151 (1961).

16. Gray, L. H., *J. Chim. Phys.*, 48, 172 (1951).

17. Chapiro, A., *Radiation Res.*, 6, 11 (1957).

18. Davison, W. H. T., Pinner, S. H., and Worrall, R., *Chem. Ind. (London)*, 1957, 1274.

19. Metz, D. J., *Advan. Chem. Series*, 1967, p. 170.

20. Magee, J. L., *Ann. Rev. Phus. Chem.*, 12, 389 (1961).

21. Schwarz, H. A., *J. Amer. Chem. Soc.*, 77, 4960 (1955).

22. Noyes, R. M., *J. Amer. Chem. Soc.*, 77, 2042 (1955).

23. Samuel, A. H. and Magee, J. L., *J. Chem. Phys.*, 21, 1080 (1953).

24. Williams, F., *J. Amer. Chem. Soc.*, 86, 3954 (1964).

25. Buchanan, J. W. and Williams, F., *J. Chem. Phys.*, 44, 4371 (1966).

26. Onsager, L., *Phys. Rev.*, 54, 554 (1938).

27. Gallivan, J. B. and Hamill, W. H., *J. Chem. Phys.*, 44, 1279, 2378 (1966).

28. Lin, J., Tsuji, K., and Williams, F., *J. Amer. Chem. Soc.*, 90, 2766 (1968).

29. Allen, A. O. and Hummel, A., *Discussions Faraday Soc.*, 36, 95 (1963).

30. Willaims, F., *Discussions Faraday Soc.*, 36, 257 (1963).

31. von Smoluchowski, M., *Z. Physik. Chem.*, 92, 129 (1917).

32. Debye, P., *Trans. Electrochem. Soc.*, 82, 265 (1942).

33. Williams, F., in *Fundamental Processes in Radiation Chemistry* (P. J. Ausloos, ed.), Interscience, 1968.

34. Lawton, E. J., Brubb, W. T., and Balwit, J. S., *J. Pol. Sci.*, 19, 335 (1956).

35. Wexler, S. and Marshall, R., *J. Amer. Chem. Soc.*, 86, 781 (1964).

36. Kebarle, P., Haynes, R. M., and Searles, S., *Advan. Chem. Series*, 58, 210 (1966).

37. Dauphin, J., Grosmangin, J., and Petit, J. C., *Intern. J. Appl. Radiation Isotopes*, 18, 285, 297 (1967).

38. Munson, N, S. B. and Field, F. H., *J. Amer. Chem. Soc.*, 87, 4242 (1965).

39. O'Donnell, J. H., McGarvey, B., and Morawetz, H., *J. Amer. Chem. Soc.*, 86, 2322 (1964).

40. Metz, D. J., Potter, R. C., and Thomas, J. K., *J. Pol. Sci.*, Pt. A, 5, 877 (1967).

41. Onsager, L., *J. Chem. Phys.*, <u>2</u>, 599 (1934).

42. Chapiro, A., *Radiation Chemistry of Polymeric Systems*, Interscience, New York, 1962.

43. Williams, F., Hayashi, K., Ueno, K., Hayashi, Ka., and Okamura, S., *Trans. Faraday Soc.*, <u>63</u>, 1501 (1967).

44. Benson, S. W. and North, A. M., *J. Amer. Chem. Soc.*, <u>84</u>, 935 (1962).

45. Schildknecht, C. E., *Polymer Processes*, Interscience, New York, 1956, p. 203.

46. Bates, T. H., Best, J. V. F., and Williams, F., *Trans. Faraday Soc.*, <u>58</u>, 192 (1962).

47. Charlesby, A. and Mories, J., *J. Pol. Sci.*, *Pt. C*, <u>4</u>, 1127 (1963).

48. Freed, S. and Saucier, K. M., *J. Amer. Chem. Soc.*, <u>74</u>. 1273 (1952).

49. Bonin, M. A., Busler, W. R., and Williams, F., *J. Amer. Chem. Soc.*, <u>84</u>, 2895 (1962).

50. Muller, J. C., *J. Chim. Phys. Physico-Chim. Biol.*, <u>65</u>(3), 567 (1968).

51. Ogata, A., Tabata, Y., and Hamaguchi, H., *Bull. Chem. Soc. Jap.*, <u>40</u>(9), 2205 (1967).

52. Tazuki, S., *Kobunshi Kagaku*, <u>24</u>, 302 (1967).

53. Chapiro, A. and Laborie, F., *C. R. Acad. Sci., Paris, Ser. C*, <u>267</u>(18), 1110 (1968).

54. Chapiro, A., *Advan. Chem. Ser.*, No. 82, 513 (1968).

55. Vlagin, I., Brit. Patent 1.126,127, Sept. (1968).

56. Kikuchi, Y., *Kogyo Kagaku Zasshi*, <u>71</u>(2), 296 (1968).

57. Vlagin, I., *Rev. Roum. Chim.*, <u>12</u>(9), 1085 (1967).

58. Hussain, F., Rashid, A., and Amjad, M., *Pols. J. Sci. Ind. Res.*, <u>11</u>(2), 134 (1968).

59. Abkin, A. D., Sheinker, A. P., and Tolstoukhova, L. P., *Vysokomol. Soedin, Ser. A*, <u>9</u>(4), 870 (1967).

60. Usmanov, K. U., Yulchibaev, A. A., and Yuldasheva, K., *Uzb. Khim. Zh.*, <u>11</u>(3), 41 (1967).

61. Yamaoka, H., Uchida, R., Hayashi, K., and Okamuna, S.,
    Kobunshi Kagaku, 24(261), 79 (1967).

62. Matheson, M. S., Auer, E. E., Bevilacqua, E. E., and Hart, E.
    J., *J. Amer. Chem. Soc.*, 73, 1700 (1951).

63. Ramford, C. H., Barb, W. G., Jenkins, A. D., and Onyon, P. F.,
    The Kinetics of Vinyl Polymerization by Radical Mechanisms,
    Butterworths, London, 1958.

64. Schulz, G. V. and Hostalka, H., *Polymer Letters*, 3, 175,
    1043 (1965).

65. Bates, T. H., Best, J. V. F., and Williams, F., *J. Chem. Soc.*,
    p. 1531 (1962).

66. Bonin, M. A., Calvert, M. L., Miller, W. C., and Williams, F.,
    Polymer Letters, 2, 143 (1964).

67. Cordischi, D., Lenzi, M., and Mele, A., *J. Pol. Sci., Pt. A*,
    3, 3421 (1965).

68. Hirota, K., Makino, K., Kuwata, K., and Meshitsuka, G., *Bull.
    Chem. Soc. Japan*, 33, 251 (1960).

69. Hirota, K. and Katayama, M., *Ann. Rept. Japan Assoc. Radiation
    Res. Polymers*, 5, 205 (1963-1964).

70. Hubmann, E., Taylor, R. B., and Williams, F., *Trans. Faraday
    Soc.*, 62, 88 (1966).

71. Bonin, M. A., Busler, W. R., and William, F., *J. Amer. Chem.
    Soc.*, 87, 199 (1965).

72. Murano, M. and Yamadera, R., *J. Pol. Sci., Pt. B*, 5(4), 338
    (1967).

73. Rozovskaya, N., Sheinker, A. P., and Abkin, A. D., *Radiats.
    Khim. Polim. Mater. Simp., Moscow*, 1964, 92 (1966).

74. Usmanov, K., Yulchibaev, A. A., and Yuldasheva, K., *Nauchn. Ti.
    Tashkentsk. Gos. Univ. No. 257*, 3 (1964).

75. Imoto, M., Otsu, T., and Horada, Y., *Makromol. Chem.*, 65, 180
    (1963).

76. Imoto, M., Otsu, T., and Nakabayashi, M., *Makromol. Chem.*, 65,
    194 (1963).

77. Tsuda, Y., *J. Pol. Sci.*, 54, 193 (1961).

78. Dolmatov, S. A. and Polak, L. S., *Neftekhimiya*, $\underline{3}$(5), 683 (1963).

79. Springer, J. M., AEC Accession No. 46337, Rept. No. TID-22244 (1965).

80. Kulikova, V. F., Savinova, I. V., Zupov, V. P., Kabanov, V. A., Polak, L. S., and Kargia, V. A., *Vysokomol. Soedin., Ser. A*, $\underline{9}$(2), 299 (1967).

81. Dolmatov, S. A. and Polak, L. S., *Radiats. Khim. Polim. Mater. Simp., Moscow*, $\underline{1964}$, 82 (1966).

82. Yamaoka, H., *Kobunshi Kagaku*, $\underline{29}$(270), 649 (1967).

83. Kearney, J., Stannett, V., and Clark, H. G., *J. Pol. Sci., Pt. C*, $\underline{1965}$ (pub. 1968), No. 16, (Pt. 6), 3441.

84. Eisler, S. L., U.S. Dept. Comm., Office Tech. Serv., P. B. Rept. No. 171,041 (1960).

85. Matsuzaki, K., Ishida, A., Osawa, Z., and Sano, K., *Kogyo Kagaka Zasshi*, $\underline{68}$(5), 855 (1965).

86. Naruse, T., Fulki, K., and Kuri, Z., *Kogyo Kagaku Zasshi*, $\underline{70}$ (4), 580 (1967).

87. Russo, S. and Munari, S., *J. Pol. Sci., Pt. B*, $\underline{5}$(9), 827 (1967).

88. Volkova, E. V., Zimakov, P. V., Fokin, A. V., Sorokin, A. D., Belikov, B. M., Balygina, L. A., Skobina, A. I., and Krashousov, L. A., *Radiats. Khim. Polim., Mater. Simp., Moscow*, $\underline{1964}$, 109 (1966).

89. Ishigure, K., Tabata, Y., and Oshima, K., *J. Macromol. Sci., Chem.*, $\underline{A5}$(2), 263 (1971).

90. Leittinger, M., Sliemen, F. A., Kircher, J. F., and Leininger, R. L., U.S. Atom. Energ. Comm. BMI 1678 (1964).

91. Miyake, H., Wada, T., Hayashi, K., and Okamura, S., *Kobunshi Kagaku*, $\underline{22}$(240), 265 (1965).

92. Czarnodola, H. and Waclaw, S., *Nukeonika*, $\underline{10}$(6), 375 (1965).

93. Ueno, K., Tsakamoto, H., Hayashi, K., and Okamura, S., *J. Pol. Sci., Pt. B*, $\underline{5}$(5), 395 (1967).

94. Fujii, S., *J. Sci. Hiroshima Univ.*, *Ser. A-II*, <u>31</u>(2), 89 (1967).

95. Takakura, K. and Hayashi, K., *Pura Appl. Chem.*, <u>12</u>, (1-4), 387 (1966).

96. Charlesby, A. and Mories, J., *Proc. Roy. Soc. (London)*, <u>A281</u>, 392 (1964).

97. Ueno, K., Yamaoka, H., Hayashi, K., and Okamura, S., *Intern. J. Appl. Radiation Isotopes*, <u>17</u>, 513 (1966).

98. Pepper, D. C., *Trans. Faraday Soc.*, <u>45</u>, 397 (1949).

99. Tabata, Y., *J. Macromol. Sci.*, *Pt. A*, <u>1</u>(3), 493 (1967).

100. Yamakita, H., *Kogyo Kagaku Zasshi*, <u>71</u>(3), 430 (1968).

101. Suganuma, F., *J. Pol. Sci.*, *Part A*, <u>6</u>(8), 2069 (1968).

102. Suganuma, F., *J. Pol. Sci.*, *Part A*, <u>6</u>(11), 3123 (1968).

103. Gotoda, M., to Japan Atomic Energ. Res. Inst., Japan 6809067, 1968.

104. Mitsui, H., *J. Pol. Sci.*, *Part A-1*, p. 2881 (1968).

105. Stelio, M., Castello, G., and Rossi, C., *J. Pol. Sci.*, *Pt. C*, <u>1965</u>, No. 16 (Pt. 7), 4149 (1968).

106. Machi, S., Kise, S., and Kagiya, T., *Kogyo Kagaku Zasshi*, <u>69</u>(9), 1892 (1966).

107. Hardy, G., Nyitrai, K., Kovacs, G., Fedonova, N., and Varga, J., Radiation Chem. Proc., Tihany Simp., Tihany, Hung., <u>1962</u>, 205 (pub. 1964).

108. Loeffelholz, M., Fr. Addn. 90,895 (1968).

109. Skobina, H. I. and Volkova, E. V., *Radiats. Khim. Polim.*, *Mater. Simp.*, *Moscow*, <u>1964</u>, 126 (pub. 1966).

110. Hayakawa, K., *Kobunshi*, <u>15</u>(174), 752 (1966).

111. Noro, Y., Fueki, K., and Kuri, Z., *Bull. Chem. Soc. Japan*, <u>41</u>(7) (1968).

112. Tsuji, K., Yoshida, H., Hayashi, K., and Okamura, S., *J. Pol. Sci.*, *Pt. B*, <u>5</u>(4), 313 (1967).

113. Abkin, A. D., Sheinker, A. P., and Yakovleva, M. K., *Vysokomolekul. Soedin.*, <u>3</u>, 1135 (1961).

114. Hinschberger, A., *Peintures, Pigments, and Vernis*, <u>43</u>(4), 262 (1967).

115. Vacherot, M., Marchal, J., and Hinschberger, A., *C. R. Acad. Sci., Paris, Ser. C*, 264(11), 962 (1967).

116. Hinschberger, A. and Marchal, J., *C. P. Acad. Sci., Paris, Ser. C*, 263(10), 667 (1966).

117. Musaev, U. N., Usmanov, K. U., Babaev, T. M., and Fatkhylaeva, M., *Izv. Vysshi Zavedi, Khim. Khim. Tekhnol.*, 1968, 11(10), 1170.

118. Matsuzaki, K., *Makromol. Chem.*, 110, 185 (1967).

119. Kotnienko, T. O., Polischuk, Y. N., Zelenchukova, T. G., and Polgakov, M. V., Radiats. Khim. Polim. Mater., Simp., Moscow, 1964, 69 (1966).

120. Blumstein, A., *J. Pol. Sci., Pt. B*, 5(8), 687 (1967).

121. Cazjkowski, J., Turska, E., and Kroh, J., *Biell. Acad. Pol. Sci. Ser. Sci. Chem.*, 14, (11-12), 861 (1966).

122. Minoura, Y., Sazuki, Y., Sakaraka, Y., and Doi, H., *J. Pol. Sci., Pt. A-1*, 4(11), 2757 (1966).

123. Hirota, K. and Takamura, F., *Bull. Chem. Soc., Japan*, 35, 1037 (1962).

124. Aso, C. and Tagami, S., *Amer. Chem. Soc., Div. Polym. Chem., Preprints*, 8(2), 906 (1967).

125. Brown, D. W., *Amer. Chem. Soc., Div. Polym. Chem. Preprints*, 6(2), 965 (1965).

126. Yamaoka, H., Williams, F., and Hayashi, K., *Trans. Faraday Soc.*, 63(2), 376 (1967).

127. Ueno, K., *J. Pol. Sci., Pat. B*, 6(1), 39 (1968).

128. Suzuki, H., *Kogyo Kayaku Zasshi*, 71(5), 746 (1968).

129. Gusilnikov, L. E., Nametkin, N. S., Polak, N. S., and Cherrysheva, T. I., *Izv. Akad. Nauk SSSR, Ser. Khim.*, 1964, (11), 2072.

130. Matsuda, T. and Fujii, S., *J. Pol. Sci., Pt. A-1*, 5(10), 2617 (1967).

131. Brown, D. W. and Wall, L. A., *J. Pol. Sci., Pat. Al*, 6(5), 1367 (1968).

132. Tabata, Y., Shibano, H., Oshima, K., and Sobue, H., *J. Pol. Sci., Pt. C*, No. 16, 2403 (1967).

133. Usmanov, K. U., Yulchibaev, A. A., and Yuldasheva, K., *Izv. Vyssh. Ucheb. Zaved., Khim. Khim. Tekhnol.*, 12(2), 197 (1969).

134. Schneider, C., Denaxas, J., and Hummel, D., *J. Pol. Sci., Pt. C*, No. 16, 2203 (1967).

135. Hardy, G., *Magev. Kim. Foly*, 73(10), 475 (1967).

136. Bulygina, L. A. and Volkova, E. V., *Radiats. Khim. Polim., Mater. Simp., Moscow*, 1964, 122 (1966).

137. Metz, D. J., *Polymerization and Condensation, Advances in Chemistry Series, No. 91*, American Chemical Society, Washington, 1969.

APPENDIX

Table 5-A summarizes the published work on radiation-induced homogeneous polymerization. The monomers investigated are listed alphabetically, followed by type of polymerization whether bulk or solution, type of radiation source, comments, name of investigator, and reference number. The coverage is reasonably complete up to and including 1968, except for the very early publications prior to about 1961.

TABLE 5-A

Radiation-Induced Homogeneous Polymerizations

Monomer	Bulk or solution	Source	Comments	Investigator and Reference
Acenaphthylene	Bulk	—	Also in solid	Muller [50]
Acetaldehyde	Bulk	^{60}Co	—	Atsushi [51]
Acrylamide	Solution	Gamma	Aqueous solution	Tazuki [52]
Acrylic acid	Solution	Gamma	Dilatometric	Chapiro [53]
Acrylonitrile	Solution	—	Anionic mechanism	Chapiro [54]
Acrylonitrile	—	—	Copolymerization	Vlagin [55]
Acrylonitrile	—	Gamma	With $SnCl_2$	Kikuchi [56]
Acrylonitrile	—	Gamma	Homopolymer	Vlagin [57]
Acrylonitrile	Solution	Gamma	Maximum yield at pH 5	Hussain [58]
Acrylonitrile	Solution	Gamma	Copolymer with styrene	Abkin [59]
Acrylonitrile	—	^{60}Co	Copolymer with vinyl chloride	Usmanov [60]

Monomer	State	Source	Notes	Reference
Acrylonitrile	Solution	Gamma	Copolymer with nitroethylene	Yamaoka [61]
Acrylonitrile	—	Gamma	Radical mechanism	Murano [72]
Acrylonitrile	Solution	^{60}Co	Copolymer with methyl acrylate	Rozovskaya [73]
Acrylonitrile	Bulk	^{60}Co	Copolymer with 2-methylfuran	Usmanov [74]
Acrylonitrile	—	Gamma	Presence of $ZnCl_2$	Imoto [75]
Acrylonitrile	—	Gamma	Radical mechanism	Imoto [76]
Acrylonitrile	Bulk	Gamma	Copolymer with methyl methacrylate	Tsuda [77]
α-deuterio-Acrylonitrile	—	Gamma	Copolymer with acrylonitrile	Murano [72]
Allyl acetate	-	^{60}Co	Presence of $ZnCl_2$	Kulikova [80]
Allyl acetate	-	^{60}Co	100% Yield	Dolmatov [78]
Allyl acrylate	Bulk	^{60}Co	Rapid polymerization	Springer [69]
Allyl alcohol	-	^{60}Co	Presence of $ZnCl_2$	Kulikova [80]
Allyl alcohol	Both	^{60}Co	Effect of hydroqyinone	Dolmatov [81]
Allyl alcohol	-	^{60}Co	100% Yield	Dolmatov [78]
Allyl bromide	-	^{60}Co	100% Yield	Dolmatov [78]

		^{60}Co		
Allyl chloride	-	-	100% Yield	Dolmatov [78]
p-Bromostyrene	-	Gamma	—	Yomaoka [82]
Butadiene	-	—	Copolymer with SO_2	Kearney [83]
Butadiene	Bulk	Gamma	Homopolymer	Eisler [84]
sec-Butyl acrylate	-	Gamma	Stereotacticity	Matsuzaki [85]
tert-Butyl acrylate	-	Gamma	Syndiotactic product	Matsuzaki [85]
Butyl vinyl ether	-	Gamma	Cationic mechanism	Naruse [86]
Carbon monoxide	-	Gamma	Copolymer with ethylene	Russo [87]
Chlorotrifluoro ethylene	Bulk	—	—	Volkova [88]
Chloroprene	-	-	Copolymer with SO_2	Kearney [83]
p-Chlorostyrene	-	Gamma	-	Yamaoka [82]
Chlorotrifluoro-ethylene	-	Gamma	Copolymer with propylene	Ishigure [89]
β-Cyanoethyl acrylate	-	-	Copolymer with ethyl vinyl sulfone	Leittinger [90]
1,3-Cyclohexadiene	-	Gamma	High rate at low temperature	Miyake [91]
Diethyl fumarate	-	Gamma	Homopolymer	Czarnodola [92]

Compound	Phase	Radiation	Notes	Reference
Diethyl maleate	-	Gamma	Homopolymer	Czarnodola [92]
1,3-Dioxolone	-	^{60}Co	Viscous oil polymer	Ueno [93]
Diethyl vinyl phosphonate	Both	Gamma	Liquid and solid state	Fujii [94]
Diketene	-	Gamma	Cationic mechanism	Takakura [95]
2,3-Dimethylbutadiene	-	-	Copolymer with SO$_2$	Kearney [83]
(Epoxyethyl) benzene	-	Gamma	Activation energy of 6.2 kcal/mole	Tabata [99]
Ethene sulfonic acid	-	Gamma	-	Yamakita [100]
Ethylene	Solution	Gamma	In various alcohols	Suganuma [101]
Ethylene	Solution	Gamma	In tert-butyl alcohol	Suganuma [102]
Ethylene	-	Gamma	Two stages	Gotoda [103]
Ethylene	-	Gamma	Effect of acetylene	Mitsui [104]
Ethylene	(Gas)	Gamma	Effect of acetylene	Stelio [105]
Ethylene	-	Gamma	Copolymer with carbon monoxide	Russo [87]
Ethylene	Solution	Gamma	Methanol solution	Machi [106]
Ethylene sulfide	-	^{60}Co	Viscous oil polymer	Ueno [93]
Ethyl vinyl sulfone	-	-	60% Yield	Leittinger [90]

Monomer				Reference
Hexadecyl methacrylate	—	Gamma	Activation energy of 2.5 kcal/mole	Hardy [107]
Hexafluoropropene	-	^{60}Co	Presence of oxygen	Loeffelholz [108]
Hexafluoropropene	Bulk	—	—	Volkova [88]
Hexafluoropropene	-	^{60}Co	93% Conversion	Skobina [109]
Hexafluoropropene	-	^{60}Co	Liquid polymer product	Springer [79]
Isobutene	-	-	Ionic mechanism	Hayakawa [110]
Isobutene	(Gas)	-	Cationic mechanism	Noro [111]
Isobutene	Solution	^{60}Co	Cationic mechanism	Tsuji [112]
Isobutene	Solution	Gamma	Copolymer with styrene	Abkin [113]
Isobutene	-	Gamma	Copolymer with chloro-trifluoroethylene	Ishigure [89]
Isoprene	Solution	-	Effect of silver nitrate	Fujioka [8]
Isoprene	-	-	Copolymer with SO$_2$	Kearney [83]
Isoprene	Bulk	Gamma	Run at -78°C	Hinschberger [114]
Isoprene	Bulk	^{60}Co	Copolymer with methyl methacrylate	Vacherot [115]
Isoprene	—	-	Both radical and cationic mechanisms	Hinschberger [116]
Methacrylic acid	-	Gamma	Copolymer with styrene	Musaev [117]

Monomer	Physical state	Radiation	Remarks	Reference
Methyl acrylate	Solution	^{60}Co	Copolymer with styrene	Rozovskaya [73]
2-Methylfuran	Bulk	^{60}Co	Copolymer with acrylonitrile	Usmanov
Methyl isopropenyl ketone	-	Gamma	Liquid phase	Matsuzaki [118]
Methyl methacrylate	Solution	Gamma	Copolymer with acrylonitrile	Abkin [59]
Methyl methacrylate	Bulk	^{60}Co	Copolymer with isoprene	Vacherot [115]
Methyl methacrylate	-	^{60}Co	Free radical mechanism	Kornienko [119]
Methyl methacrylate	-	^{60}Co	Extensive branching	Blumstein [120]
Methyl methacrylate	Solution	^{60}Co	Effect of biphenyl	Czajkowski [120]
Methyl methacrylate	-	Gamma	Copolymer with tri-butylvinyltin	Minoura [122]
Methyl methacrylate	-	Gamma	Presence of $ZnCl_2$	Imoto [75]
Methyl methacrylate	Bulk	Gamma	Copolymer with acrylonitrile	Tsuda [77]
Methyl methacrylate	-	-	Dielectric properties of polymer	Leittinger [90]
Methyl methacrylate	-	Gamma	Radical and cationic	Hirota [123]
Methyl methacrylate	Solution	Gamma	In ethyl chloride	Abkin [113]

Monomer	State	Radiation	Remarks	Reference
α-Methyl styrene	-	Gamma	—	Hirota [123]
o-Phthalaldehyde	-	-	Stereospecificity	Aso [124]
Nitroethylene	Solution	Gamma	Copolymer with acrylonitrile	Yamaoka [61]
3,3,4,4,4-Pentafluoro-1-butene	-	-	Transfer effects	Brown [125]
Nitroethylene	-	-	Anionic mechanism	Yamaoka [126]
Propylene oxide	-	^{60}Co	Viscous oil polymer	Ueno [93]
Propene	Bulk	^{60}Co	Atactic product	Springer [79]
Propene	-	-	Copolymer with chloro-trifluoroethylene	Ishigure [89]
Styrene	-	-	Cationic mechanism	Ueno [127]
Styrene	Solution	-	Dichloromethane solvent	Chapiro [54]
Styrene	Solution	Gamma	Triethylamine solvent	Abkin [59]
Styrene	-	^{60}Co	Free radical mechanism	Kornienko [119]
Styrene	Solution	^{60}Co	Copolymer with methyl acrylate	Rozovskaya [73]
Styrene	Bulk	Gamma	Copolymer with acrylonitrile	Tsuda [77]

Monomer				Reference
Styrene	Solution	Gamma	Homopolymerization	Eisler [84]
Styrene	Solution	Gamma	Ethyl chloride solvent	Abkin [113]
Sulfur dioxide	—	Gamma	Copolymer with vinyl chloride	Suzuki [128]
Sulfur dioxide	-	-	Copolymer with dienes	Kearney [83]
Triallyl methylsilane	-	^{60}Co	Powder product	Gusilnikov [129]
Triallyl phenylsilane	-	^{60}Co	Solid product	Gusilnikov [129]
Tributylvinyltin	Bulk	Gamma	Copolymer with styrene	Minoura [122]
Tetrafluoroethylene	Bulk	-	100% Conversion	Volkova [88]
Tetroxocane	-	Gamma	Cationic mechanism	Takakura [95]
1,1,2-Trichloro-1,3-butadiene	-	Gamma	Effect of DPPH	Matsuda [130]
Trifluoroethylene	—	Gamma	Copolymer with tri-fluoropropene	Brown [131]
Trifluoroethylene	Bulk	—	Homopolymer	Volkova [88]
Trifluoroethylene	-	^{60}Co	Liquid-phase polymerization	Tabata [132]
Trifluoroethyl vinyl ether	-	-	Homopolymer	Leittinger [90]

Monomer				Reference
3,3,3-trifluoropropene	-	Gamma	Copolymer with trifluoroethylene	Brown [131]
Trimethylvinyltin	—	Gamma	Copolymer with styrene	Minoura [122]
Trioxane	-	Gamma	Cationic mechanism	Takakura [95]
Vinyl acetate	Bulk	^{60}Co	Copolymer with vinyl fluoride	Usmanov [133]
Vinyl chloride	-	Gamma	Copolymer with sulfur dioxide	Suzuki [128]
Vinyl chloride	-	^{60}Co	Copolymer with acrylonitrile	Usmanov [60]
Vinyl chloride	-	^{60}Co	Copolymer with sulfur dioxide	Schneider [134]
Vinyl fluoride	Bulk	^{60}Co	Copolymer with vinyl acetate	Usmanov [133]
Vinyl fluoride	-	—	Homopolymer	Volkova [88]
Vinylene carbonate	—	Gamma	Liquid-state reaction	Hardy [135]
Vinylidene chloride	-	Gamma	Presence of $ZnCl_2$	Imoto [75]
Vinylidene	-	—	Homopolymer	Volkova [88]
Vinylidene fluoride	-	^{60}Co	Study of kinetics	Bulygina [136]

Chapter 6

RADIOLYTIC POLYMERIZATION IN THE SOLID STATE

I.   INTRODUCTION

The earliest example of radiation-activated polymerization of
solid-state monomers was provided by Schmitz and Lawton [1].   These
authors irradiated various monomers at low temperatures with 800
keV electrons.   Glycol diacrylates and dimethacrylates were found
to form polymers at temperatures below -50°C.   Solid-state polymerization of tetraethylene glycol dimethacrylate took place with
almost explosive speed.   Methyl methacrylate and methyl acrylate
monomers were also found to polymerize in the solid state.   The
authors observed, however, that less polymerization took place
slightly below the melting point than above the melting point
(liquid monomer).   It was noted that active radicals formed in the
solid state had a long life, and when irradiated solid polymers
were heated above the melting point, the trapped free radicals
initiated a very rapid polymerization reaction.   Since 1951, a
large amount of work has been carried out and published concerning
radiolytic polymerization of solid monomers.

As an initiator of solid-state polymerization, radiation has
certain advantages when compared with heat, light, or peroxides.
These advantages include the following.

1. The property of interacting with solid polymer throughout
its thickness in a rather uniform way, instead of affecting the
surface only or penetrating to only a limited depth. Chemical
catalysts are undesirable for use with solid monomers, since they
will either not enter the monomer crystal at all or else will enter
and disrupt the order of the crystal.

2. Often ionizing radiation can be used without the degrading
effects of heat, and in any case heat would tend to melt the solid
monomers and make it impossible to carry out the polymerization in
the solid phase.

3. It is possible for ionizing radiation to initiate either
free radical or ionic-type polymerization.

4. Contamination with chemical catalysts is often undesirable,
and this is avoided by the use of ionizing radiation.

There are also some intrinsic advantages to the use of solid-
state polymerization, which can be enumerated as follows.

1. If it is desirable to retain the crystal structure of the
monomer for some reason, the possibility exists that the polymer
crystal form will replicate the crystalline structure of the
monomer.

2. Unique alignments of monomer crystals may exist, which
will give higher rates of polymerization than in the liquid phase
because of steric factors.

3. Recombination of radicals would logically be slowed in the
solid state, leading to a slowing of the termination reaction and
an increase in rate of the overall polymerization reaction.

4. The effect of oxygen inhibition of polymerization may be
lessened, possibly because of reduced solubility of oxygen in solid
phase monomer.

5. Certain copolymerizations may become possible in solid
systems that are not possible in liquid, because of certain favorable spatial alignments of the molecules that may exist in the

solid crystal structure. Also, reactivity ratios are sometimes
modified, possibly in desirable directions, by copolymerization in
the solid phase.

II. THEORY OF SOLID-STATE POLYMERIZATION

There exists no comprehensive or well-developed theory of
radiolytic solid-state polymerization. However, it can be stated
that the principal effect which distinguishes solid phase reaction
from homogeneous reaction (liquid or gas phase) is the diffusion
phenomenon. Before two reactants can react, it is necessary for
them to diffuse through the medium and reach each other physically.
Such necessary diffusion of reactants is generally several orders
of magnitude slower in the solid phase than in the gas or liquid
phase.

Whereas liquid vinyl monomers will often undergo polymeriaza-
tion on standing unless prevented by special measures, solid
crystalline vinyl monomers will usually stand for long periods at
room temperature without undergoung a detectable amount of polymerization. An exception to this rule is p,p'-divinyldiphenyl,
which polymerizes to give a cross-linked polymer on standing at
room temperature [36]. Similar observations were made in the case
of 1-carboxybutadiene [37].

The usual initiators of vinyl polymerization cannot be employed to cause polymerization in solid monomer crystals, because such
initiators are not soluble in the crystal lattice of the monomer.
Ultraviolet light is known to cause the polymerization of solid
methacrylic and acrylic acid [38], but it is not feasible to use in
quantitative studies because the light is scattered by crystalline
monomers, and the solid monomer crystals may become semiopaque at
some point during the polymerization. High energy ionizing radiation such as γ-rays is not subject to such transparency difficulties

and will ordinarily pass through any reasonable thickness of solid
monomer without any appreciable decrease in intensity. Accordingly,
ionizing radiation has been used to cause the polymerization of a
wide variety of solid monomers. In some cases, it has been shown
that the irradiation only serves to introduce the initiating
species into the solid monomer, and that polymerization may take
place after removal of the monomer from the source (indirect or
postirradiation technique).

Proof of whether such reactions proceed by a free radical or
ionic mechanism is difficult to obtain, because the usual methods
of radical and ion study are not applicable to solid-state reactions. Inhibitors and solvents cannot be incorporated into the
monomer crystal in order to evaluate the effect on the reaction
kinetics and molecular weight of the polymer produced. Cases have
been observed where polymerization inhibitors did not slow down
solid-state polymerization [39, 40], but a question arises in such
cases as to whether the inhibitor was present in a separate phase.
The common radical inhibitors were found to be effective in one
case when the monomer appeared to exist in the glassy state [41].
Oxygen usually does not inhibit polymerization of crystalline
monomers, but this may only mean that the oxygen cannot diffuse
effectively through the crystal lattice [29].

Often the polymerization rate decreases rapidly as temperature
is raised above the melting point of the monomer [41-43], and this
may possibly indicate a change in mechanism, implying an ionic
process in the solid state. On the other hand, it may simply indicate a sudden shortening of the kinetic chain length in the
liquid state, owing to the more rapid diffusion and combination of
growing monomeric free radicals.

Interesting information has been obtained on the solid-state
polymerization of acrylamide by electron-spin-resonance (ESR)
studies, where it has been shown that the polymerizing crystal
contains a constant free radical concentration [3, 4, 44] and that
the number of polymer chains is about equal to the number of free

radicals [3, 31]. This implies that none of the radicals has been
"terminated" in the usual sense, and therefore radical-radical recombination takes place to a very small extent in the crystalline
monomer. This leaves the question of whether the mechanism is
ionic or free radical, but it has been postulated that initiation
involves the formation of an ion radical, which could then grow by
either an ionic or radical mechanism [31].

In the formation of polymer from crystalline acrylamide monomer, chain transfer has been found negligible and the molecular
weight of the product indicates the kinetic chain length. The rate
coefficient for chain propagation in crystalline acrylamide is
about 2.5×10^6 smaller than in solution [3]. The rate drops off
to still lower values at low monomer conversion for some unknown
reason [31].

A new and interesting field of study is based on the β- or
x-ray irradiation of urea and thiourea clathrate complexes of
various monomers. The urea or thiourea molecules in such complexes
lie in a honeycomb pattern with a long hexagonal hole in the
center. In the case of urea, the hole is of the proper size to
contain straight-chain compounds. Accordingly, the urea clathrate
complexes of butadiene, vinyl chloride, acrylonitrile, and acrolein
were found to produce polymers when irradiated [45].

In this connection, the "normal" polymerization of butadiene
in the liquid phase allows the monomer unit to be given three
possible configurations in the chain.

1,2-addition	1,4-*cis*-addition	1,4-*trans*-addition

The butadiene fits the hole of the clathrate complex in such a way
that only the 1,4-*trans*-addition takes place and forms a perfectly
stereoregular chain. Larger molecules are required to fill the

hole or channel of the thiourea clathrate complex, and some of the
monomer clathrates found to polymerize were 2,3-dimethylbutadiene,
1,3-hexadiene, vinylidene chloride, and 2,3-dichlorobutadiene [46].

It was noteworthy that 1,3-hexadiene, which does not polymerize in the liquid state, is found to polymerize in the thiourea
complex. This indicates that the spatial arrangement of the monomer molecule in the channel of the complex favors chain formation.
When a mixtrue of monomers was used, polymerization could either
lead to a copolymer when the monomers formed one phase in the
channel or a mixture of two homopolymers when the monomers segregated into separate phases in the clathrate channel [46]. Also,
although cyclooctatetraene did not homopolymerize, it did form
copolymers from a mixed thiourea clathrate containing dichloro-
butadiene.

Solid materials other than vinyl monomers have been found to
undergo radiolytic polymerization and copolymerization. For
example, solid formaldehyde when irradiated changes into polyoxymethylene [47, 48]. Early reports indicated that acetaldehyde
polymerized spontaneously at its melting point, and more recent work
involved an extensive study of its radiolytic polymerization in the
solid state [49]. Furthermore, irradiation produces a silicone
from hexamethylene cyclotrisiloxane [50] and a polyester from solid
propiolactone [51]. It is worth noting that the polymerizability
of formaldehyde drops quickly above its melting point.

One phenomenon of solid-state polymerization not observed in
liquid-state reactions is that the geometry of the monomer crystal
lattice may exert control on the nature of the reaction. Okamura
and co-workers [11, 52, 53] studied such a phenomenon in the radiolytic solid-state polymerization of trioxane, diketone, β-propiolactone, and 3,3-bis(chloromethyl)oxetane.

It was observed that such cyclic monomers undergo polymerization in the crystalline phase only, and that polymers produced from
large single crystals consist of bundles of fibers having an

orientation dependent on the crystalline pattern of the original
monomer. The degree of order in such polymeric fibers is quite
high, as indicated by sharp x-ray diffraction patterns. If the
fiber is melted, the original high melting point corresponding to
a high degree of order cannot be restored by mechanical drawing.
Several investigators have found that the polymer chains possess a
high degree of order in a direction perpendicular to the fiber axis.

The crystalline structure of both trioxane and its polymers
is hexagonal, with the dimension along the fiber undergoing a 0.5%
elongation during polymerization and the dimensions at right angles
to the fiber direction being contracted by 4% [52]. It has been
postulated that the formation of the polymer with very little
dimensional change in the fiber direction is a factor tending to
favor the reaction.

Work of this sort was carried out quite early (1930) by
Kohlschutter and Sprenger, who observed the growth of polyoxymethylene chains oriented parallel to the c axis of hexagonal
trioxane crystals [54, 55]. The polymerization took place when
formaldehyde vapor was allowed to contact trioxane crystals. However, there was no polymerization when tetraoxane was used in place
of trioxane. This constituted strong evidence that the reaction
was topochemical in nature. (A topochemical process has been
defined as one in which the geometrical arrangement of the reactant
structures exerts a strong influence on the character of the
reaction, and may even determine whether or not a reaction takes
place.)

It should be noted that the polymerization of such cyclic
monomers involves the formation of bonds of the same type as the
bonds broken, so that the energy released during reaction is very
small and should have no effect in disordering the crystal lattice.
By contrast, the conditions for producing the polymer chain within
the monomer crystal lattice are much less auspicious in the case
of vinyl monomers. In vinyl polymerization, a double bond is

opened and single bonds are formed, leading to a spatial contraction and an energy release of about 10 to 15 kcal/mole. In spite of this, the geometrical arrangement of the monomer molecules may be an important factor, as indicated in the polymerization of vinyl monomers contained in urea and thiourea clathrate complexes.

Other signs have been discovered that the geometry of the crystal lattice determines whether a reaction will take place [29, 37, 56]. Vinyl monomers of similar reactivity in the liquid state may differ widely in reactivity toward polymerization in the solid crystalline condition. For example, crystalline acrylamide polymerizes 100 times more rapidly than crystalline methylene-bis-acrylamide [24]. Solid methacrylic acid polymerizes, but solid α-chloroacrylic acid does not. Potassium acrylate polymerizes at room temperature, but sodium acrylate polymerizes significantly only above 130°C [57]. It has also been observed that amorphous calcium acrylate prepared from the dehydration of the monohydrate is much slower to polymerize than similar samples prepared from the dihydrate.

It should be possible to obtain information on the influence of the crystal lattice in polymerization from studies of the stereoisomerism of the polymer chain formed. In polymer technology, certain polymerization techniques lead to long sequences of asymmetric centers having either the same configuration (isotactic polymers) or regularly alternating configuration (syndiotactic polymers). The question arises as to what sort of chain ordering takes place, if any, in the solid-state polymerization of crystalline monomers. In a few instances an answer to this question has been obtained. The polymerization of methyl methacrylate in solid form at -100°C using the molecular beam technique and magnesium as initiator has been found to lead to an isotactic polymer [58]. Also, polymerization of crystalline methacrylic acid has been found to give a more isotactic product than the one obtained in solution polymerization [37].

Two lines of evidence must be taken into account in the interpretation of solid-state polymerization mechanisms. First, x-ray
findings show that partially polymerized monomer consists of crystalline monomer phase and an amorphous polymer phase. Second,
there are many indications that the monomer crystal lattice is
important in determining whether or not polymerization takes place.
To explain these data, it has been postulated [29] that polymerization takes place at the crystalline-amorphous interface, and that
the major part of the polymer chain is in the amorphous phase while
its growing end is located in the crystalline monomeric phase.

III. EXAMPLES OF SOLID-STATE POLYMERIZATION

A. Acrylamide

The solid-state polymerization of acrylamide has been extensively investigated [2, 3, 4-6]. Most of the work has been done at
the Brooklyn Polytechnic Institute and at the Brookhaven National
Laboratory. In this particular polymerization, the monomer structure is not replicated in the polymer which forms.

Acrylamide is a white, crystalline, water-soluble material
which melts at 86°C. Irradiation with γ-rays causes the crystals
to become yellowish and less transparent [24]. Intense irradiation
reduces the water solubility, and the insoluble residue consists of
cross-linked polyacrylamide. The polymer retains an appreciable
amount of crystallinity, even at rather high conversions. It
appears that polymer chains in the crystal have little effect on
the nearby crystalline domains, suggesting that chain propagation
takes place without any significant reorientation of monomer
molecules.

In general terms, the γ-activated polymerization of acrylamide
above room temperature is a rather fast process which proceeds to
high conversion [2]. Oxygen does not have an inhibitory effect
[25]. The molecular weight of the polymer generally increases with
conversion. At a conversion of 52.8% the number average molecular
weight is about 232,000 [26].

An induction period is observed before polymerization starts
at 5°C [2]. At much lower temperatures, no polymerization takes
place during irradiaiton. If a sample of monomer is irradiated at
low temperature and then brought up to room temperature, a very
rapid polymerization reaction is observed [25]. This implies that
free radicals capable of initiating polymerization had been produced by irradiation at low temperature.

The rapid polymerization at room temperature and only slight
alteration in the crystal structure of the monomer led some investigators to conclude that the polymer growth reaction in the
monomer crystal takes place in a preferred crystallographic
direction [27]. This would imply that the monomers are arranged
in the crystal in a head-to-tail sequence.

A plot of log conversion rate at 35°C versus log of dose rate
shows that the polymerization rate is proportional to the first
power of the dose rate [24]. In the same sequence of experiments,
the product molecular weight indicated by viscosity was *practically*
independent of dose rate and of temperature from 0° to 65°C. An
Arrhenius plot showed an energy of activation of 4.7 kcal/mole in
this temperature range. This compares with 6.7 and 4.3 kcal/mole
for liquid-state γ-activated polymerization of styrene and methyl
methacrylate, respectively.

The polymerization of solid acrylamide continues for a considerable length of time after the irradiation stops [28]. In some
cases, conversion of irradiated monomer to polymer proceeded
gradually over a period of several months. Examination of the
polymerizing microscope showed that the polymerization took place

in well-defined planes of the crystal and often started from
lattice defects.

The aftereffect (polymerization) of the monomer following irradiation of the monomer at -78°C was studied during storage
periods at various temperatures from 0 to 60°C [29]. Very little
polymerization took place during irradiation, but after irradiation
the polymerization continued for several months. A plot of percent
conversion during the aftereffect versus logarithm of time produced
a linear function that would be in accord with a bimolecular termination by combination of free radicals. The energy of activation
of the reaction was about 25 kcal/mole [29]. The molecular weight
was found to increase during the aftereffect by about the same
factor as the percent conversion.

As mentioned above, the reaction rate is proportional to the
dose rate and the molecular weight is independent of the dose rate.
In order to explain these facts, Restaino et al. [24] postulated
that the polymerization proceeds independently in each cluster of
radicals along the track of a particular ionizing particle. The
objection has been raised that if all chains were started and
terminated in independent clusters, the rates and molecular weights
should invrease in the same ratio, but such is not the case as
noted above.

Restaino et al. [24] have also raised an interesting question
as to whether the arrangement of monomer in the lattice affects the
specific rate constants of the individual steps of the polymerization reaction, or whether the energy release on addition of a monomer to a growing chain is sufficient to reduce the orientation of
some of its neighbors to an approximately liquid state.

B. Copolymerization of Acrylamide

Radiation-induced copolymerization of acrylamide with other
solid monomers has been studied as a means of elucidating the

mechanism of solid-state polymerization [59]. The experimental technique was facilitated by the fact that mixtures of acrylamide with certain other monomers remain in the solid state at room temperature.

One of the binary systems employed was a mixture of acrylamide (AA) and methacrylamide (MAA) which forms a solid solution. Another system evaluated was a mixture of acrylamide and acenaphthylene (ACN) which forms an eutectic mixture. A study was made of the relation between the structure of the monomer mixtures and the kinetic behavior and also the compositions of the polymer obtained using both in-source and postpolymerization.

The monomers were purified by recrystallization from suitable solvents. Monomer mixtures were prepared by weighing various compositions and adding them to glass ampules, cooling down, and sealing off the ampules *in vacuo* or in air as desired. In-source polymerization was performed in a cobalt-60 source at 25° ± 0.5°C. For postpolymerization studies, samples in ampules were irradiated by an electron beam from a Van de Graaff accelerator at -78°C, then postpolymerization was carried out at selected temperatures. The resulting polymers were washed free of monomer with methanol and dried in vacuum at room temperature. Phase equilibria were investigated by determining melting points of monomer mixtures with a micro melting point apparatus, and by x-ray diffraction studies employing CuK$_\alpha$ radiation. Polymer compositions were determined by high resolution NMR spectra or by elementary analysis.

The studies showed that the AA-MAA system forms a solid solution at all compositions below 75°C, and at low concentrations of AA forms a solid solution up to 110°C. The AA-ACN system forms an eutectic mixture, with the eutectic point at 83°C corresponding to about 90 mole % acrylamide. The AA-ACN system at 25°C would correspond to a physical mixture of the two solid polymers.

These physical differences in the two systems are reflected in their radiolytic copolymerization behavior at 25°C. In the AA-MAA system, the rate of polymerization increases as the AA content increases, the acceleration apparently being due to the effect of gel formation with increasing AA content. For the AA-ACN system, only homopolymers of AA were obtained as shown by infrared. While the homopolymer of ACN did not appear, the presence of the ACN monomer accelerated the homopolymerization of the acrylamide. Results from a series of various AA/MAA and AA/ACN starting ratios indicated that copolymers of 1:1 composition were obtained from the AA-MAA system, while homopolymers of AA were obtained from the AA-ACN system.

An investigation was also made of the polymerization of both systems in acetone solution at 25°C. Whereas the effect of monomer composition on polymer yield in the AA-MAA system was similar to that in the solid state, the corresponding effect was quite different for AA-ACN, with polymer yields decreasing greatly as the ACN content increases.

Postpolymerization was carried out at 50°C following preirradiation. For the AA-MAA system following preirradiation at -78°C, the postpolymerization at 50°C showed rates in accordance with the following equation originally developed by Morawetz,

$$Y = (k_p/k_t)(\ln t + \ln k_t[R_0]) \qquad (6-1)$$

where Y represents polymer yield, k_t is the rate constant for termination, k_p the constant for propagation, and $[R_0]$ is the original concentration of radicals present in the monomer mixture as a result of preirradiation. All three quantities k_p, k_t, and R_0 are influenced by the crystalline state of the monomer composition. In addition, $[R_0]$ is proportional to the preirradiation dose.

Percent conversion versus logarithm of time gives a linear plot for the AA-MAA system, in accordance with Eq. (6-1). The slope of the line obtained increases as the percent acrylamide in the composition increases. The intersection also changes as monomer concentration changes, because k_p, k_t, and R_0 are all functions of monomer composition. To summarize the effect of acrylamide concentration in the AA-MAA system, the effect of increasing AA concentration is to increase the rate of conversion regardless of whether the reaction is carried out in-source in the solid solution, in-source in acetone solution, or by postpolymerization in the solid state. For different preirradiation doses at -78°C, the slope of the percent conversion versus ln t curve remains constant for polymerization at 50°C, when the acrylamide concentration in the solid solution is held constant at 50 mole%.

Effect of postpolymerization temperature was studied at 30°, 50°, and 70°C using 50 mole% AA in the AA-MAA system and holding the amount of preirradiation dose constant at -78°C. The authors gave no detailed explanation of the temperature effect, but attributed the rate change to the variations in k_p, k_t, and the crystalline state of the system.

The copolymer compositions produced in the AA-MAA system indicated the same Q and e values for both in-source and postpolymerization reactions (Q = 1.18, e = 1.30 for AA; Q = 1.46, e = 124 for MAA).

Acenaphthylene hardly ever produced homopolymer in either solid- or liquid-phase reaction. Conversion of AA in the AA-ACN system during postpolymerization gave a linear plot versus logarithm of time. In the solid phase of the AA-ACN system, the homopolymerization of the AA was accelerated by the disordering of the monomer crystal (in-source reaction). In the corresponding postpolymerization reaction, the disordering by the ACN interfered with the efficient trapping of free radicals prior to polymerization.

C. Acetylene

Tabata *et al.* [7] carried out the radiation-induced polymerization of acetylene in the solid state. Cobalt-60 was used as a

source of γ-rays. The temperature of polymerization was -196°C.
Percent yield versus time was plotted for several dose rates. In
all cases the plot was linear, and in no case was an induction
period observed (Fig. 6-1).

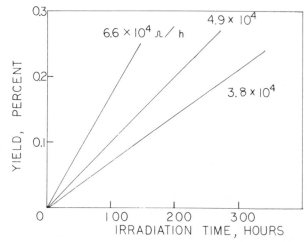

FIG. 6-1. Solid state polymerization of acetylene at -196°C.
[From Y. Tabata, B. Saito, H. Shibano, H. Sobue, and K. Oshima,
Makromol. Chem., 76, 89 (1964). Huthig und Wepf Verlag, Basel,
Switzerland.]

A plot of log grafting rate versus log dose rate indicated
dependence of rate on the first power of the dose rate. (This
contrasts with the usual dependence on the one-half power of the
dose rate in the typical free radical polymerization.) The author
concluded that the polymerization mechanism is not free radical
in character.

The effect of typical free radical scavengers on the rate of
polymerization was studied. The rate was found to be approximately
the same, regardless of whether scavengers such as DPPH or benzoquinone were present. This observation provides further evidence
that the polymerization takes place by something other than a free
radical mechanism.

The polymer resulting from the solid-state polymerization is
a rigid material of deep brown color [7]. It is insoluble in almost
all solvents, but is swelled by dimethyl formamide. The infrared
spectrum of the polymer was examined by means of the KBr-pressed
disk technique. The resulting spectrum was quite different from
that of polymer polymerized in the vapor phase [8] or in the liquid
phase [7]. From a study of its spectrum and from its insolubility
in various solvents, the author postulates that C–C double bond
linkages in the polymer from solid-phase polymerization exist
mainly in the *trans* form

$$\underset{H}{\overset{\diagdown}{}}C = C\underset{\diagdown}{\overset{H}{\diagup}}$$

The author further postulates that the two factors having the
most important bearing on the solid-state polymerization are the
monomer crystal structure and the monomer crystal electronic state.
The hypothesis is presented that the mechanism of the solid-state
polymerization is not radical or ionic in the ordinary sense, but
rather an "electronic polymerization" following a sort of collec-
tive excitation of the monomer crystal as a whole. Based on this
theory the author predicts a very small or zero activation energy
in such solid-state polymerizations at extremely low temperature.

D. Trioxane

Work on the solid-state polymerization of trioxane was carried
out in 1960 by Okamura *et al.* [9] in Japan and simultaneously by
other investigators in France and the United States. Okamura used
a dose rate of 5 x 10^4 rads/h and a polymerization temperature of
55°C. Under the experimental conditions employed, a maximum conversion of about 40% to a very fibrous polymer was obtained. A
similar procedure employing liquid monomer produced no polymerization.

The indirect postpolymerization technique was also investigated [10], involving irradiation of trioxane at 25°C, followed by aging at 55°C. The result was a highly crystalline and regular fiber of much higher molecular weight than that produced by irradiation at 55°C at the same dose. The lower molcular weight observed in the direct irradiation experiment is probably the result of polyoxymethylene chain rupture. (Independent experiments have shown that the molecular weight of polyoxymethylene can readily be reduced by irradiation.)

Aside from increased molecular weight, there are other advantages of using the postpolymerization technique. The polymerization has been found to be independent of dose rate when this technique is employed, allowing dwell time in the radiation field to be reduced to a minimum. Reduced specific viscosity of 2.5 is readily obtained, as compared to 0.5 in the direct irradiation method.

An unusual method of molecular weight control through modification of the crystal size of the trioxane can be employed in the postpolymerization technique. A regular relationship has been observed between molecular weight and particle size at a given dose. It has also been found that there is a proportionality between log of aging time and percentage conversion of monomer to polymer [11]. The molecular weight can also be controlled during aging by varying the temperature of the aging process. This latter method is somewhat limited in practice because the postpolymerization takes place only between 35°C and the melting point of the monomer crystals (64°C).

Extensive studies have been made of the structure of polyoxymethylene produced by irradiation of trioxane. The effect of the aging temperature following irradiation has been investigated carefully. x-Ray diffraction patterns of the resulting polymer indicate that it is probably a twinned crystal [12]. The melting point of the irradiation-produced polymer is about 200°C, as

compared to 180°C for commercially produced polyoxymethylenes.
This result indicates the much higher regularity possessed by the
irradiation-produced polymer. Nuclear-magnetic-resonance (NMR)
measurements on trioxane monomer indicate that molecular motion in
the solid state starts at about 40°C, which explains why irradia-
tion-induced polymerization at or near that temperature produces a
highly regular polymer [13]. This highly regular polymer can be
recrystallized from solution in the form of long whiskers. (Such
whiskers do not form when commercial polyoxymethylene is recrystallized in a similar way.)

The shape of the monomer crystal controls the shape of the
polymer crystal. The monomer can be formed as long, narrow
crystals or as short, wide crystals. Either type can be radiation
polymerized to give a polymer that replicates the crystal structure
of the monomer.

The above remarks apply to the postirradiation technique.
When polymerization is carried out by in-source irradiation at room
temperature, the resulting polymer does not have a high degree of
regularity. When using BF_3-induced polymerization at 55°C, the
final polymer is a mixture of the highly regular polyoxymethylene
and a less regular but crystalline polymer.

E. Other Cyclic Monomers

Interesting results obtained with trioxane led to the solid-
state polymerization of several other cyclic monomers. Some of
these monomeric structures are shown in Table 6-1.

For example, β-propiolactone [14] and 3,3-bis(chloromethyl)-
oxetane [11] can be irradiated in the solid state to produce
polymers. These produced polymer yields of 25% or less, as compared to much higher yields of polymer from trioxane under similar
conditions. In liquid-phase experiments, neither of these monomers
would give high molecular weight polymers.

TABLE 6-1

Cyclic Monomers That Have Been Polymerized
in the Solid State

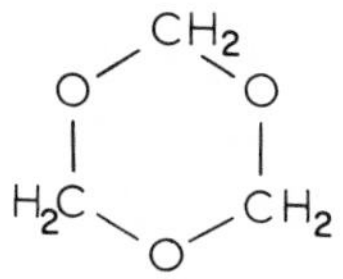

trioxane

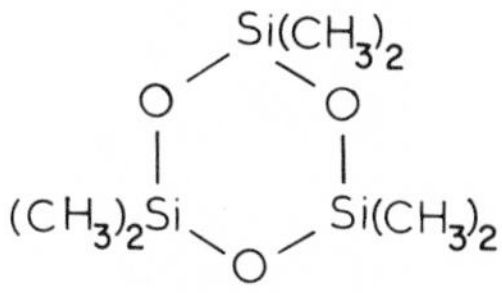

hexamethyltrisiloxane

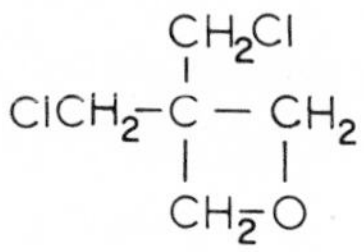

3,3-bischloromethyloxetane

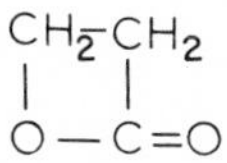

β-propiolactone

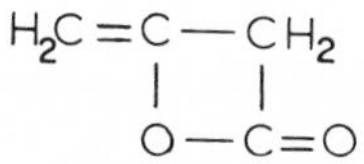

diketene

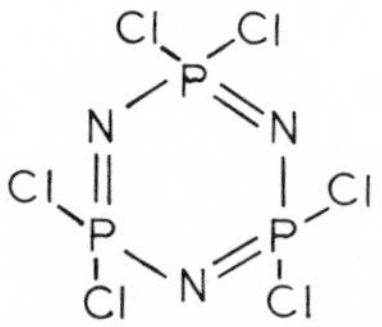

phosphonitrilic chloride

In the case of 3,3-bis(chloromethyl)oxetane, the melting point
of the irradiation-produced polymer was approximately the same as
that of polymer made by conventional chemical methods. For β-propiolactone, the irradiation-initiated polymer had a much higher
melting point than that obtained by chemical methods. The irradiation-produced polymer was much higher in molecular weight than
any polymer previously prepared by other methods. Additional
polymerization was obtained by aging in the solid state after irradiation and also by melting the monomer, resolidifying, and
irradiating again in the solid phase [15].

The polymer of β-propiolactone is polyhydracrylic ester.
After melting and recrystallizing this polymer, the melting point
dropped from 120° to 80°C, indicating a marked decrease in polymer
order. By annealing the order was increased and the melting point
became higher, but did not reach the original 120°C. The irradiation-produced polymer had reduced specific viscosity in tetra-
chloroethane in the range of 0.5 to 2.0. The polymer could be
extruded with ease and could be oriented to give a continuous fiber.
Although solid-state polymerized β-propiolactone did not have as
high a degree of regularity as radiation-polymerized trioxane, it
did have a higher degree of regularity than the polymer obtained by
conventional methods of polymerization.

Radiation-induced solid-state polymerization has also been obtained for 3,3-bis(bromomethyl)oxetane, 3-bromomethyl-3-ethyl-
oxetane, and 3-chloromethyl-3-ethyloxetane [11]. In these cases
there was no notable structural difference between the polymers
prepared by radiation activation and the polymers prepared by
conventional chemical initiation.

Most of the other monomers did not polymerize in the solid
state or else gave very low yields of polymer [16]. One interesting observation was published in connection with 3,3-bis(fluoro-
methyl)oxetane. Although this monomer did not polymerize in the
solid state near the melting point of 19°C, lowering the temperature to -78°C did result in the formation of polymer [17]. This

evidence appears to show that a lower temperature of the solid
phase may favor the propagation reaction, stabilize the initiating
site, or retard the termination step.

F. Copolymerization in the Solid State

Hayashi [18] carried out several copolymerizations in the solid
state which showed characteristics considerable dofferemt frp, those
of conventional liquid-phase copolymerizations. The names and
structural formulas of the monomers used are presented in Table 6-2.
Melting point measurements and x-ray diffraction analyses were made
for both monomer mixtures and copolymer products.

TABLE 6-2

Monomers Used in Solid Phase Radiolytic Copolymerizations

of Hayashi [18]

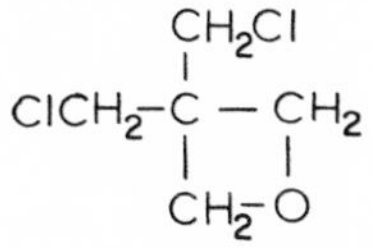

CH_2Cl $\|$ $ClCH_2-C-CH_2$ $\| \quad \|$ CH_2-O	3,3-bis(chloromethyl)oxetane (BCMO)
CH_2Cl $\|$ $C_2H_5-C-CH_2$ $\| \quad \|$ CH_2-O	3-ethyl-3-chloromethyloxetane (ECMO)
CH_2Br $\|$ $BrCH_2-C-CH_2$ $\| \quad \|$ CH_2-O	3,3-bis(bromomethyl)oxetane (BBMO)
CH_2Br $\|$ $C_2H_5-C-CH_2$ $\| \quad \|$ CH_2-O	3-ethyl-3-bromomethyloxetane (EBMO)

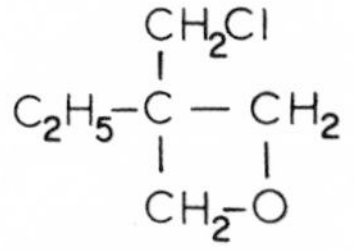

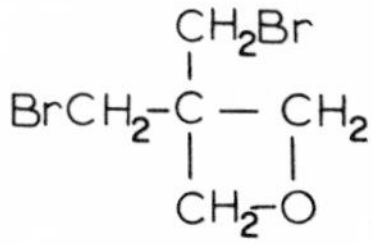

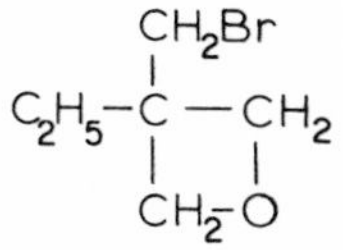

The four oxetanes were copolymerized with each other as mixed
crystals by γ-irradiation. System I was BCMO-BBMO; System II,
BCMO-ECMO; System III, ECMO-EBMO; and System IV, BBMO-EBMO.

Conditions employed can be illustrated by describing the
treatment of System I, BCMO-EBMO. Crystals obtained by shock cooling of System I were γ-irradiated at 0°C for 20.3 h at a dose rate
of 3 x 10^4 R/h. After irradiation, the resulting copolymer
crystals were washed with methanol to remove monomer.

It was observed that x-ray diagrams of the mixed crystals of
monomers were completely different from those of separate monomers,
indicating the existence of a solid solution of monomers. Pronounced x-ray diagram differences were also observed between the
copolymer and a mixture of homopolymers. The x-ray analyses were
used to plot copolymer composition against the composition of the
starting monomer mixture, as shown in Fig. 6-2. For System I, the
reactivity ratios for cationic copolymerization using BF_3-etherate
are almost the same as in the radiolytic copolymerization. However,
the reactivity ratios were quite different for Systems II, III, and
IV, as shown in Fig. 6-2 and Table 6-3.

Table 6-3 shows that copolymer composition versus starting
monomer composition gives a straight line, as shown in Fig. 6-1,
for all four cases or radiolytic copolymerization, as indicated in
the relation $r_1 = r_2 = 1$ in all four cases. This means that a
copolymer of completely random composition is produced in all four
cases, the composition of the copolymer in each case being simply
equal to the composition of the monomer mixture employed. The
author concludes that in solid-state copolymerization the physical
state seems to predominate over chemical reactivity in determining
copolymer composition.

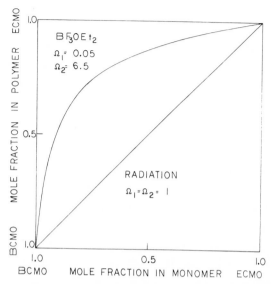

FIG. 6-2. Copolymer composition curves for copolymerization of oxetanes; (●) radiolytic copolymerization in solid state; (O) BF_3OEt_2-catalyzed copolymerization in liquid state. Source: K. Hayashi, H. Watanabe, and S. Okamura, *J. Polym. Sci.*, B1, 399 (1963); John Wiley and Sons, New York.

TABLE 6-3

Monomer Reactivity Ratios

System	By BF_3-etherate[*]		Radiolytic	
I	$r_1 = 1$	$r_2 = 1$	$r_1 = 1$	$r_2 = 1$
II	$r_1 = 0.05$	$r_2 = 6.5$	$r_1 = 1$	$r_2 = 1$
III	$r_1 = 0.7$	$r_2 = 1.6$	$r_1 = 1$	$r_2 = 1$
IV	$r_1 = 0.4$	$r_2 = 16$	$r_1 = 1$	$r_2 = 1$

[*]Using BF_3-etherate at 0.07 mole/liter at 30°C; for 0.5 h; [M] = 2 mole/liter in CH_2Cl_2.

G. Trithiane

Lando and Stannett [19] have successfully carried out the radiolytic polymerization of trithiane in the solid state. This material is the sulfur analog of trioxane. Earlier reports had stated that it did not undergo radiolytic polymerization in the solid state [20]. Lando produced the desired reaction by using high radiation doses and extensive heating at 180°C.

A solid-state reaction in which the crystal orientations of the product are correlated with the crystal orientation of the reactants has been defined as a "topotactic" reaction [21]. The polymerization of trioxane had already been found to be a topotactic process, as discussed above. Lando investigated the polymerization of trithiane to determine whether it also exhibited topotactic characteristics.

Two samples of trithiane monomer in powder form were recrystallized from chloroform, sealed in air in Pyrex tubes, and exposed to a cobalt-60 γ-radiation dose of 9.8 mrads at room temperature. One tube was opened, the contents were found to dissolve in hot chloroform, and were discarded.

The second tube was held at 195°C for 18.5 h. Extraction of this powder with hot chloroform left a 20% residue of a brown powder which did not sublime at 46°C and 50 mm pressure. (Note: Monomeric trithiane readily sublimes under such conditions.) The insoluble portion melted at 245°-260°C, and the melt could be recrystallized to form spherulites of 195°-210°C m.p. The material is stable at 210°C. (Trithiane monomer sublimes completely at 120°C and shows a melting point of 215°-216°C.) The polymeric melting point of 245°-260°C is the same as that found earlier for polythiomethylene polymerized from trithiane using cationic catalysts [22].

A crystal of trithiane which had been grown by sublimation was heated for 165 h at 180°C. An x-ray photograph of this crystal showed no change in structure from that of monomeric trithiane.

On the other hand, a crystal given a dose of 10 Mrads with γ-rays
and then heated at 180°C for 143 h showed distinct changes in
structure, as indicated by x-ray photographs. These photos showed
marked similarities to those of polythiomethylene prepared by other
methods.

H. Calcium Acrylate

In early work on the solid-state polymerization of crystalline
monomers, it had been observed that sodium and potassium acrylate
had very different rates of polymerization, indicating a difference in the favorability of the steric disposition of the reactive
groups in the crystalline structures. It was also known that in
certain cases a reaction carried out in the liquid state led to
different products from the same reaction carried out in the solid
state.

Further information on such questions was obtained by an investigation of the solid-state radiolytic polymerization of calcium
acrylate [30]. Calcium acrylate was especially useful for this
purpose, since it can be obtained in several different crystalline
or amorphous modifications. The technique employed was the irradiation of the solid-state monomer at low temperature with γ-rays,
followed by postpolymerization at higher temperature. The initial
irradiation in most cases was carried out at dry-ice temperature
(-78°C) in the presence of air. During the subsequent polymerization run, several Pyrex tubes containing monomers were placed in a
constant temperature bath. At intervals the contents of a tube were
removed, extracted with water to remove residual monomer, dried,
and weighed to determine percent conversion.

Calcium acrylate was prepared in a number of crystalline modifications, anhydrous or hydrated. Also, amorphous anhydrous
calcium acrylate was obtained by rapid dehydration of two hydrated
forms.

The postpolymerization of calcium acrylate was generally similar to that of acrylamide. The rate was highest at the start and then fell off quickly. The yields for extensive polymerization times were linear in the logarithm of polymerization time. The following general observations were made.

1. The degree of hydration does not regulate the rate of polymerization, because both anhydrous crystalline forms and the dihydrate polymerize more rapidly than the monohydrate.

2. The two anhydrous crystalline forms polymerize at quite different rates.

3. The "amorphous" salt from the monohydrate crystal polymerizes much less rapidly than "amorphous" salt prepared from the crystalline dihydrate.

An x-ray investigation showed that calcium acrylate dihydrate is monoclinic, space group $P2_1/a$, a = 14.36 Å, b = 6.59 Å, c = 11.62 Å, β = 118.9°, with four molecules per unit cell. The crystal exists in a layer structure, with the calcium ions and vinyl groups lying in sections of approximately 3 Å thickness. The acrylate bonds in one section are separated by a distance of about 7 Å from those in the next section.

An electron density map was determined from data taken with an original single crystal of calcium acrylate dihydrate after polymerization to approximately 10% conversion. The map showed the same atomic positions as the corresponding map for the original crystal.

The authors postulate that the wide difference in polymerizability of different calcium acrylate salts can only be explained by assuming that the difference in steric placement of the acrylate double bond is the dominant factor in controlling polymerization rate. If such lattice control of rate is assumed, a question arises as to the extent of polymer chain growth within the crystal lattice. A provisional answer to this question was supplied by the crystallographic data on partially polymerized crystals of calcium acrylate dihydrate, which indicated no change in the atomic positions of the unpolymerized monomer. Such data would suggest that the polymer chains grow at the surfaces of the monomer crystallites. Possibly the growing end is embedded in crystalline monomer, with the rest of the polymer forming a separate phase. This mechanism has been suggested earlier [31].

Since the lattice influences the extent of polymerization, it could conceivably affect the steric configuration of the polymer formed. For example, in calcium acrylate dihydrate the double bonds have been shown to lie in narrow sections sufficiently far apart to prevent reaction between the sections. Within each section the growing end of a given chain could probably attack several monomer units of different orientation, so that a high degree of stereoregularity would not be expected.

The author makes a further point worth noting. It may be possible that the polymerizability and product characteristics are affected only by short-range order in the vicinity of the growing chain end. For example, there is a large difference in the reactivity of the amorphous calcium acrylate derived from the monohydrate and from the dihydrate. It is notable that the product derived from the reactive dihydrate polymerized faster than that which was obtained from the inert monohydrate. This type of memory effect had been observed earlier for inorganic amorphous solids.

Work by the same author (Fadner) on barium methacrylate polymerization supplied additional evidence for the ideas and hypotheses advanced for calcium acrylate [31].

I. Glassy-State Polymerization

A number of monomers have been polymerized in the glassy state. For monomers that polymerize by a diffusion-controlled mechanism and hence polymerize more rapidly in liquid than solid phase, the glassy state usually yields more rapid polymerization than the solid state. Various anomalies have been observed in glassy-state polymerizations.

Kaetsu *et al.* [23] investigated the radiolytic polymerization of several glassy systems containing acrylamide, acrylic acid, and other components. The polymerizations were activated by γ-rays from a cobalt-60 source. Generally speaking, it was found that polymerization was very rapid at low temperatures in such systems, and that rate was very sensitive to viscosity and transition temperature of the glassy systems.

Typical mixtures used consisted of acrylamide-itaconic acid, acrylamide-malonic acid, acrylic acid-acetamide, and acrylic acid-formamide.

It was observed that irradiation at temperatures below the T_g (second transition point) caused no in-source polymerization, but by warming above T_g after irradiation, an almost explosively rapid polymerization took place and was completed in a few minutes, the temperature rising to 80°-160°C owing to heat of polymerization. Irradiation above T_g caused in-source polymerization to be completed quickly to nearly 100% conversion. These facts indicate that polymerization does not take place below T_g, but does occur above T_g as a result of molecular diffusion.

The differences between in-source and postpolymerization can be listed as follows: (1) The polymerization heat peak of differential thermal analysis is very small for in-source polymerization; (2) a large temperature rise is found in the postpolymerization; (3) polymer resulting from in-source polymerization is closely cross-lined, whereas that obtained by postpolymerization is not so highly cross-linked; and (4) during postpolymerization, crystallization of nonpolymerizable components occurs rapidly. For in-source polymerization, no crystallization occurs, and a more transparent appearance is retained owing to the closely cross-linked structure including nonpolymerizable substances.

In-source polymerization of several glassy systems was carried

out at -78°C. Polymerizations were very rapid and often showed
autoacceleration, owing to the high viscosity of the monomeric systems and consequent decrease in termination rate. At temperatures
high above T_s, the polymerization rate decreases with temperature
in the "normal" fashion. However, the rate increases with decreasing temperature near T_s, owing to the decrease of bimolecular
termination as viscosity increases. Near T_g the rate decreases
again, so that conversion rate appears to go through a peak in the
general neighborhood of T_s. Thus, the odd effects of temperature
can be explained on the bais of the relation between viscosity and
molecular diffusion.

Several glassy systems were studied by the postpolymerization
technique, including acrylamide-malonic acid, acrylamide-succinic
acid-acetamide, and acrylamide-itaconic acid. Almost 100% conversions are obtained in the acrylamide systems following rapid postpolymerization. The tempertures of the reaction mixtures rise 40°-
100°C above the T_g as a result of the heat of the explosive polymerization, which therefore cannot be followed by the usual
weighing methods.

Equations were developed which allow the rapid polymerization
rates to be estimated from temperature-increase rates. These
results indicated a proportionality between polymerization rate and
irradiation dose, suggesting a lack of the usual bimolecular termination and possible inactivation by the occlusion of propagating
chains.

To obtain more information on the mechanism of glassy-state
polymerization, ESR spectra were measured for acrylamide-malonic
acid and acrylamide-itaconic acid. The spectrum of the acrylamide-
malonic acid system irradiated at -196°C consisted of a quintet
spectrum and a triplet spectrum assigned to $CH_3-\overset{\bullet}{C}H$ $(CONH_2)$ and
$\text{\Large\wedge\wedge} CH_2\overset{\bullet}{C}H$ $(CONH_2)$, respectively. When the temperature was raised,
the quintet spectrum had disappeared at -129°C, and at -82°C the
spectrum had changed to that of a peroxy radical, $\text{\Large\wedge\wedge} CH_2CH(-O-O\bullet)CONH_2$.

The peroxy radical spectrum changed into the triplet spectrum of the propagating radical, $-CH_2\overset{\bullet}{C}H(CONH_2)$, at about -60°C, that is, a little above the glass transition temperature (T_g). The author states that postpolymerization began with the peroxy radical at this temperature, in agreement with the results of the polymerization experiments. The spectrum of the propagating radical disappeared sharply at about -49°C.

Similar ESR studies were made of the acrylamide-itaconic acid system. In this case the ESR spectrum seemed to be produced by the superposition of the spectra of two kinds of propagating radicals, $\text{\Large\textbackslash\textbackslash\textbackslash} CH_2\overset{\bullet}{C}H(CONH_2)$ and $\text{\Large\textbackslash\textbackslash\textbackslash} CH_2\overset{\bullet}{C}(COOH)\,CH_2COOH$. These radicals survived to a slightly higher temperature than did the propagating radical in the acrylamide-malonic acid system.

The degree of polymerization was found to be high, in agreement with the low termination rate in glassy systems. Viscosity data for in-source polymers were used to compute Huggins' constant, k', often regarded as an index of the degree of branching. It was found that k' of polymer produced between T_s and T_g is extremely large (3.0 to 3.4), whereas k' of polymer produced above T_s has the normal value. The author postulates that at a temperature between T_g and T_s active species are formed in the polymer chain by irradiation, and these begin to propagate at a very high rate owing to the high viscosity and consequent decrease in termination rate. Under these circumstances the branching might well develop to an unusually high degree.

In the case of postpolymerization in glassy systems, the degree of polymerization increases with dosage and polymerization rate, and k' decreases with increasing dosage.

J. Other Monomers

1. Hexamethylcyclotrisiloxane

This material was the first cyclic monomer to be polymerized in the solid state by means of irradiation. The structure of the

polymer did not replicate the monomer crystal nor was a useful
polymer produced. The polymer was found to be heavily cross-linked
[32, 33].

Irradiation caused little or no polymerization above the melting point. It may be that the initiating species is stable only in
the solid state. In fact, the initiating species may be slightly
unstable even in the solid state. Marans [34] found no polymerization when the monomer is irradiated at 25°C and then heated to
slightly below the melting point (m.p. 62°C).

2. Tetraoxane

Tetraoxane is the eight-membered ring corresponding to the six-
membered ring of trioxane (Table 6-1). Investigation of its solid-
state polymerization revealed the following [35].

1. Higher yields were obtained than for trioxane, with the
employment of aging temperatures over 100°C in the solid state.

2. For results comparable to trioxane, shorter aging times
could be employed.

3. The resulting polymer was not as well ordered a crystal
as the one from trioxane.

IV. SUMMARY OF CHARACTERISTICS OF RADIATION-INDUCED
SOLID-STATE POLYMERIZATION

No general laws have been formulated concerning the characteristics of solid-state polymerization, and it is possible that no
concise summary can be given because of the broad nature of the
subject and the wide variety of features presented by individual
examples of solid-state polymerizations. Certain observations can
be made that apply to most solid-state polymerizations, or at least

to a large number of experimental examples, and these can be summarized as follows:

1. *The mechanism* of radiation-induced solid-state polymerization can be ionic or free radical or a combination of both. The
rules to determine what type of mechanism will apply are no more
definite than those for liquid-phase polymerization. Irradiation
may form ions that are capable of initiating polymerization, but
such ions are sensitive to small amounts of impurities that may react with them and prevent them from acting as initiators, in close
analogy with what happens in liquid-phase polymerization.

2. *Slowness of diffusion* in the solid phase is an important
factor, and it hinders access of monomer to the growing radicals
and may hinder even more strongly the termination reaction which
takes place by radical recombination. The slowness of diffusion in
the crystal structure of the monomer is also important in shielding
and prolonging the lifetime of free radicals formed by irradiation
and trapped in the solid polymer. If the shielding is too good,
diffusion of monomer to the trapped radicals is also prevented, and
the polymerization reaction is slowed. Generally speaking, free
radicals trapped in solids are more shielded and have much longer
lifetimes than radicals formed in liquid-phase monomers.

3. *Monomer crystal structure* can either promote or hinder
polymerization, and both effects have been observed. Monomer crystal structure may or may not influence polymer crystal structure.
Cases are known where the polymer crystal structure replicates the
crystal structure of the monomer. Such a reaction that is influenced by the geometric arrangement of the reactants has been called
a "topochemical" reaction.

4. *The effect of inhibitors* or additives presents special
problems in evaluation, because the additive may not be soluble in
the solid monomer. If it does dissolve in the solid monomer, it
may disrupt the monomer crystalline structure and change the reaction character by this disruption.

5. *The effect of oxygen* may differ from its effect in liquid-phase polymerization, possibly because of the slowness of oxygen diffusion into solid monomers. Oxygen may thus have a small or reduced effect in its inhibition of solid-phase polymerization. Slow oxygen diffusion may allow preirradiation to be carried out in air, followed by a successful postpolymerization reaction.

6. *Copolymerization reactivity ratios* may differ from those observed in the liquid-phase. If two solid monomers form separate solid phases when mixed together, they are apt to polymerize to form two homopolymers. If the two monomers form a solid solution, the reactivity ratios may still be changed because of the effect of crystal structure orientation that may favor or disfavor additon of one monomer to another or to itself. In many experimental cases involving a solid solution of 1:1 mole ratio, the resulting co-polymer contains monomers in a 1:1 mole ratio. In some cases the reactivity ratios may be the same in solid- as in liquid-phase copolymerization.

REFERENCES

1. Schmitz, J. V. and Lawton, E. J., *Science*, 113, 718 (1951).
2. Mesrobian, R. B., Ander, P., Ballantine, D. S., and Dienes, G. J., *J. Chem. Phys.*, 22, 565 (1954).
3. Morawetz, H. and Fadner, T. A., *Makromol. Chem.*, 34, 162 (1959).
4. Baysal, B., Adler, G., Ballantine, D., and Colombo, P., *J. Pol. Sci.*, 44, 117 (1960).
5. Baysal, B., Adler, G., Ballantine, D., and Glines, A., *Polymer Letters*, 1, 257 (1963).

6. Adler, G. and Reams, W., *J. Chem. Phys.*, 32, 1698 (1960).

7. Tabata, Y., Saito, B., Shibano, H., Sobue, H., and Oshima, K., *Makromol. Chem.*, 76, 89 (1964).

8. Jones, A. R., *J. Chem. Phys.*, 32, 953 (1960).

9. Okamura, S., Hayashi, K., Nakamura, Y., *Radiation and Isotopes (Japan)*, 3, 416 (1960).

10. Hayashi, K., Watanabe, H., and Okamura, S., *Polymer Letters*, 1, 397 (1963).

11. Hayashi, K. and Okamura, S., *Makromol. Chem.*, 47, 230 (1961).

12. Curazzolo, G., Leghissa, S., and Mammi, M., *Makromol. Chem.*, 60, 171 (1963).

13. Komaki, A. and Matsumoto, T., *Polymer Letters*, 1, 671 (1961).

14. Okamura, S., Hayashi, K., Kitanishi, Y., and Nishii, M., *Isotopes and Radiation (Japan)*, 3, 510 (1960).

15. David, C., Van der Panen, J., Provost, E., and Ligotti, A., *Polymer*, 4, 391 (1963).

16. Nakashio, S., Kondo, M., Tsuchita, H., and Yamada, M., *Makromol. Chem.*, 52, 79 (1962).

17. Hayashi, K., Nishii, M., Moses, K., Shimizer, A., and Okamura, S., *American Chem. Soc., Reprints Div. Pol. Chem.*, 5, No. 2, 951 (1964).

18. Hayashi, K., *J. Pol. Sci.*, Pt. B, 1, 397 (1963).

19. Lando, J. B. and Stannett, V., *J. Pol. Sci., Pt. B*, 2 375 (1964).

20. Hayashi, K. and Okamura, S., in *Proceedings of the International Symposium on Radiation-Induced Polymerization, Battelle, Nov. 29-30, 1962*, USAEC Publication T1D7643, 1962, p. 163.

21. Lotgering, F. K., *J. Inorg. Nucl. Chem.*, 9, 113 (1959)

22. Gipstein, E., Willisch, F., and Swerting, O., *Polymer Letters*, 1, 237 (1963).

23. Kaetsu, I., Tsuji, K., Hayashi, K., and Okamura, S., *J. Pol. Sci., Pt. A-1*, 5, 1899 (1967).

24. Restaino, A. J., Mesrobian, R. B., Ballantine, D. S., Morawetz, H., Davis, G. J., and Metz, D. J., *J. Amer. Chem. Soc.*, 78, 2939 (1956).

25. Henglein, A. and Schulz, Z., *Naturforsch.*, 9B, 617 (1954).

26. Ballantine, D. S., BNL Report 294 (T-50), 1954.

27. Schulz, R., Henglein, A., von Steinwehs, H. E., and Bombauer, H., *Angew. Chem.*, 67, 232 (1955).

28. Adler, G., Ballantine, D. S., and Baysal, B., *J. Pol. Sci.*, 48, 195 (1960).

29. Fadner, T. A. and Morawetz, H., *J. Pol. Sci.*, 45, 475 (1960).

30. Lando, J. B. and Morawetz, H., *J. Pol. Sci.*, Pt. C, No. 4, 789 (1963).

31. Fadner, T. A. and Morawetz, H., *J. Pol. Sci.*, 45, 475 (1960).

32. Adler, G., *J. Chem. Phys.*, 31, 848 (1959).

33. Adicoff, A. and Burlant, W., *J. Pol. Sci.*, 27, 269 (1958).

34. Marans, N. S., *Pol. Eng. Sci.*, P. 14 (1966).

35. Hayashi, K., Ochi., H., Nishii, M., Miyake, Y., and Okamura, S., *Pol. Letters*, 1, 427 (1963).

36. Valyi, I., Janssen, A. G., and Mark, H., *J. Phys. Chem.*, 49, 461 (1945).

37. Morawetz, H. and Rubin, I. D., *J. Pol. Sci.*, 57, 686 (1962).

38. Bamford, C. H., Jenkins, A. D., and Ward, J. C., *J. Pol. Sci* , 48, 37 (1960).

39. Chapiro, A. and Stannett, V., *J. Chim. Phys.*, 57, 35 (1960).

40. Sobue, H. and Tabata, Y., *J. Pol. Sci.*, 43, 459 (1960).

41. Amagi, Y. and Chapiro, A., *J. Chem. Phys.*, 59, 537 (1962).

42. Bensasson, R. and Marx, R., *J. Pol. Sci.*, 48, 53 (1960).

43. Chen, C. S. H., *J. Pol. Sci.*, 58, 389 (1962).

44. Adler, G., Ballantine, D., and Baysal, B., *J. Pol. Sci.*, 48, 195 (1960).

45. White, D. M., *J. Amer. Chem. Soc.*, 82, 5678 (1960).

46. Brown, J. F. and White, D. H., *J. Amer. Chem. Soc.*, 82, 5671 (1960).

47. Chachaty, C., Magat, M., and Terhinassian, L., *J. Pol. Sci.*, 48, 139 (1960).

48. Tsuda, Y., *J. Pol. Sci.*, 49, 369 (1961).

49. Pshezhetskii, V. S., Kargin, V. A., and Bakh, N. A., *Vysokomol. Soedin.*, 3, 925 (1961).

50. Burlant, W. and Taylor, C., *J. Pol. Sci.*, 41, 547 (1959).

51. David, C., Gosselain, P. A., and Mussa, G., *Bull. Soc. Chim.*
     Belg., 70, 583 (1961).

52. Hayashi, K., Kitanishi, Y., Nishii, M., and Okamura, S.,
     Makromol. Chem., 47, 237 (1961).

53. Okamura, S., Hayashi, K., and Kitanishi, Y., *J. Pol. Sci.*, 58,
     927 (1962).

54. Kohlschutter, H. W., *Ann. Chem. Liebigs*, 482, 75 (1930).

55. Kohlschutter, H. W. and Sprenger, L., *Z. Phys. Chem.*, B16, 284
     (1932).

56. Duling, N. and Price, C. C., *J. Amer. Chem. Soc.*, 84, 578
     (1962)

57. Morawetz, H. and Rubin, I. D., *J. Pol. Sci.*, 57, 669 (1962).

58. Kargin, V. A., Kabanov, V. A., and Zubov, V. P., *Vyskomol*
     Soedin., 2, 303 (1960).

59. Nishii, M., Tsukamoto, H., Hayashi, K., and Okamura, S., *J.*
     Appl. Pol. Sci., 11, 1117 (1967).

60. Houilleres du Bossin du Nordet du Pas de Calais, Brit. Patents
     871, 298; 889, 170; and 901, 200.

61. Finkelsthein, E. I., Gorbatov, E. Y., Chernyak, I. V., and
     Abkin, A. D., *Vysokomol. Soedin., Ser. B*, 10(6), 397 (1968).

62. Swiderski, J., *Rocz. Chem.*, 42(7-8), 1299 (1968).

63. Hardy, G., Varga, J., Nagy, G., Varga, E., and Sara, L., *Magg.*
     Kem. Foly., 73(1), 43 (1967).

64. Adler, G., *J. Pol. Sci.*, *Pt. C*, No. 16 (Pt. 2), 1211 (1967).

65. Hardy, G. and Nagy. L., *J. Pol. Sci., Pt. C*, No. 16 (Pt. 5),
     2667 (1967).

66. Charlesby, A., Garratt, P. G., and Morris, J., *Nature*, 196.
     574 (1962).

67. Zurakowska-Orszagh, J., *Polish Acad. Sci., Inst. Ncul. Res.*,
     Dept. No. 307/XVII (1962).

68. Zurakowska-Orszagh, J., *Chim. Ind. (Paris)*, 93(4), 404 (1965).

69. Chapiro, A. and Sommerlatte, T., *C. R. Acad. Sci., Paris,*
     Ser. C, 264(23), 1825 (1967).

70. Tabata, Y., Oshima, K., and Sobue, H., *J. Fac. Univ. Tokyo,* *Ser. A*, 1, 46 (1963).

71. Tabata, Y., *Kobunshi*, 1968, 17(197), 758.

72. Herz, J. H., Stannett, V., and Turner, R. T., *Makromol. Chem.* 110, 173 (1967).

73. Ishigure, K., Tabata, Y., and Oshima, A., *J. Macromol. Sci.,* *Pt. A*, 1(4), 591 (1967).

74. Neth. Appl. 6,611,640 (1967).

75. Chapiro, A. and Cordias, P., *J. Chem. Phys.*, 64(2), 334 (1967).

76. Chapiro, A., U.S. At. Energ. Comm. Reprot TID-7643, 136, 1962.

77. Ito, M. and Kuri, Z., *Kogyo Kagaku Zasshi*, 70(1), 109 (1967).

78. Bowden, M. J., O'Donnell, J. H., and Sothman, R. D., *Makromol. Chem.*, No. 122, 186 (1969).

79. Ivanov, V. S., Bezhan, I. P., and Levando, L. K., *Vestn. Leningr. Univ.*, 20(10), *Ser. Fiz. i. Khim.*, No. 2, 157 (1965).

80. Morawetz, H., U.S. At. Energ. Comm. Report NYO-1715-16, 1966.

81. Gusakovskaya, I. G. and Goldanskii, V. I., *Vysokomol. Soedin. Ser. B*, 9(5) 390 (1967).

82. Zurakowska-Orszagh, J., *J. Pol. Sci., Pt. C*, No. 16 (Pt. 7), 4219 (1965).

83. Hardy, G., Nyitrai, K., Vorga, J., Hovacs, G., and Federova, N., *J. Pol. Sci., Pt. C*, No. 4, 923 (1963).

84. Fujii, S., Yamakita, H., and Matsuda, T., *Kobunshi Kagaku*, 23, 369 (1966).

85. Kagiya, T., Izu, M., and Fukui, K., Polymer Letters, 2, 779 (1964).

86. Miura, M., Aoyagi, T., and Hinai, T., *Kogyo Kagaku Zasshi*, 67(3), 485 (1964).

87. Kitanishi, Y., Hayashi, K., and Okamura, S., *Radiation and Isotopes*, 3, 346 (1960).

88. Tabata, Y., Suzuki, T., and Oshima, K., *J. Macromol. Chem.*, 1(4), 817 (1966).

89. Okamura, S., Hayashi, K., Kitauishi, Y., Watanabe, H., and Nishii, M., *Nippoa Isotope Kaigi Hobunshii*, 4, 309 (1966).

90. Ito, A., *Japan Chem. Quart.*, 4(4), 10-15 (1968).

91. Okamura, S., Hayashi, K., and Mori, K., AEC-Translation 6361, p. 174.

92. Hardy, G., Nyitrai, K., Kovacs, G., Federova, N., and Varga, J., *Radiation Chim. Proc., Tihany Simp., Tihany, Hung.*, 1962, 205 (1964).

93. Hardy, G. and Nyitrai, K., *Eur. Pol. J.*, 5(1), 133 (1969)

94. Milyutinskaya, R. I. and Bagdasaryan, K. S., *Kinet. Katal.*, 8(3), 677 (1967).

95. Ishida, S. and Saito, S., *J. Pol. Sci., Pt. A-1*, 5(4), 689 (1967).

96. Jager, P. and Waight, E. S., *J. Pol. Sci., Pt. A*, 1, 1909 (1963).

97. Lipscomb, N. T. and Weber, E. C., *J. Pol. Sci., Pt. A-1*, 5(4), 779 (1967).

98. Hardy, G. and Nyitrai, K., *Magy. Kem. Foly.*, 72(12), 517 (1966).

99. Tabata, Y., Kimura, H., and Sobue, H., *Polymer Letters*, 2, 23 (1964).

100. Caglioti, V., Cordischi, D., and Mele, A., *Nature*, 195, 492 (1962).

101. Fujiwara, K., Hayashi, K., and Okamura, S., *Nippon Hoshasen Kobunshi Kenkyu Kyokai Nempo*, 4, 183 (1962).

102. Kuri, Z. and Ito, M., *Kogyo Kagaku Zasshi*, 69(5), 1066 (1966).

103. Sobue, H., Tabata, Y., and Shibano, H., AEC Translation 6316, p. 469.

104. Tabata, Y., Shibano, H., Oshima, K., and Sobue, H., *J. Pol. Sci., Pt. C*, No. 16, 2403 (1967).

105. Hayashi, K., Ochi, H., Mishu, M., Miyake, Y., and Okamura, S., *Polymer Letters*, 1, 427 (1963).

106. Matsuda, T. and Fujii, S., *J. Pol. Sci. Pt. A-1*, 5(10), 2617 (1967).

107. Nishii, M., Hayashi, K., and Okamura, S., *Nippon Hoshasen Kobunshi Kenku Kyokai Nempo*, 7, 159 (1965-1966).

108. Magat, M., *Polymer*, 3, 449 (1962).

109. Goldecki, Z., Karvalk, J., Pekala, W., and Kroh, J., *Bull. Acad. Pol. Sci., Ser. Sci. Chim.*, 15(5), 209 (1967).

110. Parrish, C. F., Trinles, W. A., and Burton, C. K., *J. Pol. Sci., Pt. A-1*, 5(10), 2557 (1967).

111. Hardy, G., *Magy. Kim. Foly.*, 73(2), 55 (1967).

APPENDIX

Table 6-A summarizes the literature on radiation-induced solid-state polymerizations from about 1951 through 1970. The names of the monomers are listed alphabetically, followed by melting point, irradiation temperature, investigator, reference, and comments. Under "comments" is given information of special interest for each monomer, such as type of radiation source employed, yield figures, effect of oxygen, effect of added inhibitors, effect of electric field or pressure, effect of crystallinity, type of mechanism, and name of any monomer employed as a comonomer in copolymerization. A brief mention under "comments" often indicates the main subject or theme of the particular reference.

TABLE 6-A

Radiation-Induced Solid-State Polymerizations

Monomer	Melting point, °C	Irradiation temp., °C	Comments	Investigator and Reference
Acenaphthalene	92	25	Copolymerized with acrylamide	Nishii [59]
Acetaldehyde	-123	-196	8% Yield	———— [60]
Acetonitrile	-41	-45	0.4% Yield	Marans [34]
3-Acetoxymethyl, 3-chloromethyloxetane	35	4	1% Yield	Marans [34]
Acrolein	—	-269	Gamma source	Finkelsthein [61]
Acrylamide	85	-78	100% Yield	Morawetz [3]
Acrylamide	85	—	Effect of oxygen	Swiderski [62]
Acrylamide	85	–	Effect of benzoquinone	Hardy [63]
Acrylamide	85	–	Solid solution in propionamide	Adler [64]
Acrylamide	85	–	Copolymerization with methacrylamide	Nishii [59]

Acrylamide	85	–	Copolymerization with acrylic acid	Hardy [65]
Acrylamide	85	-196	Effect of electric field	Charlesby [66]
Acrylamide	85	–	^{60}Co Source	Zurakowska-Orszagh [67]
Acrylamide	85	–	Copolymerization with maleic anhydride	Zurakowska-Orszagh [68]
Acrylic acid	12	-18	Violent polymerization	Mesrobian [2]
Acrylic acid	12	-16	Effect on stereo-regularity	Chapiro [69]
Acrylic acid	12	–	Eutectic with acrylamide	Hardy [65]
Acrylonitrile	-82	-100	15% Yield	Tabata [70]
Acrylonitrile	-82	-196	———	Tabata [71]
Acrylonitrile	-82	-120	Copolymer with β-propiolactone	Herz [72]
Acrylonitrile	-82	-196	Copolymer with methacrylonitrile	Ishigure [73]
Acrylonitrile	-82	0	10,000 Torr pressure	[74]
Acrylonitrile	-82	-94	Copolymer with styrene	Chapiro and Cordias [75]

Monomer			Conditions	Reference
Acrylonitrile	-82	-196	Effect of electric field	Charlesby [66]
Acrylonitrile	-82	-196	^{60}Co Source	Tsuda [48]
Acrylonitrile	-82	-	Gamma or electron beam	Chapiro [76]
N-tert-amylacrylamide	-	-	Effect of oxygen	Swiderski [62]
Allyl alcohol	-129	-196	Copolymer with SO_2	Ito [77]
Allyl chloride	-	-196	^{60}Co Source	Tsude [48]
Barium acrylate	-	35	75% Yield	Restaino [24]
Barium methacrylate	-	15	Gamma source	Bowden [78]
Biphenylene dimaleimide	-	260	^{60}Co Source	Ivanov [79]
3,3-bis(Bromomethyl)-oxetane	24	0	Effect of crystallinity on yield	Nakashio [16]
3-Bromomethyl 3-Ethyloxetane	4	—	Formed polymer	Hayashi [18]
Butadiene-1-carboxylic acid	75	-	Gamma source	Morawetz [80]
Butane dioldimethacrylate	—	-107	Gamma source	Gusakovskaya [81]

N-tert-Butylacryl-amide	–	–	Effect of oxygen	Swiderski [62]
N-tert-Butylacryl-amide	–	–	Copolymer with maleic anhydride	Zurakowska-Orszagh [82]
Cetyl methacrylate	–	–	Polymer formed	Hardy [83]
β-Chloroethyl vinyl phosphonate	–	-78	Free radical mechanism	Fujii [84]
3,3-bis(Chloromethyl)-oxetane	20	0	25% Yield	Nakashio [16]
3,3-bis(Chloromethyl)-oxetane	20	0	85% Yield	Kagiya [85]
3,3-bis(Chloromethyl)-oxetane	20	–	Effect of benzo-quinone	Hardy [83]
3-Chloromethyl, 3-cyanomethyl-oxetane	8	-50	2.4% Yield	Marans [34]
3-Chloromethyl 3-ethyloxetane	-3	–	Formed polymer	Hayashi [18]
3-Chloromethyl 3-ethyloxetane	-3	-86	Effect of temperature	Miura [86]
N-Cyclohexyl Acrylamide	–	–	Effect of oxygen	Swiderski [62]

Monomer				Reference
Diketene	-7	-78	7% Yield	Kitanishi [87]
Diketene	-7	-196	Effect of pressure	Tabata [88]
Diketene	-7	—	Gamma or electron beam	Okamura [89]
Diketene	-7	–	Ionic mechanism	Okamura [53]
Dimethyl fumarate	—	–	Copolymer with acrylamide	Zurakowska-Orszagh [68]
3-Fluoromethyl-3-ethyloxetane	-	-	Effect of temperature	Miura [86]
3,3-bis(Fluoromethyl)-oxetane	19	-78	1.4% Yield	Hayashi [17]
Formaldehyde	-118	-196	40% Yield	Chachaty [47]
Formaldehyde	-118	-196	Gamma source	Ito [90]
Formaldehyde	-118	-196	^{60}Co Source	Tsuda [48]
Glyoxal	15	-196	Polymer formed	Okamura [91]
Hexadecyl methacrylate	8	-	Polymer formed	Hardy [92]
Hexadecyl vinyl ether	-	-	Polymer formed	Hardy [93]

Monomer			Description	Reference
N-tert-Heptyl acrylamide	-	-	Effect of oxygen	Swiderski [62]
N-tert-Hexyl acrylamide	-	-	Effect of oxygen	Swiderski [62]
Isobutyl vinyl ether	-	-196	^{60}Co Source	Tsuda [48]
Isobutylene	-141	-196	Amine inhibition	Milyutinskaya [94]
Isobutylene	-141	-192	1.6% Yield	David [15]
Itaconic anhydride	-	-	First-order reaction	Ishida [95]
Lithium acrylate	-	-78	24% Yield	Morawetz [57]
Maleic anhydride			Copolymer with N-tert butylacrylamide	Zurakowska-Orszagh [82]
Maleic anhydride	-	-	Copolymer with N-tert pentylacrylamide	Zurakowska-Orszagh [68]
Methacrylic acid	16	-18	100% Yield	Restaino [24]
Methacrylamide	110	30	0.7% Yield/h	Restaino [24]
Methacrylamide	110	-	Effect of oxygen	Swiderski [62]
Methacrylamide	110	25	Copolymer with acrylamide	Nishii [59]
Methacrylamide	110	-	^{60}Co Source	Zurakowska-Orszagh [67]

Monomer				Reference
Methacrylamide	110	20	^{60}Co Source	Jager [96]
Methacrylonitrile	-	-	Attempted copolymerization with acrylonitrile	Ishigure [73]
N,N-Methylene bis-(acrylamide)	-	30	1.9% Yield/h	Restaino [24]
Methyl methacrylate	-	-65	First-order respect to polymer	Lipscomb [97]
α-Methyl styrene	-	-196	^{60}Co Source	Tsuda [48]
N-*tert*-Nonyl acrylamide	-	-	Effect of oxygen	Swiderski [62]
5-Norbornene 2-carboxylic acid	40	30	Rate highest at 30°C	Hardy [98]
Octamethylcyclotetra siloxane	15	14	7% Yield	Tabata [99]
N-*tert*-Octyl acrylamide	-	-	Effect of oxygen	Swiderski [62]
N-*tert*-Pentyl acrylamide	-	-	Copolymer with maleic anhydride	Zurakowska-Orszagh [68]
Phosphonitrilic chloride trimer	114	100	10% Yield	Caglioti [100]

Monomer				
Potassium acrylate	360	-78	2% Yield	Morawetz [57]
Potassium methacrylate	—	-78	3.5% Yield	Morawetz [57]
β-Propiolactone	-33	-78	15% Yield	Hayashi [52]
β-Propiolactone	-33	-120	Copolymer with styrene	Herz [72]
β-Propiolactone	-33	—	Gamma and electron beam	Okamura [89]
β-Propiolactone	-33	-	Ionic mechanism	Okamura [53]
Sodium acrylate	-	-78	13% Yield	Morawetz [57]
Sodium methacrylate	-	-78	8% Yield	Morawetz [57]
Sorbic acid	-	-	Polymer formed	Fujiwara [101]
Styrene	-30	-	Copolymer with β-propiolactone	Herz [72]
Styrene	-30	-94	Copolymer with acrylonitrile	Chapiro [75]
Styrene	-30	-	Free radical mechanism	Kuri [102]
Styrene	-30	-	Gamma or electron beam	Chapiro [76]
Sulfur dioxide	-	-196	Ionic and Radical mechanism	Ito [77]

Sulfur dioxide	Copolymer with styrene	Kuri [102]	–	–
Tetrafluoro-ethylene	Polymer formed	Sobue [103]	-140	–
Tetrafluoro-ethylene	Effect of crystal-linity	Tabata [104]	-140	–
Tetraoxane	99% Yield	Hayashi [105]	112	-78
Trichlorobuta-diene	Effect of DPPH	Matsuda [106]	-49	-196
Trioxane	Effect of benzo-quinone	Hardy [63]	64	—
Trioxane	Effect of crystal form	Nishii [107]	64	–
Trioxane	Gamma source	Ito [90]	64	–
Trioxane	Gamma or electron beam	Okamura [89]	64	–
Trithiane	Effect of crystal-linity	Lando [19]	216	180
Vinyl acetate	Copolymer with acrylonitrile	Ishigure [73]	-93	-196

Vinyl acetate	-93	-196	^{60}Co Source	Tsuda [48]
Vinyl caprolactam	—	—	Polymer formed	Magat [108]
Vinylcarbazole	65	30	1.1% Yield/h	Restaino [24]
Vinylcarbazole	65	—	Gamma source	Galdecki [109]
Vinylcarbazole	65	—	14% Yield	Parrish [110]
Vinyl stearate	34	26	20% Yield	Restaino [24]
N-Vinyl succin-imide	—	—	Formed polymer	Magat [108]
N-Vinyl succin-imide	—	—	Gamma source	Hardy [111]
N-Vinyl succin-imide	—	—	Polymer formed	Hardy [92]

Chapter 7

IRRADIATION OF POLYMERS: CROSS-LINKING VERSUS SCISSION

I.   INTRODUCTION

A.   General Discussion

Following the first nuclear chain reaction in 1942, research on the irradiation of polymers was prompted by the search for plastics materials able to withstand the ionizing radiations of nuclear reactors.   It was soon found that the effects of radiation are not always detrimental, and that some plastics are toughened and made infusible by moderate doses of radiation.   Further studies had the objective of learning how to encourage the beneficial reactions at the expense of the detrimental, that is, to accelerate chain cross-linking and retard chain scission so as to enable the vulcanizing of rubber and the toughening of plastics.

In one of the early investigations, Farmer [10] observed that irradiation of polystyrene caused an increase in electrical conductance.   Winogradof [11] found that x-ray irradiation decreases the tensile strength of cellulose acetate fibers, and Day [12] studied the appearance of paramagnetic resonance absorption in irradiated polymethyl methacrylate.   The burst of world-wide interest in polymer radiation chemistry did not come until 1952, when Charlesby [13] carried out an extensive series of investigations and clearly showed

that radiation could change polyethylene into a cross-linked, insoluble, heat-resistant material. There followed a large number of
papers reporting in detail the effects of ionizing radiation on a
variety of polymers. Research activity is still on the increase in
this field at present, spurred by the now proved fact that desirable modifications of many polymers can be induced by radiation.

Small chemical changes produced by moderate radiation doses
can cause large changes in the physical properties of polymers. A
polymeric substance becomes essentially one large molecule when an
average of about one cross-link unit per molecule has been produced
by irradiation. Since the polymer molecule is commonly made up of
several thousand monomer units, this result corresponds to less
than 0.1% chemical change, and even smaller changes are sufficient
to considerably modify the viscosity of the polymer or its solutions. In the case of polyethylene, a dose of 2×10^6 rads
increases the softening point from about 90° to over 250°C and
greatly decreases the solubility of the polymer in its usual solvents.

The effectiveness of radiation in modifying polymers is influenced by the presence of oxygen or additives, type of radiation
used, degree of polymer crystallinity, and presence of solvent.
Other factors affecting radiation-induced reactions include non-
localization of absorbed energy, energy transfer, electron trapping
and recapture, and abstraction of bound hydrogen by hydrogen atoms.

Charlesby [14] and Lawton *et al.* [15] observed that polymers
may either cross-link or degrade depending on their chemical nature.
Both processes take place simultaneously, and the classification in
Table 7-1 merely indicates the process that appears to predominate.
There has been some disagreement in the development of Table 7-1,
possibly due to the presence at times of additives and plasticizers.
(Note that polyvinyl chloride appears in both Groups I and II.)
Also, a change in irradiation conditions may at times shift a
polymer from one group to the other.

TABLE 7-1

Predominant Processes in Irradiated Polymers

Group I. Cross-linking	Group II. Scission
Polyethylene	Polyisobutylene
Polypropylene	Poly-α-methylstyrene
Polystyrene	Polymethacrylates
Polyacrylates	Polymethacrylamide
Polyacrylamide	Polyvinylidene chloride
Polyvinyl chloride	Cellulose
Polyamides	Cellulose acetate
Polyesters	Polytetrafluoroethylene
Polyvinylpyrrolidone	Polytrifluorochloroethylene
Natural rubber	Polymethacrylic acid
Synthetic rubbers (except	Polyvinyl chloride
polyisobutylene)	Poly-α-methacrylonitrile
Polysiloxanes	Polyethylene terephthalate
Polyvinylalcohol	
Polyacrolein	
Polyacrylic acid	
Polyvinyl alkyl ethers	
Polyvinyl methyl ketone	
Polymethylene	
Chlorinated polyethylene	
Chlorosulfonated polyethylene	
Polyacrylonitrile	
Sulfonated polystyrene	
Polyethylene oxide	

The net result of cross-linking is that the molecular weight
of the polymer increases with increasing dose until a three-
dimensional network is formed where each polymer chain is linked
to one other chain on the average. The reaction can be indicated

diagrammatically.

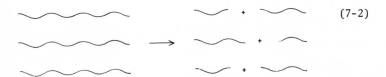

(7-1)

The final structure represents a polymer that will not melt unless
the temperature is raised considerably above its normal melting
point (and possibly not then, depending on dose size). The process
resembles the vulcanization of rubber. A cross-linked polymer will
dissolve only partially in one of its usual solvents. When an
attempt is made to dissolve the polymer, a swollen gel fraction
is left as an insoluble residue, the size of the gel fraction
increasing with radiation dose size.

When scission predominates in an irradiated polymer, the
molecular weight decreases as dose increases. The final product
may be a low molecular weight liquid in some cases. Radiation-
induced degradation at room temperature is generally not a chain
reaction and yields are low. The reaction can be indicated as
follows.

(7-2)

Polytetrafluoroethylene degrades rapidly when irradiated in the
presence of oxygen, and finally loses its strength and becomes
powderlike. Polymethyl methacrylate resists large doses at room
temperature, although it will ultimately crack and come apart.
When it is heated during irradiation, evolution of gas swells it

to a foamy mass. The composition of the gas is analogous to that
produced by the irradiation of low molecular weight esters.

Several theories have been advanced to explain why scission
predominates in some polymers and cross-linking in others. It has
been observed [16] that polymers with scission predominating have
the following structure:

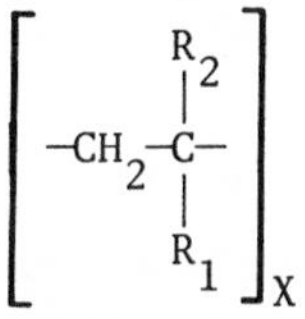

$$\left[-CH_2-\overset{\overset{\textstyle R_2}{|}}{\underset{\underset{\textstyle R_1}{|}}{C}}- \right]_X$$

the implication being that R_1 and R_2 produce a steric strain which
weakens the bonds of the backbone chain. Also, most of the Group
II polymers have low heats of polymerization, which correlates with
a tendency to form monomer on pyrolysis and undergo scission during
irradiation. Table 7-2 illustrates a fair correlation between heat
of polymerization and tendency to degrade to monomer.

TABLE 7-2

Cross-linking Versus Scission[a]

Polymer	Predominant effect of radiation	Heat of polymerization, kcal/mole	Monomer yield on pyrolysis, wt%
Polyethylene	Cross-linking	22	0.025
Polypropylene	Cross-linking	16.5	2
Polymethyl acrylate	Cross-linking	19	2
Polystyrene	Cross-linking	17	40
Polyisobutylene	Scission	10	20
Polymethyl methacrylate	Scission	13	100
Poly-α-methylstyrene	Scission	9	100

[a]Taken from Refs. [17-19].

Any comprehensive theory would need to take account of what
steps would follow the initial scission,

$$-CH_2-\underset{\underset{R_1}{|}}{\overset{\overset{R_2}{|}}{C}}-CH_2-\underset{\underset{R_1}{|}}{\overset{\overset{R_2}{|}}{C}}- \;\rightarrow\; -CH_2-\underset{\underset{R_1}{|}}{\overset{\overset{R_2}{|}}{C}}\cdot \;+\; \cdot CH_2-\underset{\underset{R_1}{|}}{\overset{\overset{R_2}{|}}{C}}- \qquad (7\text{-}3)$$

where the extent of scission would be influenced by the tendency of
the resulting radicals to (1) recombine with disproportionation or
become inactivated by hydrogen abstraction from another molecule,
or (2) recombine and couple. Whereas recombination would seem to
be favored in solid polymers where the radicals remain near each
other, there does not appear to be any theoretical basis for
determining whether recombination or disproportionation would predominate in specific cases.

One factor that must be considered in polymer chain scission
is deduced from studies on the radiolysis of low molecular weight
linear and branched chain hydrocarbons. Generally speaking, chain
scission appears to be favored over cross-linking in branched chain
hydrocarbons. For example, the yield of dimer formation in
branched alkanes is much smaller than the yield in the corresponding linear alkanes. Only small amounts of dimer are formed by the
irradiation of 2,2-dimethylbutane, a substance that contains a
tetrasubstituted carbon atom. Also, the proportion of C–C bond
scissions with respect to total fragmentation increases with
branching and reaches 100% in neopentane.

The cross-linking mechanism may possibly involve the production of polymer radicals at neighboring sites on adjacent chains,
accompanied by the loss of molecular hydrogen as follows.

$$-CH_2CH_2CH_2 \;\rightarrow\; -CH_2\overset{\cdot}{C}HCH_2- \;+\; H\cdot \qquad (7\text{-}4)$$

$$H\cdot \;+\; -CH_2CH_2CH_2- \;\rightarrow\; -CH_2\overset{\cdot}{C}HCH_2- \;+\; H_2 \qquad (7\text{-}5)$$

$$-CH_2\overset{\bullet}{C}HCH_2- \ + \ -CH_2\overset{\bullet}{C}HCH_2- \ \rightarrow$$

$$-CH_2-CH-CH_2-$$
$$|$$
$$-CH_2-CH-CH_2- \tag{7-6}$$

Charlesby [20] has objected to this reaction sequence on the grounds that it would not allow significant protection by small concentrations of additive, while actually the gelation dose is doubled by 1% of a suitable additive in a solid polymer and by a somewhat lower concentration in a liquid polymer. The alternative would be to assume that polymer radicals are formed at random and migrate to each other before combining. This explanation is also somewhat unsatisfactory [20], since it would imply complete protection by an additive concentration that normally far exceeds the radical concentration, whereas only partial protection is noted experimentally.

Although some investigators have postulated an ionic mechanism for cross-linking, Charlesby [20] argues against this hypothesis for the following reasons.

1. Cross-linking takes place in dilute aqueous solutions where there is small chance that two polymer ion molecules will be in the proper position for linking during the short life span of an ion.

2. Polymer radicals have been observed in irradiated solid polymer by ESR spectroscopy and in aqueous solutions (pulse radiolysis) at concentrations similar to the resulting cross-linking density.

3. Radical scavengers in suitable concentration reduce the cross-linking density in both solid and liquid polymers.

Charlesby [20] has suggested that the cross-linking sequence is somewhat as follows. A polymer radical and a hydrogen atom are first formed. A fraction of these hydrogen atoms abstracts in the close vicinity, forming secondary polymer radicals. The resulting

polymer radicals combine readily, with small chance of interference
by additives. If hydrogen does not abstract in the first few
collisions, it may migrate for some distance and become thermalized
before abstracting a second hydrogen and forming a second polymer
radical. Such dispersed radicals (as well as thermalized hydrogen)
may well react with a suitable protective additive that is present
at low concentration.

Another difficulty is to explain the high mobility of polymer
radicals, even in the solid phase. One way to account for the
apparent migration of polymer radicals is by hydrogen addition to
certain radicals and hydrogen abstraction elsewhere from polymer,
the process being accelerated by the presence of hydrogen gas.
This is in line with ESR data indicating that the radical concentration in polyethylene decreases more quickly in the presence of
hydrogen [21].

There are other chemical effects of radiation on polymers
aside from cross-linking and degradation, including the production
or removal of unsaturation, color formation, and the evolution of
gases such as hydrogen, methane, ethane, carbon monoxide, and
carbon dioxide. Crystalline polymers lose at least a fraction of
their crystallinity. Polyethylene becomes yellow or brown, polyvinyl chloride green, and polymethyl methacrylate brownish pink.
Surface fluorescence is produced on polyethylene, polymethyl
methacrylate, nylon, and polyvinyl chloride [24].

The physical effect of chemical cross-linking is small as long
as only a few molecules are involved, but increases rapidly as
cross-linking density increases, and the gel point is reached when
a three-dimensional network first extends throughout the system.
At the gel point some of the material becomes insoluble in solvents
which dissolved the original polymer. Taking note that two monomer
units must be cross-linked to form one cross-link, it can be shown
that the dose r in rads to reach the gel point is expressed by

$$r = \frac{6.023 \times 10^{23} \times 100}{2 \times 6.24 \times 10^{13} \times G \times M_w} \qquad (7\text{-}7)$$

where G is the number of cross-links produced per 100 eV absorbed and M_w is the weight-average molecular weight. In the special case where all original moleules are of the same size, gel formation starts when there is one cross-link for each two molecules originally present. When the original molecules have a distribution such that the weight-average molecular weight is twice the number-average molecular weight, gel formation starts when there is one corss-link for every two weight-average molecules initially present.

A process known as "end-linking" should also be considered. When scission occurs in the main chain of a molecule, at least one of the fragments may link to the main chain of a neighboring molecule to give a branched molecule of a higher molecular weight (see Fig. 7-1). The physical properties of an end-linked polymer would be essentially identical to those of a cross-linked polymer, as shown by an examination of Fig. 7-1.

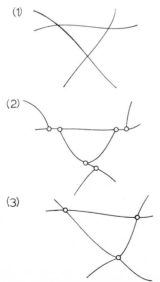

FIG. 7-1. Initiation of gelation by end-linking and cross-linking; (1) Starting condition — three separate molecules. (2) End-linking — one molecule broken twice, another once. (3) Cross-linking.

If chain scission takes place simultaneously with cross-linking, the dose required to reach the gel point is increased. The amount of gel expressed as a fraction of total weight will not continue to increase with dose above the gel point but will approach a certain finite value at infinite dose size. If the gel produced by irradiation is purified by extraction and subjected to additional irradiation, soluble polymer will be formed, the amount of soluble polymer being determined by the ratio of scission to cross-linking.

Other equations relating radiation-induced cross-linking and/or scission to the physical properties of polymers have been developed. Physical methods are required for the complete investigation of cross-linking and scission effects in large molecules. Unvulcanized rubbery polymers become tough and similar to vulcanized rubber after irradiation. The equilibrium modulus, which is zero for an uncross-linked rubber, becomes finite at a minimum total dose dependent on the type of chain structure and inversely related to the molecular weight. The equilibrium modulus has been estimated from statistical theory to be [22]

$$\tau = (\rho RT/M_c)(1 - 2M_c/M)(\alpha - 1/\alpha^2) \tag{7-8}$$

where τ is the tensile force per unit original cross-sectional area, ρ equals density of the polymer, M_c equals average molecular weight between cross-links, α equals ratio of extended to initial length, and R is the gas constant. This equation holds well at small deformations, but fails at large elongations [23]. At constant α, τ will increase approximately linearly with dose as M_c is reduced by formation of cross-links. Equation (7-8) can be employed to estimate the number of cross-links produced by irradiation, if the conditions of measurement are carefully controlled so that the equilibrium modulus is actually measured.

A reduction in τ evidently indicates a predominance of scission over cross-linking, whereas an increase in τ means predominant cross-linking. Studies of tensile strength and breaking elongation give less clear indications of radiation effects. When cross-linking is the major effect, tensile strength may increase to a maximum and then decrease, followed by a final increase at high cross-linking density when the polymer becomes brittle. In other cases, a steady increase in cross-linking density may be accom-panied by a continuous decrease in tensile strength. Breaking elongation is difficult to interpret in a fundamental way.

Measurements of modulus and tensile strength of plastics are more difficult to interpret than for rubbers, because for plastics the measurements are made at a greater distance from equilibrium conditions. With plastics it is preferable to make the measure-ments well above the second-order transition or, when the polymer is normally crystalline, above the first-order transition. How-ever, nonequilibrium measurements correspond more closely to actual service conditions and must be employed when commercial engineering data are required.

The rate of scission may be measured independently of changes in cross-linking by using the stress relaxation method of Tobolsky *et al.* [25, 26]. In this method, the sample is stretched to a constant elongation and the retractive force τ is measured from time to time. Under equilibrium conditions, τ decreases as the chains undergo scission. The following relation has been derived for g, the number of scissions introduced per gram:

$$g = -(1/M_c)\ln(\tau/\tau_0) \qquad (7-9)$$

where τ_0 is the starting tension. This technique has been employed in the investigation of the scission of rubbers by radiation. When used properly, there should be no significant change in sample dimensions during irradiation.

The equilibrium modulus first rises above zero at a certain dose, and at the same dose an insoluble gel fraction first begins to appear. At the gel point and beyond, the polymer is essentially

infusible and no longer becomes semiliquid at the softening point.
All these property changes are caused by the formation of an infinite cross-linked network of polymer chains.   If chain scission
predominates over cross-linking, the eventual effect of irradiation
is a weakening and embrittlement of the polymer.

The radiation-induced production of double bonds makes irradiated polymers more prone to oxidize.   Hence, some of the
improvement in high temperature resistance of polyethylene may be
offset by greater ease of oxidation.   Use of the proper antioxidant
is therefore important in an irradiated polymer.

Any treatment which affects the chemical composition of polymers will also affect the electrical properties.   Degradation of
polyvinyl chloride releases HCl within the polymer, which increases
the conductance, dielectric constant, and power factor considerably.
In line with this fact, Sisman and Bopp [27] observed that exposure
of polyvinyl chloride-acetate to pile radiation reduced the volume
resistivity from 10^{14} to less than 10^6 ohm cm.

A similar effect is produced in polymers having a structure
such that ionizable molecules cannot be split out.   Farmer [28]
observed that the volume resistivity of polystyrene (normally 10^{20}
ohm cm) may be reduced by a factor of 10^8 or more by exposure to
4000 rep of x-rays.   The increased conductance lasts for several
days after the cessation of irradiation, falling off gradually in
an exponential manner.   The phenomenon is caused by the formation
of ions and electrons during irradiation that persist for a considerable period of time.   Similar effects have been found in many
other polymers including polyethylene, polymethyl methacrylate, and
polytetrafluoroethylene.

The cross-linking and degradation reactions noted in polymers
irradiated in the pure state are also observed in polymers irradiated in solution.   The earliest studies on the irradiation of
polymer solutions were by Alexander and Fox [29], who irradiated
aqueous solutions of polymethacrylic acid, and by Wall and Magat

[30], who irradiated polystyrene dissolved in organic solvents.   A
solvent has several important effects on the degradation of a polymer.   For example, it allows increased mobility of the polymer,
and radicals produced from the solvent may react with the polymer
or with polymer radicals.   Energy transfer between solvent and
polymer may also take place.

Radiolysis of solutions of polymethyl methacrylate and polyisobutylene causes degradation at all concentrations, whereas
radiolysis of solutions of polystyrene, polyvinyl chloride, polyvinyl alcohol, and rubber produces cross-linking under certain
conditions.   With some water-soluble polymers such as polyvinylpyrrolidone, the minimum dose for gel formation exhibits a complex
dependence on polymer concentration.   As the concentration drops
to about 1%, the gelation dose also decreases even though the
polymer molecules are farther apart.   This has been explained as
being due to the higher contribution made to polymer radical formation by the radiolytic products of water.   At still lower concentrations, the gelation dose increases greatly, until below a
certain critical concentration no network is formed even at the
highest doses.

Polymer solutions are useful for investigating radiation
protection afforded by various additives.   The increase in gelation
dose is an indication of the protection afforded and can be employed to evaluate the effects of polymer concentration, additive
concentration, oxygen, and pH.

Some of the most efficient protective compounds are those
containing the sulfhydryl group.   Also, the benzene ring provides
a considerable amount of radiation protection to molecules containing it.   For example, solid polystyrene absorbs about 2000 eV
for each cross-link formed, whereas most cross-linking polymers
absorb only 20 to 30 eV per cross-link.   Also, when styrene is
added to other monomers before polymerization, the resulting polymer will be composed of styrene (S) and other (X) units,

–X–X–X–S–X–X–S–X–X–S–S–X–

The styrene units in the chain are found to assist in protecting the
polymer from radiation damage. By changing the S/X ratio, it has
been noted that a benzene ring offers protection for about four
carbon atoms along the chain [31]. Addition of a small percentage
of aniline or thiourea has been found to reduce the decomposition
of polymethyl methacrylate by a factor of 4 [32], probably by
energy transfer to the additive.

    Additional information on fundamental questions of mechanism
may be obtained by comparing the effects of α-, β-, and γ-radiation.
Alpha-radiation differs from both β and γ in producing a much
denser distribution of reactive species along the α-particle track.
Accordingly, protective agents are far less effective with α-
radiation than with β and γ, because of the high concentration of
radicals and other reactive species present in the dense α-track.
By combining diffusion theory with data on the effectiveness of
radical scavengers in eliminating certain products, the diameter
of the α-track can be estimated.

    In the case of both α- and β-radiation, very high doses are
needed to destroy crystallinity in polyethylene. This is true even
though the energy released in the path of each α-particle through
the polymer corresponds to an estimated rise of 10^4°C in temperature, after correcting for the energy used in ionization and excitation [33]. However, there is small indication of local heating
and loss of crystallinity because of the radiation. Long-term
heating of the polyethylene to a few degrees above the melting
point is far more efficient in producing loss of crystallinity.
Charlesby [20] has presented some pertinent comments concerning
the effect of radiation on polyethylene crystals, as revealed by
studies using the electron microscope. Large single crystals were
employed, having each polymer molecule folded back and forth to
give parallel chains. Based on theory, for a random distribution
of cross-links the radiation dose required for incipient gelation

is one cross-link unit (0.5 cross-links) per weight average molecule. Internal links between different units in the same molecule
are ineffective in causing gelation. In the study on large single
crystals, it was discovered that the gelation dose is 10 times as
large as for the same polymer grown under conventional conditions.
Since the radical concentration was no different from the usual
value as shown by ESR, it was deduced that most of the links formed
in the single crystal were internal links which had no effect on
solubility. It was also found that if the single crystals were
under pressure during growth, the gelation dose was greatly reduced. This was explained by postulating that pressure would
bring the surfaces of adjacent crystals closer together and cause
an interleaving of the surface loops.

The presence of oxygen changes the effect of radiation on
polymers. With thin polymer films, oxygen diffuses in and causes
an effect throughout the film. For thicker films, the oxidation
reaction may take place only at the surface of the film. When
polyethylene is irradiated in oxygen, cross-linking still takes
place, as shown by gel fraction determination. However, the nature
of the mechanism has changed considerably, because even prolonged
irradiation does not render the polymer infusible. Possibly the
links consist of peroxide bridges between the molecules, which
decompose when the polymer is heated. In addition, water and other
groups are formed including carbonyl, carboxyl, and hydroxyl. Some
of the reactions involved may be the following:

$$-CH_2-CH_2-\overset{\bullet}{C}H-CH_2- + O_2 \rightarrow -CH_2-CH_2-\overset{\overset{\textstyle O-O\bullet}{\textstyle |}}{C}H-CH_2- \qquad (7\text{-}10)$$

$$-CH_2-CH_2-\overset{\overset{\textstyle O-O\bullet}{\textstyle |}}{C}H-CH_2- \rightarrow -CH_2-CH_2-CO-CH_2- + \bullet OH \qquad (7\text{-}11)$$

$$-CH_2-CH_2-CH_2-CH_2- + \bullet OH \rightarrow H_2O + -CH_2-CH_2-\overset{\bullet}{C}H-CH_2- \qquad (7\text{-}12)$$

B.   Statistical Theory of Cross-linking

For cross-linking polymers, irradiation results in the formation of an insoluble network or gel, which makes up a certain
fraction (gel fraction) of the initial sample weight. When cross-
linking takes place at random between all monomeric units, a relation between gel fraction and radiation dose can be derived in
terms of the initial molecular weight distribution and the rate of
formation of cross-links [59, 60]. Chain scission may also take
place, but if this is also random it can be taken into account in
the theoretical equations [58, 61, 62]. Frequently the analysis
is facilitated by the fact that the initial chains are linear and
the extent of cross-linking and chain scission are proportional to
total radiation dose.

In developing the theory, it is assumed that all mers in the
system are equally likely to take part in cross-linking with any
other mer in the system. Stockmayer [63] has shown that the gel
point is reached when

$$q_g \frac{1}{(\bar{P}_w)_0} \qquad\qquad\qquad\qquad (7\text{-}13)$$

where q is the fraction of mers taking part in cross-linking, q_g
is the value of q at the gel point, and $(\bar{P}_w)_0$ is the weight-average
degree of polymerization in the original system. If cross-linking
is accompanied by random scission of main chain bonds, $(\bar{P}_w)_0$ in
Eq. (7-13) must be replaced by $\bar{P}_w$ for the initial chains as modified by chain scission. The relation between the displacement of
the gel point and p, the fraction of initial bonds broken, is a
function of the distribution and structure of the initial molecules.

When a polymer is cross-linked by irradiation, q and p are
usually proportional to the dose, R, and the reactions are generally
random and independent so that

$$q = q_0 R \qquad\qquad\qquad\qquad (7\text{-}14)$$

$$p = p_0 R \tag{7-15}$$

The molecular weight changes in the pregel region are therefore dependent on p_0, q_0, and the initial molecular weight distribution. Charlesby [9] showed that for the case of the most probable initial distribution,

$$\frac{1}{\overline{P}_n} = \frac{1}{(\overline{P}_n)_0} + \left(p_0 - \frac{q_0}{2} \right) R \tag{7-16}$$

$$\frac{1}{\overline{P}_w} = \frac{1}{(\overline{P}_w)_0} + \left(\frac{p_0}{2} - q_0 \right) R \tag{7-17}$$

although Eq. (7-16) holds for any initial molecular weight distribution. Equation (7-17) applies strictly to only the most probable initial distribution, but Inokuti and Dole [64] have demonstrated that $1/\overline{P}_w$ versus R should be approximately linear even for broader or narrower initial distributions.

The postgel solubility behavior of polymers with most probable distributions is described by the Charlesby-Pinner function [61],

$$S + S^{\frac{1}{2}} = \frac{p_0}{q_0} + \frac{2}{q_0 (\overline{P}_w)_0 R} \tag{7-18}$$

where S is the fraction of poolymer which is extractable (soluble) at radiation dose R. Hence, the gel-point dosage R_g is

$$R_g = \frac{1}{(\overline{P}_w)_0 \left(q_0 - \dfrac{p_0}{2} \right)} \tag{7-19}$$

Equation (7-16) can be expressed

$$\frac{1}{\overline{P}_n} = \frac{1}{(\overline{P}_n)_0} \left[1 - \frac{(q_0 - 2p_0)}{2(2q_0 - p_0)} \frac{R}{R_g} \right] \qquad (7\text{-}20)$$

and Eq. (7-17) can be written

$$\frac{1}{\overline{P}_w} = \frac{1}{(\overline{P}_w)_0} \left(1 - \frac{R}{R_g} \right) \qquad (7\text{-}21)$$

Hence, postgel measurements will yield the ratio p_0/q_0, and the location of the gel-point R_g for a sample of known molecular weight provides $q_0 - (p_0/2)$. The variation of P_n with dose [Eq. (7-16)] supplies an independent measure of q_0 and p_0 in the pregel region, thus testing the assumption that only random cross-linking and scission take place and that the parameters q_0 and p_0 are independent of dose.

Equation (7-18) has been shown valid for a number of cross-linking polymers [58, 65, 66]. Equations (7-20) and (7-21) have been employed to analyze several pregel polymers. Equation (7-19) has been verified by the observation of a constant $R_g (\overline{P}_w)_0$ product in silicones over a factor of 100 in molecular weight [67] and in polyethylene over a factor of 10 [68].

Most of the above comments relate to the "pregel" system and are mainly based on experimental measurements made prior to gela-tion. Much information can be obtained after the start of gelation, by considering the relative amounts of sol (soluble portion) and gel (insoluble portion). If only cross-linking takes place, the solubility curve depends only on the initial distribution of molecular weight. Equation (7-22) applies to primary chains with a large number of mers per chain [60],

$$1 - X = \int_{0}^{\infty} W(n)e^{-\alpha Xn} \, dn \qquad (7\text{-}22)$$

where α equals cross-link density, the fractional number of mers in the system involved in cross-links, $1 - X$ is the fraction (by weight) of the polymer which is soluble, and $W(n)$ is the original molecular weight distribution, the fractional weight of polymer with degree of polymerization n in the uncross-linked polymer.

For reference purposes in the subsequent discussion, the number-, weight-, and z-average degree of polymerization are defined as follows.

$$\overline{DP}_n = \left(\int_{0}^{\infty} \frac{W(n)}{n} \, dn \right)^{-1} \qquad (7\text{-}23)$$

$$\overline{DP}_w = \int_{0}^{\infty} nW(n) \, dn \qquad (7\text{-}24)$$

$$\overline{DP}_z = \frac{\int_{0}^{\infty} n^2 W(n) \, dn}{\int_{0}^{\infty} nW(n) \, dn} \qquad (7\text{-}25)$$

Ionizing radiation generally produces cross-links in direct proportion to the total dose, $\alpha = kR$, and also main chain scission in proportion to total dose. Hence, it is convenient to employ two parameters, k as already defined and β as a measure of number of main chain scissions produced per cross-linked unit formed. These parameters are assumed to be characteristic constants for each polymer and independent of molecular weight and distribution.

When chain scission takes place simultaneously with cross-linking, the solubility behavior is changed by what is considered as a continuous alteration of the primary molecular weight distribution. This leads to a modified solubility equation for linear primary chains [62, 69],

$$1 - X = \frac{X^2 w(s) + \beta^2 + 2X\beta \, 1/s \cdot \int_0^s w(\lambda) \, d\lambda}{(X + \beta)^2} \qquad (7\text{-}26)$$

where $w(s) = \int_0^\infty W(n) e^{-sn} \, dn$, and $s = \alpha(X + \beta)$. Inokuti [62] employed (7-26) to compute theoretical gel curves for various values of the chain scission parameter.

The typical features of the Charlesby-Pinner curve [61] are presented in Fig. (7-2), where $(S + S^{\frac{1}{2}})$ is plotted against the reciprocal of radiation dose. Below a critical dosage R_g the polymer remains completely soluble, but beyond the gel-point dosage (R_g) the solubility decreases with slope S_0. For large R, the $(S + S^{\frac{1}{2}})$ function approaches a limiting value I_∞ with a limiting slope S_∞.

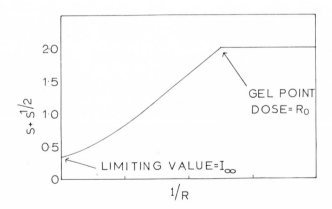

FIG. 7-2. Typical features of the Charlesby-Pinner plot, $(S + S^{\frac{1}{2}})$ versus 1/R.

The behavior of the gel curve near I_∞ may be obtained directly from Eq. (7-26), employing the assumption that α is linear in R [58],

$$I_\infty = \lim_{R \to \infty} [S + S^{\frac{1}{2}}] = \beta \tag{7-27}$$

$$S_\infty = \lim_{R \to \infty} \left[\frac{d}{d(1/R)}(S + S^{\frac{1}{2}}) \right] = \frac{\overline{DP}_w}{\overline{DP}_n} \rho \tag{7-28}$$

The averages $\overline{DP}_w$ and $\overline{DP}_n$ pertain to the original polymer sample, and ρ is the dosage at which gelation would have taken place in the absence of chain scission. Charlesby and Pinner [61] derived equations equivalent to (7-27) and (7-28) and demonstrated that they are generally true even for originally nonlinear chains. When chain scission occurs, the actual gel point dose R_g is greater than ρ. Also Stockmayer [63] has shown that ρ is related to the original $\overline{DP}_w$ of the sample by

$$\alpha(\text{gel point}) = \frac{1}{\overline{DP}_w} \tag{7-29}$$

so that

$$\rho = \frac{1}{k\overline{DP}_w} \tag{7-30}$$

It is possible to compute ρ approximately from the properties of the gel curve in the neighborhood of the gel point. At the gel point, $X = 0$ and Eq. (7-26) yields

$$1 - \beta/2 = \frac{1}{\alpha\beta} \int_0^{\alpha\beta} w(\lambda) \, d\lambda \tag{7-31}$$

Differentiation of Eq. (7-26) gives the rate of increase of gel
with cross-link density at the gel point.

$$\left(\frac{dX}{d\alpha}\right)_{x=0} = \frac{2\alpha(1 - \beta/2) - w(\alpha\beta)}{3\beta w(\alpha\beta) + (\beta - 1)} \tag{7-32}$$

If the primary distribution is not too broad and if β is small,
$w(\alpha\beta)$ can be expanded in powers of β in Eqs. (7-31) and (7-32).
The coefficients in this expansion series will depend on the moments
of the primary distribution, and these can be expressed as the
various $\overline{DP}$'s of the system. By simplifying and discarding terms of
order β^2 and above, one obtains [58]

$$\frac{1}{\rho} = \frac{1}{R_g} + \frac{\beta}{S_0} \tag{7-33}$$

$$\left(\frac{S_0}{3\rho}\right)\frac{\overline{DP}_z}{\overline{DP}_w} = 1 + \beta \frac{R_g}{S_0}\left(\frac{3}{2}\frac{\overline{DP}_{z+1}}{\overline{DP}_z} - 2\right) \tag{7-34}$$

where

$$S_0 = \left\{\frac{d}{d(1/R)}\left[(1 - X) + (1 - X)^{\frac{1}{2}}\right]\right\}_{x=0} \tag{7-35}$$

Equations (7-33) and (7-34) become exact expressions for $\beta = 0$,
and they are also exact for any β value in samples having the "most
probable" distribution: $W(n) = a^2 ne^{-an}$, $\overline{DP}_w/\overline{DP}_n = 2$, $\overline{DP}_z/\overline{DP}_w = 3/2$,
and $\overline{DP}_{z+1}/\overline{DP}_z = 4/3$. Therefore, Eqs. (7-33) and (7-34) are essen-
tially first-order corrections near the gel point for small amounts
of chain scission when the initial distribution differs from the
most probable distribution.

The above theoretical expressions are useful in determining characteristics of the molecular weight distribution from the gel curve behavior. For example, Eqs. (7-27), (7-28), (7-33), and (7-34) supply a means of calculating the two radiation parameters, ρ and β and the two distribution parameters, $\overline{DP}_w/\overline{DP}_n$ and $\overline{DP}_z/\overline{DP}_w$, from the four gel curve properties, R_g, S_0, I_∞, and S_∞.

II. TYPES OF CROSS-LINKING[*]

In order to present a systematic discussion of research to date on the radiation-induced cross-linking of the polymers, it is convenient to classify the work done under four main headings

1. *A/B Cross-linking*: the linking of polymer A to polymer B, sometimes called "cocross-linking" or "intercross-linking."

2. *A/A Cross-linking*: the linking of polymer A itself.

3. *A/M/A Cross-linking*: the linking of polymer A to itself by the addition of a monomer M or other small organic molecule.

4. *A/I/A Cross-linking*: cases where polymer A appears to be linked to itself by adding small particles of inorganic materials such as metal, S_iO_2, glass, etc. (The evidence for and against chemical linkage to the particles will not be reviewed at this point.)

The early research was mainly concerned with the A/A cross-linking of a polymer to itself, and the studies up to 1961 were

---

*In the attempt to compress subjects of principal interest into a single chapter, some selectivity must be employed. In the remainder of the chapter greater emphasis is placed on cross-linking with less emphasis on chain scission.

reviewed extensively in Chapiro's book [34], including work on
polyethylene, polypropylene, polystyrene, polyisoprene, polybutadiene, polyacrylates, polyvinyl chloride, polyorgano-siloxanes,
polyamides, polyesters, polyvinyl alcohol, polyvinyl acetate, poly-
acrylonitrile, polyvinyl methyl ether, fluorinated elastomers, and
polyelectrolytes.

Recently, interest has been shifting to the other three types
listed above, including cross-linking of polymer A to some other
polymer B, linking of polymer A to itself through an added molecule
such as allyl methacrylate, and modification of polymer properties
through irradiation of polymers containing small particles of
inorganic fillers.

An examination of the classification into four cross-linking
types immediately suggests additional areas of considerable interest
where essentially no results have been published. Examples would
be the A/M/B type involving the cross-linking of polymer A to
polymer B by the addition of a monomer or small organic molecule,
the A/I/B type involving the modification of polymer A/polymer B
mixtures by the addition of inorganic particles followed by irradiation, and the A/P/B type where polymers are linked by replacing
the linking monomer molecule with a low concentration of a "linking
polymer" molecule.

Additional details and examples of the four cross-linking
classifications investigated to date are presented in the following
sections.

A.   Cross-linking of Polymer A to Polymer B

Various techniques have been designed for the cross-linking of
one polymer to another, but all involve bringing the two types of
polymeric chains into close contact. These may be arbitrarily
divided into categories, according to whether physical contact
methods or solution methods are employed:

1. Physical Contact Methods

a. Surface Contact. Films of polymer A and polymer B are irradiated while their surfaces are brought into close contact by any suitable method. Possible methods include the application of one or more of the following: pressure, heat, or ultrasonic vibration.

b. Mill Roll Mixing. Another technique is to mix polymer A and polymer B mechanically on hot mill rolls followed by irradiation. The product can then be dissolved in a common solvent for viscosity studies, if desired.

2. Solution Methods

a. Solvent Solution of A and B. When common solvents for polymers A and B are available, a solution containing A and B can be prepared and irradiated. If desired, the solution can then be evaporated to yield a cross-linked film of A and B.

b. Melt Solution of A and B. If melted polymer A is soluble in melted polymer B, a homogeneous melt of A and B may be prepared and irradiated.

c. Irradiation of Cast Film. A variant of the above methods is to cast a film from a mixed solution of polymers A and B, followed by irradiation of the cast film. This method has proved successful in some A/B combinations [39] where other techniques did not work.

d. Film of A in Solution of B. An approach that has been found successful is to irradiate a film of A having a surface coating of a solution of polymer B. Because of the slow diffusion of molecules of polymeric size, this usually results in the grafting of a surface layer of polymer B on the film of polymer A. (Surface grafting of *monomeric* B on a film of polymer A is often difficult to attain because of the rapid diffusion of monomeric B into the film.)

Each of the above methods may be successful for particular A/B polymer pairs and unsuccessful for others. The theory has not been sufficiently developed to reliably predict whether a given technique will work for polymers A and B, or even whether A and B can be cocross-linked under any conceivable conditions.

Henglein [86, 87] has made extensive studies of the theory and practice of cocross-linking polymers A and B in solution (method 2,a above). The following theoretical points were emphasized:

1. When two polymers, A and B, are irradiated in homogeneous solution, two types of free macroradicals can potentially be formed; combination of the two different macroradicals corresponds to the formation of a cross-link between A and B (cocross-linking).

2. If the rate constants for combination of the appropriate macroradicals are indicated by k_{AA}, and k_{BB}, and k_{AB}, cocross-linking will be favored when $k_{AB} > k_{AA}$ and $k_{AB} > k_{BB}$ but will not be favored when $k_{AA} > k_{AB}$ and $k_{BB} > k_{AB}$.

3. Even if macroradical A does not combine with macroradical B, cross-linked chains of A can become physically entangled with cross-linked chains of B, forming what Henglein calls a "mixed-network." He refers to a network of A and B formed by linking of A and B radicals as a "co-network."

4. At low degrees of cross-linking, cocross-linking will be indicated by the appearnace in the solution of polymer molecules of the type that normally result from the radiation-induced grafting of one monomer on the other polymer, such as,

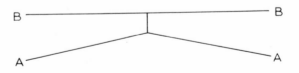

5. Cocross-linking will also be indicated by the dependence
of the rate of cross-linking on polymer concentration. In each
solvent, there is a "critical concentration" for A and also for B
below which scission predominates and gelation cannot be induced no
matter how large the radiation dose. However, if cocross-linking
occurs, both A and B can be present at less than critical concentration and yet gel formation takes place, showing that both A and
B are involved in forming the total network (gelation).

In addition to the above five characteristics of cocross-
linking, a sixth concept due to Chapiro [34] should be mentioned.

6. Chapiro postulated that cocross-linking would be expected
to take place if both A and B are of the cross-linking type
(Table 7-1). If B was of the cross-linking type and A of the
scission type, one would expect the formation of grafted structures,
such as

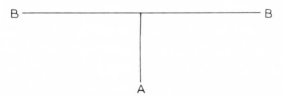

If neither A nor B was of the cross-linking type, the yield of cocross-linking would be expected to be very low. However, no
experimental check of these postulates for a large number of A/B
polymer pairs has been made.

The above six concepts provide a qualitative but not quantitative theory of cocross-linking. At least they serve as a basis on
which further research can build in this field of great practical
and theoretical signigicance.

The principles of cocross-linking in solution are perhaps best
illustrated by Henglein's work on polyvinylpyrrolidone and polyacrylamide. Separate water solutions of the two polymers were
first irradiated by γ-rays from ^{60}Co at an exposure dose rate of

about 7×10^4 R/h. The effects of irradiation were studied at
several different concentrations of each polymer. As polymer concentration decreases, more solvent radicals and less macroradicals
will be formed. However, the simultaneous scission of the main
chain by direct action of the radiation will still take place with
unchanged efficiency. The result is that the p_0/q_0 ratio (density
of scission to density of cross-linking) will increase as polymer
concentration decreases. Hence, a critical concentration should
exist below which gelation is not possible ($p_0/q_0 > 2$), no matter
how large the dose. Henglein [86] estimated the critical concentration in organic solvents to be about 10% by weight.

Water is a much more efficient solvent for the cross-linking
of a polymer, and the critical concentration in water is about 1%
by weight. The data are generally presented as a plot of the gel
dose versus polymer concentration, which passes through a minimum
slightly above the critical concentration. The addition of solvent
to a polymer first leads to a decrease in gel dose, since the
mobility of the macroradicals in recombining is increased as
viscosity falls. This effect is counterbalanced and surpassed as
concentration drops, owing to the decrease in intermolecular cross-
linking and the increase in p_0/q_0.

The work on a water solution of polyvinylpyrrolidone (PVP)
alone showed a critical concentration of 0.28 wt% [87], whereas
the corresponding critical concentration for polyacrylamide (PAM)
alone was 0.25 wt%. Hence, a solution containing 0.27 PVP and also
0.24% PAM should not gel on irradiation, unless cocross-linking
between the two polymers takes place. As a matter of fact, a solution containing 0.20% PVP and 0.12% PAM did exhibit gelation on
irradiation, indicating clearly that cocross-linking was taking
place between the two polymers.

Examples of the other experimental methods listed above for
causing cocross-linking between A and B can also be cited. Method
1,a was employed in a commercial application for bonding of polyethylene film to polyester film by irradiation with ultraviolet

light [88]. Although ultraviolet was employed in this particular
case, the principle is the same as with ionizing radiation, the
latter having the advantage of greater depth of penetration plus
the property of passing through opaque materials.

Matsuda *et al.* [89] studied γ-ray-induced effects involving
cocross-linking of two water-soluble polymers. The polymers
evaluated included polyvinyl alcohol (PVA), polyvinylpyrrolidone
(PVP), polyacrylamide (PAM), methylcellulose (MC), and polymethacrylic acid (PMA). Two polymers at a time were dissolved in water
at less than the critical concentrations for gelation. Irradiation
caused cocross-linking between two polymers if both were of the
cross-linking type. On the other hand, gel formation was retarded
if one of the two was of the degradative type. Cocross-linking of
PVA/PVP was studied by measuring the viscosity of irradiated solutions and the solubility of the irradiated polymers in methanol.
In the case of PVA/PVP and PVA/PAM, linear relations were observed
between the gelation dose and the amount of PVP or PAM. The network formation of cross-linking-type polymers was retarded by the
presence of methyl cellulose.

Hori *et al.* [90] found γ-irradiation helpful in the homogeneous
mill roll blending of high- and low-density polyethylene. The
specific heat/temperature plots of the blends were studied. The
unirradiated blend had two specific heat peaks. The specific heat
plot showed only one peak for the blend irradiated in the molten
state. By irradiating in the molten state, both polymers were
blended homogeneously.

Blokh *et al.* [91] investigated the γ-irradiation of natural
rubber blends with polyethylene, and butadiene/styrene rubber blends
with polyethylene. The blends showed much higher cross-linking
densities under similar conditions than did the pure polymers. The
covulcanization obtained between rubbers and polymers was suggested
as a means of obtaining artificial leather.

Since the above examples of cocross-linking are of comparatively recent origin, it may be worth mentioning some of the earlier

work for the sake of completeness. Irradiation of a film cast from
a solution of polymethyl methacrylate and polystyrene produced
cocross-linking of the two polymers [92], whereas irradiation of a
mechanical mixture of the two did not. Irradiation of mixtures of
polyethylene with another polymer caused cocross-linking in several
cases including rubber [93], polystyrene [94], polyisobutylene
[94], and nylon [94]. Polyvinyl alcohol was cocross-linked to the
surface of polyethylene film by bringing the lattter in contact
with a solution of polyvinyl alcohol and irradiating the system
with a beam of high energy electrons [95].

B. Cross-linking of Polymer A to Itself

1. Polyethylene and Related Polymers

In order to discuss the effect of radiation on polyethylene,
it is necessary to describe the structure of polyethylene. For
example, polyethylene contains many more methyl groups than would
correspond to the chain ends of unbranched molecules. From infrared
studies, it has been concluded that polyethylene contains both *long
branches* and *short branches* attached to the polymer backbone. The
long branches are assumed to form by chain transfer from growing
polymer radicals to other polymer chains. Crystallinity is related
inversely to the number of short branches, which may form by
intrachain transfer of the growing polymer chain to itself, a
process called "backbiting." The short-chain branches will be
mostly butyl groups, if a transient six-membered ring is the preferred configuration during the backbiting step.

Therefore, the polyethylene chain may contain any or all of
the following structures in varying proportions,

$$\text{WW} \!-\!(\text{CH}_2\text{CH}_2)_x\!-\!\text{CH}\!-\!\text{CH}\!-\!(\text{CH}_2\text{CH}_2)_y\!-\!\underset{\underset{\text{CH}_2\text{CH}_2\text{CH}_2\text{CH}_3}{|}}{\overset{\overset{\text{CH}_2\text{CH}_2\text{CH}_2\text{CH}_3}{|}}{C}}\!-\!(\text{CH}_2\text{CH}_2)_z\!-\!\text{WW}$$

short branch, tetrafunctional

CH$_2$
CH$_2$ CH$_2$CH$_2$CH$_2$CH$_3$
CH$_2$ short branch,
CH$_2$ trifunctional
long
branch

Infrared studies have aided in identifying various types of unsaturation in polyethylene. For polyethylene polymerized at high pressure, the total unsaturation is made up of roughly 60% vinylidene, $RR'C=CH_2$, 20% *trans*-vinylene, $RHC=CHR'$, and 20% vinyl, $RCH=CH_2$.

Polyethylene polymerized at low pressure using Ziegler catalysts has much less short- and long-chain branching. This structure corresponds to a high degree of crystallinity, high density, and high softening temperature, as compared to the analogous properties for "high pressure" polyethylene.

The radiation chemistry of polyethylene has been studied for many years by a large number of investigators. The major chemical effects of the irradiation of polyethylene can be summarized as follows.

a. <u>Formation of Cross-links</u>. This is one of the most important chemical effects and accounts for many of the observed changes in physical properties. These include changes in elastic modulus, density, transparency, and melting behavior. The disappearance of

the crystalline regions is related to the production of internal
strain in the material by the formation of cross-links.

    b. Evolution of Gas. Irradiation causes the evolution of
gas, mainly hydrogen, but containing a small percentage of hydrocarbons. The hydrocarbon fraction contains mainly C_2, C_3, and C_4
compounds, probably explained by the breaking off of side chains
[35].

    c. Decay of Vinyl and Vinylidene Unsaturation. Some of the
earliest investigators found that any vinyl and vinylidene unsaturation disappears rapidly during irradiation. Early studies
involved the effect of temperature on G values for vinyl and vinylidene decay.

    d. Formation of *trans*-Vinylene Unsaturation. An important
reaction noted in irradiated polyethylene (or paraffins) is the
formation of *trans*-vinylene (t-Vl) unsaturation. This product is
less influenced by temperature and the presence of scavengers than
is cross-linking. The concentration of *trans*-vinylene may continue
to rise or fall for some time after irradiation has ceased,
probably as a result of reactions of radicals trapped in the polymer.

    e. Formation of Radicals and Other Reactive Intermediates.
The evidence indicates that irradiation of polyethylene leads to the
formation of a host of reactive species, including positive ions,
trapped electrons, electronically excited groups, alkyl and allyl
free radicals, atomic hydrogen, and carbanions.

    Dole and associates [36,38] have worked extensively on the
identification and reactions of such reactive intermediates. The
principal experimental tools employed were ultraviolet spectroscopy
and ESR spectroscopy. Although many ion-molecule reactions have
been postulated to explain unsaturation and cross-linking in polyethylene, there is very little direct evidence for such reactions.
Dole [37] has suggested that the production of *trans*-vinylene
unsaturation in polyethylene may take place by

$$\overset{+}{-CH_2-CH_2-} \rightarrow H_2 + \overset{+}{-CH=CH-} \tag{7-36}$$
$$(trans)$$

a concept that is supported by the small effect of temperature on $\underline{G(\underline{t}-Vl)}$, and by the fact that reaction (7-36) should be exothermic to the extent of about -0.145 eV [36].

Dole argues that at large doses, free radicals should be plentiful enough to be significant in trapping electrons. The reaction

$$R\cdot + e^- \rightarrow R:^- \tag{7-37}$$

to form a carbanion should be exothermic to the extent of about 1 eV, as estimated from the electron affinity of the methyl free radical, which is about 1.08 eV.

Atomic hydrogen must exist in irradiated polyethylene, because radicals such as $R\cdot$ are known to cause the exchange of deuterium between D_2 and polyethylene after irradiation, possibly by a chain reaction [38] of the following sort.

$$R\cdot + D_2 \rightarrow RD + D\cdot \tag{7-38}$$

$$D\cdot + RH \rightarrow R\cdot + HD \tag{7-39}$$

The evidence implied that the D atoms did not migrate very far after formation before abstracting a hydrogen as indicated.

There is considerable evidence for the existence of the secondary alkyl radical $-CH_2\overset{\cdot}{C}HCH_2-$ in irradiated polyethylene [40-42].

The allyl free radical has been identified in irradiated polyethylene. It may exist in uncombined form for several months after irradiation [37, 43]. The presence of such allyl radicals is most noticeable in high density polyethylene, and it may be the cause of

the slow room temperature oxidation of this material as indicated
by the formation of carbonyl groups absorbing at 1725 cm^{-1} in the
infrared. Some question has been raised as to whether the allyl
absorption peak at 258 nm corresponds to a terminal allyl free
radical, $-\overset{\bullet}{C}HCH=CH_2$, or the chain allyl free radical, $-CH_2\overset{\bullet}{C}HCH=CHCH_2-$

Polyenyl free radicals have also been identified and studied
in irradiated polyethylene. Relatively stable polyenes such as
$-CH_2(-CH=CH)_n-CH_2-$ with n values up to 5 and reactive polyenyl
structures such as $-\overset{\bullet}{C}H(CH^-=CH)_n-CH_2-$ have been identified and investigated by ultraviolet spectroscopy [44]. The shift of λ max of
polyenyl free radicals to longer wavelengths with increaseing n is
linear with n. Dole and Bodily [45] showed that during irradiation
at liquid nitrogen temperature few allyl free radicals were formed,
but the number increased on warming the sample to room temperature
after λ-ray irradiation at -196°C. Several investigators found
that the concentration of allyl free radicals is reduced by irradiating the sample with ultraviolet light at -196°C. By contrast, the
dienyl and trienyl free radicals increase both on irradiation with
ultraviolet and on warming to room temperature. The esr spectra of
the polyenyl free radicals tend to overlap each other, so in this
case ultraviolet spectroscopy with its separate polyenyl radical
absorption peaks provides a useful complement to ESR spectroscopy.

Dole obtained some interesting results in infrared studies of
polyethylene samples irradiated at -196°C, heated to room temperature, and then cooled again to liquid nitrogen temperature. There
was an increase in the 968 cm^{-1} band (vinylene group) on heating,
in addition to an increase in the 942 cm^{-1} band (allyl free radical).
It was suggested that these changes were caused by decay reactions
of the alkyl free radicals. The vinylene group possibly resulted
from the recombination of alkyl free radicals on adjacent carbon
atoms on the same chain after migration of either or both of the
free radical centers to neighboring carbon atoms. It was further
postulated that the allyl free radical was formed by the migration
of a free radical to a carbon atom next to a vinylene or vinyl

double bond. Since no change in vinyl absorbance was caused by heating, the increase in alkyl free radicals must have resulted to some extent from processes other than the decay of vinyl groups, although vinyl group decay may possibly contribute to allyl radical formation as suggested by Auerbach [46].

Using ultraviolet absorption Bodily and Dole [47] measured the rate of diene formation in high density γ-irradiated polyethylene at -196°C, and obtained a G(diene) value of 0.20. The increase in diene concentration was linear with dose, whereas the allyl and dienyl free radicals as well as the conjugated trienes grew with increasing rates as the irradiation continued. The triene was formed at a rate proportional to the square of the dose, suggesting that triene production requires the prior presence of some other group. It was postulated that the triene was formed by adding a vinylene group in a conjugated position to a conjugated diene. This hypothesis was confirmed by rate calculations showing that the rate of triene formation computed from diene concentration agrees with the experimentally observed value. The higher polyene products could hardly be detected at -196°C. The appearance of the higher polyenes at room temperature but hardly at all at liquid nitrogen temperature shows that they may be formed by the migration of free radicals during or after irradiation.

Dole found that the allyl radical is produced at low concentration at -196°C. Heating of the sample to room temperature after irradiation caused the allyl concentration to increase considerably. The allyl absorption at 258 nm could be reduced greatly by irradiation with ultraviolet light.

Bodily and Dole summarized several possible steps of the degradation that were in accord with the experimental observations [47]. The allyl radical structure formed during γ-irradiation at -196°C is probably

$$-CH_2\dot{C}HCH{=}CH-$$
I

which is more stable than

$$-\overset{\bullet}{C}HCH_2CH{=}CH-$$

II

by about 0.70 eV, as indicated by gas-phase data. Therefore, conversion of I into II is an endothermic process. Absorption of a light photon at 258 nm of 4.80 eV at -196°C would provide ample energy to convert I into II. On warming to room temperature, II would presumably revert to I with an evolution of about 0.70 eV.

In the conversion of I to II and in the reverse reaction, the transfer of the hydrogen may take place intermolecularly rather than intramolecularly. Dole has suggested two possible mechanisms, one involving carbanions and the other involving carbonium ions. The carbonium mechanism is

$$-CH_2\overset{\bullet}{C}HCH{=}CH- \; + \; -CH_2\overset{+}{C}HCH_2 \underset{heat}{\overset{h\nu}{\rightleftharpoons}}$$

$$-CH_2\overset{+}{C}\,HCH{=}CH- \; + \; -CH_2\overset{\bullet}{C}HCH_2- \qquad (7\text{-}40)$$

For reaction (7-40) to occur, the ionization potential of the allyl radical must be greater than that of the alkyl free radical, and such does appear to be the case from the data of Kiser [48].   In addition, reaction (7-40) satisfies both ultraviolet and ESR spectral observations.

While the concentration of allyl free radicals decreases during uv irradiation, the concentrations of dienyl and trienyl free radicals continue to grow with both uv irradiation at -196°C and heating to room tmperature. Possibly, absorption of uv by a diene group, for example, causes migration of a hydrogen atom to the allyl free radical in this way forming a dienyl free radical and changing the allyl free radical into a vinylene group. An alternative possibility is that on heating a free radical center of an alkyl group could migrate to a carbon atom next to a diene group, thus forming a dienyl free radical.

A plot of allyl free radical absorption at 258 nm versus time showed that its concentration does not grow linearly with time, but rather increases with the dose (for high density polyethylene γ-irradiated at -196°C). Comparison of this observed growth rate with a computed rate based on the allyl radical arising from a random juxtaposition of a vinylene group and an alkyl radical indicates that the allyl radical was formed at a higher rate than one would have predicted statistically. Furthermore, the rate of increase in allyl radicals on heating to room temperature after irradiation indicates that the allyl radical was not produced by random placement of an alkyl radical adjacent to a vinylene group.

Growth plots for dienyl and trienyl free radicals (measured by absorption at 285 and 323 nm, respectively) showed that neither is linear with dose, and that both probably require the formation of a prior product. Dienyl and trienyl free radicals probably arise from migration of alkyl free radical centers to positions adjacent to dienes or trienes, respectively.

In later work, Kang et al. [49] determined the temperature coefficient over the range 35°–120°C of the following processes that take place during the γ-ray radiolysis of high density polyethylene: hydrogen evolution, vinyl decay, vinylene double bond and conjugated diene formation, formation of cross-links, and chain scission. The value of $G(H_2)$ increased gradually from 3.68 to 4.11 as temperature increased from 35° to 120°C. A large increase in $G(H_2)$ took place on passing to the liquid state at 140°C. At each of the temperatures studied, $G(H_2)$ decreased slightly as dose increased.

A first-order decay plot of vinyl (Vi) unsaturation, log [vinyl] versus dose, gave linear curves at the various temperatures from 35° to 120°C, with the first-order constant k_1 equal to 1.3×10^{-21} g eV^{-1} at 35°C.

Linear curves at various temperatures resulted from changes in trans-vinylene (t-Vl) concentration plotted according to the zero-order growth and first-order decay observed earlier by Dole et al. [50]. Growth and decay constants were independent of temperature

over the range 35°–120°C. The first-order decay constant was computed to be 0.56 x 10^{-21} g eV^{-1} at room temperature. (Earlier
results indicated the constant to be 0.64 x 10^{-21} g eV^{-1} in the
liquid state at 142°C.) The G value for vinylene formation extrapolated to zero dose was the same at all temperatures from 35° to
120°C.

The gel fraction was plotted against dose at different temperatures to determine the effect of temperature on gel point dose,
r_g. The value of r_g declined from 2.65 mrads at 35°C to 2.20 mrads
at 120°C. The Charlesby-Pinner plot, $S + S^{\frac{1}{2}}$ versus reciprocal
dose, $1/r$, where S is the sol fraction, proved less convenient in
determining r_g because of extrapolation difficulties. In order to
calculate G values for cross-linking and scission, it was first
necessary to obtain a theoretical function involving both chain
scission and cross-linking yields that was consistent with the
gel-sol data. A theoretical treatment was carried out based on the
original work of Saito [51] and using Wesslau's [52] molecular
weight distribution function, which is believed to give a good
description of the molecular weight distribution in linear polyethylene.

Although details cannot be given here, it was found that a
plot of [X]/r versus r was linear with positive slope at each temperature studied, where [X] equals cross-links per milliliter. It
was observed that [X]/r x 10^6 is almost exactly equal to G(X) when r
is expressed in mrads. Therefore, the increase in [X]/r with dose
demonstrates that the G value for cross-linking increases with dose.
Values were computed for $\lambda = G(S)/G(X)$, where G(S) is the yield of
chain scission which is very probably constant as dose varies [53].
Extrapolation to zero dose showed that $G_0(X)$ increased from 0.96
to 1.08 as temperature increased from 35° to 120°C, while $G_0(S)$
increased from 0.192 to 0.259 as temperature underwent the same
increase.

At higher doses, the experimentally determined gel fractions
were higher than predicted theoretically, which was attributed to

the increase in G(X) with dose. From a mechanistic viewpoint, G(X) might be expected to increase with dose if vinylene double bonds decay to form cross-links. Assuming that no unmeasured (intra-molecular) cross-links are formed, and that no residual free radicals remain, the following material balance equation results,

$$G(H_2) = G(X) + G(t-Vl) + 2 \, G(diene) \qquad (7-41)$$

when it is also assumed that trienes and higher polyenes are negligible at the doses employed. The right and left side of Eq. (7-41) were compared at 35° and 120°C and at zero dose and maximum dose (27 mrads). The agreement was almost perfect in most of these cases, the *worst* agreement occurring at zero dose and 35°C, where the right and left sides of the equation were equal to 3.9 and 3.7, respectively. Such an excellent material balance indicates that we now have a quantitative knowledge of the main chemical events taking place during the γ-ray radiolysis of polyethylene. Processes having little or no effect on intermolecular cross-linking include the decay of vinyl groups, chain scission, and intramolecular cross-linking.

It is of interest to compare radiolysis of polyethylene with that of polyethylene oxide (PEO), the latter having a chain structure with one atom of oxygen between each pair of $-(CH_2-CH_2)-$ units. Polyethylene oxide has a molecular weight of 10^5 to 10^7 and is prepared by the polymerization of ethylene oxide in the presence of heavy metal catalysts [1]. The solution viscosity of aqueous or organic solutions of PEO decreases during long-term storage, a change that is promoted by the presence of strong acids, oxidizing agents, and heavy metal ions [2]. Mechanical agitation also causes degradation of such solutions, so that care must be taken in pre-paring them to prevent significant viscosity decreases.

Pearson [3] irradiated polyethylene oxide with radiation from cobalt-60 and observed cross-linking in air at low doses (1 Mrad) with chain degradation predominating at higher doses. Changes in

intrinsic viscosity were studied as a function of molecular weight
and dose size. Salovey and Dummont [4] investigated the cross-
linking of low molecular weight polyethylene oxide and found that
electron irradiation *in vacuo* causes both degradation and cross-
linking with a ratio of chain scissions to cross-links of 0.6.
Nitta *et al.*[5, 6] observed that polyethylene oxide is cross-linked
when exposed to cobalt-60 radiation or electron beams. His study
of main-chain scission and hydrogen evolution. Crouzet and Marchal
[7] investigated the irradiation of polyethylene oxide solutions
with radiation from cobalt-60 and postulated two mechanisms, one
leading to cross-linking and the other resulting in oxidative degradation accompanied by chain scission and molecular weight
decrease.

One of the more recent studies was made by King [1], who found
that the γ-ray or electron-beam irradiation of PEO produced changes
in solution properties, molecular weight, and molecular weight
distribution. The relative significance of cross-linking and degradation was observed to be dependent on dose rate, oxygen concentration, and particle size of the solid polymer irradiated.

The 4000-Ci cobalt-60 source included 12 pencils of 6-inch
active length arranged in holders in a 6-in. diam. circle. For
electron-beam irradiations, the vertically mounted Van de Graaff
accelerator was adjusted to deliver a current of 250 μA at 2 MeV.
In a typical experiment, powdered PEO was delivered from a vibrating
feeder to a belt moving at 90 in./min, which carried the resin
through the electron beam from the accelerator. After irradiation,
aqueous solutions of the irradiated resins were prepared and viscosities were measured with a Brookfield viscometer.

The other property measured was pituitousness (stringiness),
determined by lowering a duNuoy ring into the solution, raising it,
and measuring the time required for the solution to break all
contact with the ring.

It was found that large changes in viscosity were caused by

rather small doses. For example, with a resin sample of 600,000 initial molecular weight, the viscosity of a 5% aqueous solution dropped from 1620 to 23 cp after a dose of 0.5 Mrad absorbed by the solid, powdered PEO. The size of the viscosity reduction becomes less as the initial molecular weight drops. For irradiations performed *in vacuo* the decrease in viscosity was much less than for the corresponding irradiation carried out in air. Long-term irradiation in the absence of air, however, forms insoluble (cross-linked) products.

There is a considerable difference in the effect of electron and γ-irradiation. In causing viscosity reduction, a γ-dose rate of only 0.1 Mrad/h is about three times as effective as electron dose rates in the 200 to 4000 Mrad/h range. King has postulated that greater oxygen availability may be responsible for the greater efficiency of the low dose rate irradiation.

The rate of viscosity decrease slows greatly as irradiation continues. For example, the viscosity of PEO of 200,000 molecular weight drops from 680 to 12 cp after absorbing a dose of 0.5 Mrad. Increasing the dose to 4.6 Mrads causes only a small decrease to 8 cp. Further irradiation has a rather small effect on the viscosity.

Shear stability of irradiated PEO was evaluated by plotting percent viscosity retention (of aqueous solutions) versus agitation time with a ten-blade stirrer turning at 1080 rpm. Doses of 0.1 t0 0.4 Mrad gave large increases in shear stability, as indicated by little or no drop in solution viscosity after 2 h of agitation.

Effect of irradiation on solution stringiness was evaluated by plotting pituitousness against total dose absorbed. A dose of only 0.5 Mrad for PEO of 3,000,000 molecular weight formed a product which gave an aqueous solution having essentially no stringiness.

King used the radiolysis of diethyl ether, studied by Ng [8], as a model for the chemical reactions occurring in PEO radiolysis. The latter obtained the following relative probabilities for bond

cleavage in diethyl ether: C $-$ H, G = 0.34; C $-$ O, G = 1.6;
C $-$ C, G = 0.28. Cleavage of α-C $-$ H bonds led to the radical
$CH_3 - \overset{\cdot}{C}H - O - CH_2 - CH_3$, a radical which was about 40 times as
abundant as the corresponding radical formed by loss of a β-hydro-
gen. One of the major products was 2,3-diethoxybutane formed by
radical combination.

$$2 \ CH_3-\overset{\cdot}{C}H\!-\!O\!-\!O\!-\!CH_2\!-\!CH_3 \rightarrow CH_3\!-\!\underset{\underset{\underset{CH_3}{|}}{\underset{CH_2}{|}}{\overset{|}{\underset{O}{|}}C}} - \underset{\underset{\underset{CH_3}{|}}{\underset{CH_2}{|}}{\overset{|}{\underset{O}{|}}C}}\!-\!CH_3 \tag{7-42}$$

Cleavage of the C $-$ O bond formed alkoxy and alkyl radicals, which
took part in several reactions. Abstraction was more important
than disproportionation for the alkoxy radical, as shown by an
8/1 ratio of ethanol to acetaldehyde in the products.

   By analogy with diethylether, King postulated that the relative
amounts of various bonds broken in PEO should be as follows:
C $-$ H, G = 2.7; C $-$ O,G = 6.4; C $-$ C, G = 0.6. Some of the steps
postulated by King were

$$-CH_2\!-\!CH_2\!-\!O\!-\!CH_2\!-\!CH_2\!- \rightarrow -CH_2\!-\!\overset{\cdot}{C}H\!-\!O\!- \ + \ -CH\!\!=\!\!CH\!-\!O\!-$$

$$+ \ H\!\cdot \ + \ H_2 \tag{7-43}$$

$$-CH_2\!-\!CH_2\!-\!O\!-\!CH_2\!-\!CH_2\!- \rightarrow -CH_2\!-\!CH_2\!\cdot \ + \ \cdot O\!-\!CH_2\!-\!CH_2\!- \tag{7-44}$$

$$-CH_2\!-\!CH_2\!-\!O\!- \rightarrow -CH_2\!\cdot \ + \ \cdot CH_2\!-\!O\!- \tag{7-45}$$

King refers to the first radical on the right side of Eq. (7-43)
as the "backbone radical." Hydrogen atoms may react by abstraction
to form more polymer backbond radicals.

$$-CH_2-CH_2-O- + H\cdot \rightarrow -CH_2-\overset{\cdot}{C}H-O- + H_2 \qquad (7\text{-}46)$$

A doublet in the ESR spectrum was observed by Nitta *et al.*[5, 6] and assigned to this backbone polymer radical. The combination of such backbone radicals would possibly account for the cross-linking ot two PEO chains.

The radicals formed above may disappear by disproportionation

$$-O-CH_2-CH_2\cdot + \cdot O-CH_2CH_2-O- \rightarrow$$
$$-O-CH_2-CH_3 + OHC-CH_2-O- \qquad (7\text{-}47)$$

$$-O-CH_2-CH_2\cdot + \cdot O-CH_2-CH_2-O- \rightarrow$$
$$-O-CH{=}CH_2 + HO-CH_2-CH_2-O- \qquad (7\text{-}48)$$

or possibly undergo abstraction followed by disproportionation;

$$-O-CH_2-CH_2-O\cdot + -CH_2-CH_2-O- \rightarrow$$
$$-O-CH_2-CH_2-OH + -CH_2-\overset{\cdot}{C}H-O- \qquad (7\text{-}49)$$

$$-CH_2-\overset{\cdot}{C}H-O- + \cdot CH_2-CH_2-O- \rightarrow$$
$$-O-CH{=}CH-O- + CH_3-CH_2-O- \qquad (7\text{-}50)$$

The ratio of scission to cross-linking was found to range from 0.7 to 0.95 by the method of Charlesby [9], implying that cross-linking and degradation take place with approximately equal probability. The only radical observed in the ESR spectrum at room temperature is the backbone radical, indicating that radicals formed by C – O and C – C cleavage disappear quickly. Since these radicals do not exist long enough for oxygen to diffuse to their

site, it appears probable that the backbone radical is the important one in oxidative degradation. The following reactions were
postulated for this radical.

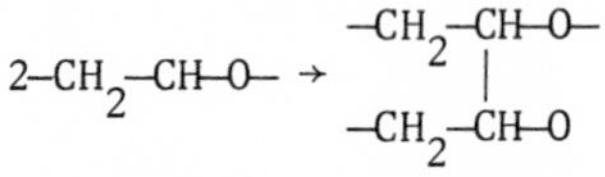

$$2-CH_2-CH-O- \rightarrow \quad \begin{array}{c} -CH_2-CH-O- \\ | \\ -CH_2-CH-O \end{array} \qquad (7\text{-}51)$$

$$2 \ -CH_2-\overset{\bullet}{C}H-O- \rightarrow -CH_2-CHO + \cdot CH_2-CH_2-O- \qquad (7\text{-}52)$$

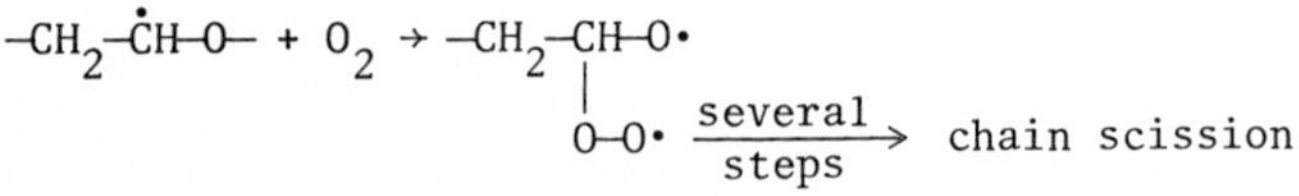

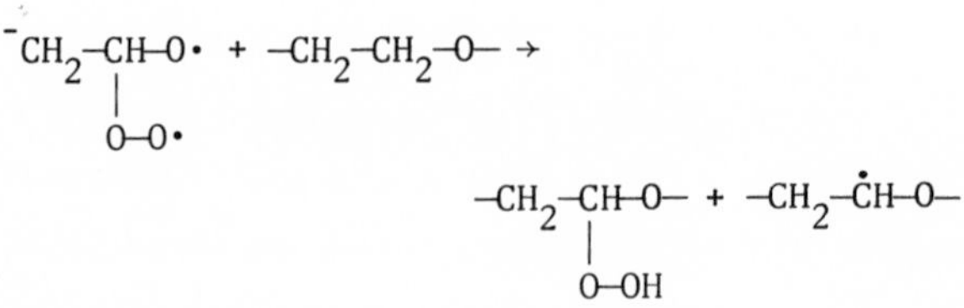

$$-CH_2-\overset{\bullet}{C}H-O- + O_2 \rightarrow \begin{array}{c} -CH_2-CH-O\cdot \\ | \\ O-O\cdot \end{array} \xrightarrow[\text{steps}]{\text{several}} \text{chain scission} \qquad (7\text{-}53)$$

$$\begin{array}{c} ^-CH_2-CH-O\cdot \\ | \\ O-O\cdot \end{array} + -CH_2-CH_2-O- \rightarrow$$

$$-CH_2-CH-O- + -CH_2-\overset{\bullet}{C}H-O- \qquad (7\text{-}54)$$
$$| $$
$$O-OH$$

Reaction (7-51) is cross-linking, whereas reaction (7-52) is a rearrangement leading to aldehyde production. When oxygen is
present, reaction (7-53) competes with cross-linking. Reaction
(7-54) indicates that the peroxy radical can form a hydroperoxide,
the latter possibly decomposing with resulting chain scission. The
existence of unstable perocides and hydroperoxides in the irradiated PEO is in agreement with the observed rapid drop and eventual
leveling off in the viscosity of solutions prepared in the 24 h
following solid polymer irradiation.

Viscosity-average molecular weights were measured for a series
of PEO samples, and from these the number-average molecular weights
were computed using relations derived by Bailey et al. [54]. From
the number-average molecular weights, estimates of G values for
chain scission ranging from 185 to 250 were computed. Those
figures are rough estimates only, because of uncertainties about

the molecular weight distribution. The estimated G values were computed for the low dose rate cobalt-60 irradiations. The high viscosity and shear susceptibility of dilute solutions are probably attributable to the high molecular weight fraction of the PEO. The extremely large molecules are the ones most likely to be affected by irradiation. This would cause a narrowing of the molecular weight distribution into one having a more nearly random distribution.

2. Polystyrene

The structure of polystyrene is believed to consist entirely of a head-to-tail arrangement of the monomeric units. The number of branches is essentially negligible, in view of the very low chain transfer constant for polystyrene with the growing polymer radical.

The study of polystyrene irradiation is of special interest because of the presence of benzene rings, which are known to have a protective action against the effects of irradiation. In actuality, polystyrene is one of the most stable polymers with respect to radiation, and very large doses are required to produce any significant changes in properties. Some of the early work in this field was done by Wall and Brown [55], who irradiated polystyrene with γ-rays from cobalt-60 and analyzed the evolved gas in a mass spectrometer. The gas evolved was almost pure hydrogen, with a $G(H_2)$ value equal to approximately 0.026. Such a value is of the same order of magnitude as the hydrogen yield in the radiolysis of benzene. Wall studied viscosity versus irradiation time in the pregel range for polystyrene irradiated *in vacuo* and found a steady viscosity increase with dose until the gel point was readied, at which point the viscosity rose sharply providing good definition of the gelation dose.

Charlesby [56] irradiated polystyrene in a reactor and investigated the relation between the radiation dose and the sol fraction after the gel point. The results indicated an initially random molecular weight distribution and a p_o/q_o ratio of 0 to 0.2, showing scission to be relatively unimportant. Swelling measurements [56] indicated that the cross-links are introduced at random and their number is approximately proportional to dose.

When polystyrene is irradiated in air at relatively low dose rates, oxidation takes place and chain scission may preodminate over cross-linking. For example, Feng and Kennedy [57] found that the intrinsic viscosity of polystyrene irradiated in air decreased markedly. After large doses, a yellow color was formed in the polymer having a maximum absorption at 3400 Å. The polymer had become soluble in methanol, indicating the possible presence of hydroxyl and carboxyl groups.

Graessley [58] studied the cross-linking of polystyrene, using γ-rays from a cobalt-60 souce. Useful information on the molecular weight distribution of polystyrene samples was obtained from the effects of cross-linking on the solubility properties.

Using special techniques, samples of most probable distribution, narrow distribution, and broad distribution were prepared. The $\overline{DP}_w$ of the samples was obtained by the Zimm method using a Brice-Phoenix light scattering photometer. The $\overline{DP}_w/\overline{DP}_n$ and $\overline{DP}_{z+1}/\overline{DP}_z$ were obtained by sedimentation techniques.

After irradiation, the samples were extracted exhaustively in benzene at 35°C, the insoluble residue representing the gel fraction. Charlesby-Pinner plots of the resulting data were drawn. The solubility data for all MPD (most probable distribution) samples gave linear plots, which is consistent with the requirement that distributions of the form $W(n) = a^2ne^{-an}$ have solubility curves of the type

$$(1 - X) + (1 - X)^{\frac{1}{2}} = 2\rho (1/R) + \beta \qquad (7\text{-}55)$$

The values of ρ and β are obtained from the slope and intercept of this curve, respectively. The corresponding curves for the N samples (narrow distribution) and B samples (broad distribution) were not linear but exhibited characteristic shapes. The $\overline{DP}_w/\overline{DP}_n$ and $\overline{DP}_z/\overline{DP}_w$ parameters for these samples were computed from R_0, S_0, I_∞, and S_∞ by the use of Eqs. (7-28) and (7-34). The average value of the chain scission parameter β was computed to be 0.35, whereas the value of k was found to be 0.53×10^{-11} rads.

Values of $\overline{DP}_w/\overline{DP}_n$ from the gel curve and from sedimentation studies were in close agreement, as were the values of $\overline{DP}_z/\overline{DP}_w$, showing that information on the molecular weight distribution of a linear polymer can be obtained from its cross-linking/solubility behavior regardless of its initial distribution (provided the chain scission parameter is small).

Alberino and Graessley [70] also drew some interesting conclusions concerning the pregel behavior of radiation-cross-linked polystyrene. In this work, the radiation-cross-linking properties were determined for samples ranging in $\overline{M}_w$ from 50,000 to 3,000,000. Values for $\overline{M}_n$ and $\overline{M}_w$ of the samples as prepared were measured by the osmotic pressure technique and light scattering, respectively. Values of $\overline{M}_w/\overline{M}_n$ were about 1.5 or less, indicating that the distributions were somewhat narrower than the most probable distribution ($\overline{M}_w/\overline{M}_n = 2.0$). There was no particular trend in distribution breadth with average molecular weight or method of preparation.

Plots of $1/\overline{P}_w$ versus dose were found to be linear for pregel samples, as were plots of $(S + S^{\frac{1}{2}})$ versus $1/R$ for postgel samples (Charlesby-Pinner plots). Values of R_g were obtained from pregel measurements by extrapolating $1/\overline{P}_w$ to 0, and from postgel measurements by extrapolating to $S + S^{\frac{1}{2}} = 2$. The agreement between R_g values obtained by the two methods was excellent.

The behavior of some of the samples was in agreement with the theory of random cross-linking, but others showed slight deviations. For some, the plot of $1/\overline{P}_w$ versus R gave a straight line, and for these samples the change from $(\overline{P}_n)_0$ to the value at the gel point $(\overline{P}_n)_g$ was about 0.10, which is approximately the change predicted by Eq. (7-20) using the value $p_0/q_0 = 0.35$ from the postgel behavior. For other polymer samples, the plot of $1/\overline{P}_w$ versus R gave

a curve. A correlation study with the preparation techniques
showed that straight-line plots were for samples prepared by bulk
polymerization, whereas the curved plots were for samples polymerized in ethylbenzene solution.

A more detailed analysis of the deviations from the theory of
random cross-linking by the solution-polymerized samples led to
the conclusion that coupling reactions involving the end groups
alone could have led to the observed effects. If approximately
12% of the chains in the solution polymers had reactive end groups
which could somehow couple randomly and rapidly with other end
groups in the system during irradiation, the observed results would
be obtained. Alberino and Graessley [70] have summarized the
difficulties involved in visualizing the type of reaction that
meets these requirements.

C. Cross-Linking of a Polymer to Itself by Added Monomer

Radiation cross-linking in the presence of an added polyfunctional monomer such as allyl methacrylate (AMA) has been studied
for polyvinyl chloride [74, 75], cellulose acetate [76], polymethyl
methacrylate [77], and other polymers [78]. The technique is of
considerable theoretical and practical interest because it permits
cross-linking to take place at reduced doses compared to those
required in the absence of polyfunctional monomer. It also affords
an efficient means of cross-linking polymers not normally cross-
linked by radiation. The following summarizes the recent work of
this type done with polyethylene, nylon, polyvinyl alcohol, and
polypropylene.

1. Polyethylene

Odian and Bernstein [72] made a detailed study of the radiation cross-linking of low density polyethylene in the presence of
AMA and other polyfunctional monomers. The polyethylene samples
were equilibrium swollen with monomer at 25°C and then irradiated

with γ-rays from a ^{60}Co source or with electrons from a 1.5-MeV
Van de Graaf generator. The former was used for irradiations to
12 Mrads, and the latter for irradiations above 12 Mrads. Gel
contents and swelling ratios were determined by standard techniques
for the irradiated samples. Tensile strength and heat resistance
measurements were also made on some of the irradiated samples.

The results showed that the incipient gelation dose is reduced from 0.5 Mrad for pure polyethylene to 0.02 Mrad for polyethylene containing 4.5% AMA. The G (cross-link) value for polyethylene was computed from an equation of Charlesby's [9],

$$G = 4.8 \times 10^5 / R_g \overline{M}_w \qquad\qquad (7\text{-}56)$$

where R_g is the dose for incipient gelation and $\overline{M}_w$ is the weight-
average molecular weight of the original polymer. The G (cross-link)
values for pure PE and AMA-containing PE were found to be 1.9 and 48,
respectively.

The cross-linking of monomer-free PE had been shown by early
investigators to be dependent on total dose, with a first-order
dependence on radiation dose rate. A plot of percent gel versus
dose was made for the PE/AMA system, and it was found that this
cross-linking reaction is also dependent on total dose, with a
first-order dependence on dose rate.

Linear plots of log weight swelling ratio versus log dose were
obtained for both pure PE and PE/AMA. The weight swelling ratio
(d_m) was defined as

$$d_m = W_{WG} / W_G \qquad\qquad (7\text{-}57)$$

where W_G is the dry weight of the gel sample used and W_{WG} is the
weight of the wet (swollen) gel sample. It was found that at the
same dose the PE/AMA system gives gels with lower swelling ratio
than the pure PE. Also, in order to obtain a gel of a specific
swelling ratio, lower radiation doses were employed in the PE/AMA
system.

The quantity $(S + S^{\frac{1}{2}})$ was plotted against reciprocal dose
$(1/R)$ and extrapolated to infinite dose to obtain the intercept.
The intercept represents the scission/cross-linking ratio p_0/q_0
[Eq. (7-18)]. The plot gave a value of $p_0/q_0 = 0.20$ for both pure
PE and PE/AMA. Hence, the presence of 4.5% AMA strongly affected
the course of the cross-linking reaction at low doses (<3 Mrads),
but did not change the ultimate effect of the irradiation.

Physical strength tests showed that the PE/AMA system was
superior to the monomer-free polyethylene in tensile strength at
all dose levels. Heat-aging studies at 188°C showed PE/AMA irradiated to 1.2 Mrads equal to pure PE samples irradiated to 30
Mrads.

Odian postulated that the cross-linking of PE/AMA samples took
place in two steps: (1) rapid initial polymerization of the AMA,
and (2) reaction of the polymerized AMA with polyethylene chains,
tying the latter into a three-dimensional cross-linked network.
The polymerized AMA is probably a rather low molecular weight graft
and homopolymer bearing pendant allyl groups. Possible ways for
attaching this structure ot the PE chains include: (1) The pendant
allyl double bonds may react with free radicals on neighboring PE
chains, causing cross-linking of the chains.   (2) A large fraction
of the pendant allyl groups would probably be converted to allyl
radicals by irradiation. Coupling of these allyl radicals to PE
chain radicals would contribute to the total cross-linking.
(3) These allyl radicals could abstract H atoms from the PE chains,
thus increasing the G value for PE radical production, resulting in
an increase in G (cross-links). It was observed that added mono-
functional monomers such as styrene or methyl methacrylate did not
increase the cross-linking of the polyethylene.

2.   Nylon-66 and Polyvinylalcohol

Bernstein et al. [71] studied the γ-induced cross-linking of
nylon-66 and polyvinylalcohol in the presence of small added per-
centages of AMA. In both polymers, cross-linking was promoted by

the presence of the AMA and did not take place in its absence.
In the experimental procedure, strips of polymer were swollen in
methanol/water/monomer, flushed with nitrogen, sealed, and irradiated at 0.02 to 0.06 Mrad/h using ^{60}Co.

The results were similar to those obtained in the studies on
polyethylene [72, 73]. The incipient radiation dose was reduced,
and increased gel fractions were obtained per unit dose when AMA
monomer was present. Water was found to expedite radiation cross-
linking for both polymers. Gel fractions of 40% were obtained by
irradiating nylon-66 to 3 Mrads, and gel fractions of 70% by irradiating polyvinyl alcohol to the same dose. It was possible to
regulate the level of AMA incorporation and total gel fraction by
varying the water/monomer ratio or the total dose. The process
probably involved radiation-induced radical formation on the polymer backbone either directly or indirectly, followed by radical
addition to the AMA. The grafted chains would grow in length and
presumably interact to form a three-dimensional cross-linked
network.

This approach allowed a substantial degree of cross-linking
without much chain scission at lower doses. It was claimed as the
first successful radiation cross-linking at low dose of a polyamide
and polyol.

3. Polypropylene

Several studies [79-80] have shown that polypropylene undergoes both cross-linking and chain scission on irradiation, the
relative extent of the two processes apparently depending on the
degree of crystallinity. Stereoregular polypropylene cross-links
less efficiently than polyethylene. Benderly and Bernstein [81]
studied the irradiation of stereoregular polypropylene and found
that the incorporation of AMA prior to irradiation expedites
cross-linking, just as it does for polyethylene.

Sheets of polypropylene were irradiated with γ-rays from ^{60}Co
for dose levels up to 5.0 Mrads. Higher doses were obtained with

a 2-MeV van de Graaf accelerator, which imparted 2 Mrads per pass. For tensile testing, ^{60}Co irradiations were performed on dumbbell-shaped specimens. All disks for dielectric testing were prepared subsequent to irradiation.

The results showed that stereoregular polypropylene can be cross-linked and thermally upgraded by irradiation in nitrogen when AMA is incorporated prior to irradiation. The creep compliance of the cross-linked samples was determined as a function of time under load. Compliance at 185°C was found to decrease with increasing dose, at a constant testing load and monomer level. Irradiated polymer without AMA was found to melt and flow after 1 h at 214°C, whereas irradiated polymer containing AMA retained its form, supplying additional evidence of cross-linking in the AMA-containing polymer.

Tensile strength and gel fraction size of monomer-cross-linked polypropylene increased with dose up to 5 Mrads. Irradiation to 5 Mrads of polypropylene/AMA increased the tensile strength by about 30-35%. Incipient gelation took place at less than 0.3 Mrad; above 5 Mrads, 80% gel resulted. In the absence of AMA, irradiation did not produce similar changes in the polypropylene. In fact, both ^{60}Co irradiation in nitrogen (to 5 Mrads) and machine irradiation in air (to 37 Mrads) produced a decrease in tensile and no gelation.

D.   Cross-Linking of Polymers Containing Added
Small Paricles

Many investigators have observed that improved elastic modulus and tensile strength of polymers can be obtained by the incorporation of small inorganic particles (fillers) followed by irradiation. The technique has the advantage that the degree of reaction can be readily controlled in a quantitative way merely by changing the radiation dose.

Although many authors have attributed such results to the formation of chemical linkages (grafting) between the particles and the polymer chains, Charlesby [20] has questioned this viewpoint. Charlesby studied the increase in Young's modulus of silicone gums containing finely divided carbon black or silica powder as a function of radiation dose. Based on this work, he saw no reason to postulate chemical linkage between the particles and the polymer chains. Rather, he hypothesized that the increase in modulus was caused by the physical presence, within each polymer network loop, of the particles that reduced the maximum extension of which each chain was capable by their physical presence within the available volume. It was further observed that the effectiveness of a filler as a reinforcing agent was determined primarily by particle size and only to a small extent by its nature (whether carbon black, silica, surface treated, or not).

Other investigators have expressed the opinion that irradiation results in the formation of chemical linkages between the polymer chains and filler particles. The effects observed are well illustrated by the work on polyethylene, polypropylene, and epoxy resins.

1. Polyethylene

Raff *et al.* [82] found that blends of polyethylene with reinforcing fillers such as glass fiber, aluminum oxide, and carbon black, after being subjected to γ-radiation, exhibited improvements in physical properties. The improvement was attributed to radiaation-induced grafting of the polyethylene to the inorganic fillers, and to cross-linking of the polyethylene.

The cobalt-60 source was used to irradiate specimens contained in an argon atmosphere in glass vials. The dose rate was about 1 Mrad/h, and total dose was between 10^7 and 10^8 rads. This was chosen as the optimum range for promoting radiation-induced polymer grafting, while holding radiation damage at a minimum.

Gamma-irradiation was found to increase the physical strength of both unfilled and reinforced low density polyethylene (LDPE) and high density polyethylene (HDPE). Irradiation of the HDPE/ aluminum oxide samples to 10^7 rads resulted in the recovery of most of the initial loss of tensile strength of the pure polymer when filler was incorporated, as well as considerably greater flexural strength and improved stiffness.

Blends of LDPE and carbon black increased considerably in tensile strength on irradiation. Carbon black, not as most fillers, did not produce an initial decrease in tensile strength when blended with either HDPE or LDPE. Hence, irradiated samples containing carbon black attained the highest recorded values of tensile strength (up to 560 kg/sq cm).

With a 60% glass fiber content in LDPE, specimens irradiated to 8 x 10^7 rads attained the same stiffness as samples made of unfilled HDPE, but their strength values were lower. Stiffness values are higher for HDPE/glass fiber blends, and they increase even further after irradiation.

Creep resistance studies were carried out by attaching fixed loads to each type of material at 70°F and measuring amount of creep as a function of time. It was found that superior creep resistance of PE blends results from irradiation. The results were interpreted in terms of three chemical reactions initiated by γ-radiation: (1) degradation, (2) cross-linking, and (3) polymer grafting to filler. Under the conditions employed of exposure to radiation in an inert atmosphere, it was assumed that oxidative degradation was negligible. The complete list of irradiation runs included preliminary experiments on irradiated PE-containing sodium alumino-silicate, calcium carbonate, magnesium oxide, silica, and silica gel. Of this latter group, magnesium oxide and silica gel were superior to the other fillers, as regards improvement of properties by irradiation.

2. Polypropylene

Isotactic (stereoregular) polypropylene has a tensile strength of about 5000 psi and has utility in a number of applications. In the ionic polymerization of propylene to produce the stereoregular product, a by-product of less crystallinity is produced. Depending on conditions, the by-product polymer has a tensile varying from 100 to 1400 psi and little utility. Bernstein [83] irradiated the by-product polymer plus silica filler in an attempt to improve its properties. Addition of AMA was also evaluated, and combinations of PP/AMA/SiO$_2$ were found to give the best results.

Irradiation was carried out using γ-rays from a ^{60}Co source. Three types of by-product polyethylene were employed: (1) a semi-crystalline, intermediate molecular weight by-product, (2) a slightly crystalline (10%) low molecular weight commercially available grade, and (3) a very high molecular weight noncrystalline polymer. Radiation-induced cross-linking of all three types took place in the presence of AMA monomer. Tensile strength improvements were produced, and when silica filler was also present the tensile strength increase was even greater. In comparing the three types of polypropylene studied, it was found that the amount of tensile strength improvement under conditions of constant radiation dose, monomer level, and reinforcing agent level depended on the initial degree of crystallinity.

The semicrystalline (ca. 33%) polypropylene of intermediate molecular weight underwent a tensile increase from 275 to 1100 psi after 3.3 Mrads in the presence of AMA and filler. The low molecular weight type of 10% crystallinity showed only a small increase in tensile from the same treatment. The very high molecular weight noncrystalline polymer was improved from 100 to 500 psi in tensile strength. It was found that conversion of the silica hydroxyl groups or vinyl groups prior to blending produces further improvement in the tensile strength of the polypropylene/AMA system after

irradiation. The semicrystalline polypropylene exhibited 10 to
15% greater tensile than that imparted by untreated silica. The
vinyl groups on the silica probably aid in incorporating the silica
into the cross-links, thus increasing the breaking strength of the
polymer. If AMA is not incorporated, irradiation decreases the
tensile strength of the polypropylene whether or not it contains
silica.

Improvement in tensile generally occurred as the dose was
increased until the total dose reached a limiting value, above
which thenwas little or no additional improvement. For the semicrystalline polypropylene, the limiting dose value was reached
in the 5 to 10 Mrads range.

3. Epoxy Resins

Some researchers have reported evidence that when epoxy resin
in contact with metal is irradiated bonds are formed between the
metal and the resin. For example, Khimchenko *et al.* [84] studied
the curing reactions of bisphenol A phenoxy resins containing
colloidal metal particles during irradiation with γ-rays from
cobalt-60. This composition cured at room temperature when irradiated, and the presence of the metal particles altered the final
stage of the curing reaction. It was postulated that curing in the
presence of the metal particles produced a three-dimensional
strucutre.

In another study involving metal, Raff and Subramanian [85]
found that the strength of joints between stainless steel or copper
and an epoxy resin could be increased up to 300% by γ-irradiation.
Metal inserts of copper or steel were embedded in an epoxy compound,
which was then cured by heating to 90°C for 0.5 h or more, followed
by irradiation. The metal inserts were cleaned by treating with
KBr/H_3PO_4 or other chemical mixtures prior to embedding in the
epoxy resin. The initial bond strength prior to irradiation was
a function of the survace cleaning treatment, but in all cases irradiation produced a further increase in bond strength between

resin and metal. For example, stainless-steel/epoxy bonds started at a strength of 8.4 kg/cm^2 and increased to 21 kg/cm^2 after irradiating to a dose of 4 x 10^7 rads. Copper/epoxy bonds had an initial strength of 11 kg/cm^2, which increased to 32 kg/cm^2 after a dose of 6 x 10^7 rads.

The authors postulated that irradiation caused changes in the polymer/metal interfaces, leading to atomic displacements in the metal and free radicals in the polymer. This type of work on polymer/metal adhesion induced by irradiation opens up a wide field for further research.

REFERENCES

1. King, P. A., in *Irradiation of Polymers, Advances in Chemistry Series*, 66, American Chemical Society, 1967.

2. McGary, C. W., Jr., *J. Polymer Sci.*, 467, 51 (1960).

3. Pearson, R. W., *Radioisotopes in Scientific Research, Proceedings of the First UNESCO International Conference, Paris, 1957*, Vol. 1, Pergamon Press, New York, 1958, p. 151.

4. Salovey, R. and Dummont, F. R., *J. Polymer Sci., Pt. A*, 1, 2155 (1963).

5. Nitta, I., Onishi, S., and Nakajima, Y., *Annual Report of Japanese Association for Radiation Research on Polymers*, Vol. 3, 1961 AEC-tr-6372, p. 437.

6. Nitta, I., Onishi, S., and Fujimoto, E., *Annual Report of Japanese Association for Radiation Research on Polymers*, Vol. 1, 1958 AEC-tr-6231, p. 320.

7. Crouzet, C. and Marchal, J., *J. Polymer Sci.*, 59, 317 (1962).

8. Ng, M. K. M., and Freeman, G. R., *J. Amer. Chem. Soc.*, $\underline{87}$, 1935 (1965).

9. Charlesby, A., *Atomic Radiation and Polymers*, Pergamon Press, New York, 1960.

10. Farmer, F. T., *Nature*, $\underline{150}$, 521 (1942).

11. Winogradof, N. N., *Nature*, $\underline{165}$, 72 (1950).

12. Day, M. J. and Stein, G., *Nature*, $\underline{168}$, 644 (1951).

13. Charlesby, A., *Proc. Roy. Soc. (London)*, $\underline{A215}$, 187 (1952).

14. Charlesby, A., *Nature*, $\underline{171}$, 167 (1953).

15. Lawton, E. J., Bueche, A. M., and Balwit, J. S., *Nature*, $\underline{172}$, 76 (1953).

16. Miller, A. A., Lawton, E. J., and Balwit, J. S., *J. Pol. Sci.*, $\underline{14}$, 503 (1954).

17. Simha, R., *Trans. N.Y. Acad. Sci.*, $\underline{14}$, 151 (1952).

18. Wall, L. A. and Florin, R. E., *Paper at 126th Meeting Amer. Chem. Soc.*, New York, Sept. 12, 1954.

19. Roberts, D. E., *J. Res. Nat. Bur. Std.*, $\underline{44}$, 221 (1950).

20. Charlesby, A., in *Advances in Chemistry*, Vol. 66, American Chemical Society, 1967.

21. Ormerod, M. G., *Phil. Mag.*, $\underline{12}$, 118 (1965).

22. Flory, P. J., *Chem. Rev.*, $\underline{35}$, 51 (1944).

23. Bovey, F. A., *The Effects of Ionizing Radiation on Natural and Synthetic High Polymers*, Interscience, New York, 1958.

24. Sun, K. H., *Modern Plastics*, $\underline{32}$, No. 1, 236 (1954).

25. Tobolsky, A. V., Metz, D. J., and Mesrobian, R. B., *J. Amer. Chem. Soc.*, $\underline{72}$, 1942 (1950).

26. Tobolsky, A. V., Prettyman, I. B., and Dillon, J. H., *J. Appl. Phys.*, $\underline{15}$, 380 (1944).

27. Sisman, O. and Bopp, C. D., ORNL-928 (1951); *Nuclear Sci. Abst.*, $\underline{8}$, No. 2792.

28. Farmer, F. T., *Nature*, $\underline{150}$, 521 (1942).

29. Alexander, P. and Fox, M., *Nature*, $\underline{169}$, 572 (1952).

30. Wall, L. A. and Magat, M., *J. Chim. Phys.*, $\underline{50}$, 308 (1953).

31. Alexander, P. and Charlesby, A., *Proc. Roy. Soc. (London)*, *Ser. A*, $\underline{230}$, 136 (1955).

32. Alexander, P., Charlesby, A., and Ross, M., *Proc. Roy. Soc. (London), Ser. A*, 223, 392 (1954).

33. Blackburn, R., Charlesby, A., and Woods, R. J., *European Polymer J.*, 1, 161 (1965).

34. Chapiro, A., *Radiation Chemistry of Polymeric Systems*, John Wiley, New York, 1962.

35. Lawton, E. J., Zemany, P. D., and Balwit, J. S., *J. Amer. Chem. Soc.*, 76, 3437 (1954).

36. Dole, M. and Bodily, D. M., in *Advances in Chemistry*, Vol. 66, American Chemical Society, 1967.

37. Dole, M., *Mechanisms of Chemical Effects in Irradiated Polymers* (R. A. V. Raff and K. W. Doak, eds.), Vol. I, Chap. 16, Interscience, New York, 1965.

38. Dole, M. and Cracco, F., *J. Phys. Chem.*, 66, 193 (1962).

39. Ballantine, D. S., Colombo, P., Glines, A., Manowitz, B., and Metz, D. J., BNL Report 414 (T-81), 1956.

40. Lawton, E. J., Balwit, J. S., and Powell, R. S., *J. Chem. Phys.*, 33, 395, 405 (1960).

41. Libby, D. and Ormerod, M. G., *Phys. Chem. Solids*, 18, 316 (1961).

42. Smaller, B. and Matheson, M. S., *J. Chem. Phys.*, 28, 1169 (1958).

43. Onishi, S., Sugimoto, S., and Nitta, I., *J. Chem. Phys.*, 37, 1283 (1962).

44. Fallgatter, M. B. and Dole, M., *J. Phys. Chem.*, 68, 1988 (1964).

45. Dole, M. and Bodily, D. M., *Third Intern. Conf. on Radiation Res., Cortino d'Ampezzo, Italy*, June, 1966; J. Chem. Phys., 45, 1428, 1433 (1966).

46. Auerbach, I., *Polymer*, 8, 63 (1967).

47. Bodily, D. M. and Dole, M., *J. Chem. Phys.*, 45, 1433 (1966).

48. Kiser, R. W., *Introduction to Mass Spectrometry and Its Applications*, Prentice-Hall, 1966, p. 319.

49. Kang, H. Y., Saito, O., and Dole, M., *J. Amer. Chem. Soc.*, 89, 1980 (1967).

50. Dole, M., Milner, D. C., and Williams, T. F., *J. Amer. Chem. Soc.*, 80, 1580 (1958).

51. Saito, O., *J. Phys. Soc. Japan*, 13, 198, 1451 (1958).

52. Wesslau, H., *Makromol. Chem.*, 20, 111 (1956).

53. Yu, H. and Wall, L. A., *J. Phys. Chem.*, 69, 2072 (1965).

54. Bailey, F. E., Kucera, J. L., and Imhof, L. G., *J. Pol. Sci.*, 32, 517 (1958).

55. Wall, L. A. and Brown, D. W., *J. Phys. Chem.*, 61, 129 (1957).

56. Charlesby, A., *J. Pol. Sci.*, 11, 513, 521 (1953).

57. Feng, P. Y. and Kennedy, J. W., *J. Amer. Chem. Soc.*, 77, 847 (1955).

58. Graessley, W. W., *J. Phys. Chem.*, 68, 2258 (1964).

59. Charlesby, A., *Proc. Roy. Soc.*, A222, 542 (1954).

60. Baskett, A. C., *Int. Symp. Macromol. Chem., Milan and Turin,* 1954; *Ric. Sci. Suppl.*, p. A379 (1955).

61. Charlesby, A. and Pinner, S. H., *Proc. Roy. Soc.*, A244, 367 (1959).

62. Inokuti, M., *J. Chem. Phys.*, 38, 2999 (1963).

63. Stockmayer, W. H., *J. Chem. Phys.*, 12, 125 (1944).

64. Inokuti, M. and Dole, M., *J. Chem. Phys.*, 38, 3006 (1963).

65. Schultz, A. R. and Bovey, F. A., *J. Pol. Sci.*, 22, 485 (1956).

66. Glover, L. C. and Lyons, B. J., *Amer. Chem. Soc., Pol. Chem. Preprints*, 9, 243 (1968).

67. Charlesby, A., *Proc. Roy. Soc.*, A230, 120 (1955).

68. Kitamaru, R., Mandelkern, R., and Faton, J. G., *J. Pol. Sci., Pt. B*, 2, 511 (1964).

69. Small, D. A., *J. Pol. Sci.*, 18, 431 (1955).

70. Alberino, L. M. and Graessley, W. W., *J. Phys. Chem.*, 72(12), 4229 (1968).

71. Bernstein, B. S., Odian, G., Orban, G., and Tirelli, S., *J. Pol. Sci., Pt. A*, 3, 3405 (1965).

72. Odian, G. and Bernstein, B. S., *J. Pol. Sci., Pt. A*, 2, 2835 (1964).

73. Odian, G. and Bernstein, B. S., *J. Pol. Sci., Pt. B*, 2, 819 (1964).

74. Miller, A. A., *J. Appl. Pol. Sci.*, 5, 388 (1961).

75. Pinner, S. H., *Nature*, 183, 1108 (1959).

76. Pinner, S. H., Greenwood, T. T., and Lloyd, D. G., *Nature*, 184, 1303 (1959).

77. Pinner, S. H. and Wycherly, V., *J. Appl. Pol. Sci.*, 3, 388 (1960).

78. Lyons, B. J., *Nature*, 185, 604 (1960).

79. Schnabel, W. and Dole, M., *J. Phys. Chem.*, 67, 295 (1963).

80. Salovey, R. and Dummont, F. R., *J. Pol. Sci.*, *Pt. A*, 1, 2155 (1963).

81. Benderly, A. A. and Bernstein, B. S., *J. Appl. Pol. Sci.*, 13, 505 (1969).

82. Raff, R. A. V., Schecter, D. J., and Subramanian, R. V., *Modern Plastics*, 74. September (1971).

83. Bernstein, B. S. and Lee, J., *F. and E. C. Prod. Res. and Dev.*, 6(4), 211 (1967).

84. Khimchenko, Y. I., Meleshevich, A. P., Ulberg, Z. R., and Natanson, E. M., *Fiz.-Khim. Mekh. Liofilnost Dispersnykh Sist.*, p. 230 (1968).

85. Raff, R. A. V. and Subramanian, R. V., *J. Appl. Pol. Sci.*, 15, 1535 (1971).

86. Henglein, A., *Radiat. Res. Proc. Int. Congr., 3rd*, 1966 (G. Silini, ed.), 1967, p. 316.

87. Henglein, A., *Makromol. Chemie*, 32, 226 (1959).

88. *Modern Packaging*, p. 214, April (1970).

89 Matsuda, T., Lin, C., and Hayakawa, K., *Kobunshi Kagaku*, 18, 492 (1961).

90. Hori, S., Iida, S., Sakami, H., and Adachi, T., *Nagoya Kogyo Gijutsu Shikensho Hokoku*, 11, 48 (1962).

91. Blokh, G. A., Zhurko, V. A., Denisenko, E. V., Tripenyuk, Z. A., and Belonsova, A. P., *Radiats. Khim. Polim. Mater. Simp., Moscow*, 1964, 321 (1966).

92. Ballantine, D. S., Colombo, P., Glines, A., Manowitz, B., and Metz, D. J., BNL Report 414 (7-81), 1956.

93.   Charlesby, A. and Pinner, S. H., Brit. Pat. 805,897 (1958).

94.   Wippler, C., French Pat. 1,166,793 (1958).

95.   Miller, A. A., French Pat. 1,168,117 (1958).

Chapter 8

EMULSION POLYMERIZATION; POLYESTERS;
THERMOSETTING SYSTEMS IN GENERAL

The purpose of this chapter is to collect and present available information concerning the radiation chemistry of three
important categories concerning which little has been published:
(1) emulsion polymerization systems, (2) polyester systems, and
(3) thermosetting systems in general.

A number of monomers have been polymerized by irradiation
in emulsion form, and several significant advantages over thermal
initiation have been discovered, but little has been done in the
way of commercialization beyond some pilot plant scale research.
Polyester resin coatings have been cured by radiation, and this
application of radiation has been commercialized in recent years.
Little progress has been made on radiation curing of thermosetting
and condensation systems in general, except for a few exceptions
to be described in the following pages.

I. EMULSION POLYMERIZATION

A. Introduction

There are a number of reasons for studying radiation-induced
emulsion polymerization. In the first place, classic methods of

initiating emulsion polymerization by the use of peroxides or
hydroperoxides produce free radical concentrations that vary with
temperature, and often change throughout the course of the reaction.
High energy irradiation offers a source of free radical initiation
which essentially does not vary with temperature or time. Hence,
comparative experiments can be carried out at different temperatures with a relatively constant concentration of free radicals.
Such a constant source of free radicals is a distinct advantage in
kinetic studies. Other possible advantages include ease of control,
less pH change than often found with catalysts, no combined or free
catalyst fragments such as sulfate ions, and lack of interference
on changing emulsifiers such as, for example, to a cationic system.

Other reasons for studying radiation-induced emulsion polymerization involve differences between this method and the more
conventional methods and the special information that can be derived by investigating these differences. One remarkable difference
is the greater stability of the emulsion during radiation initiation, attributed to the formation of a negative charge on the latex
particles by the radiation.

The generally accepted theory of emulsion polymerization was
developed by Harkins [1] and given quantitative treatment by Smith
and Ewart [2]. There are four principal ingreaients in an emulsion
polymerization: the water medium, the monomer which is immiscible
with water, an emulsifier of the oil-in-water type, and a source of
free radicals (initiator). These ingredients are stirred together
to produce an emulsion of the monomer in water.

The emulsifier plays an important part in emulsion polymerization. At the beginning of the reaction it forms the outer layer of
spherical aggregates known as *micelles*, which are perhaps 50 to
100 Å in diameter and contain liquid monomer on the inside. Most
of the monomer is dispersed in droplets of 1 to 10 μm in diameter.

It has been shown that essentially no polymer is formed in the
monomer droplets. Very slow polymerization can take place in the

homogeneous phase (water), but this can only account for a very
small fraction of the polymer formed. The initiator decomposes in
the water phase to produce the primary radicals, which slowly add
monomer units until the radical/monomer adduct reaches such a
length that it possesses surface-active properties, at which point
it diffuses to a monomer/water interface. The interface reached
by the radical/monomer adduct is more probably that of a micelle
than that of a monomer droplet, simply because the preponderance of
surface present is that of the tiny micelles, which are present at
a concentration of about 10^{18} micelles/ml. Once in the micelle,
the radical propagates by addition of monomer at a high rate,
changing the monomer-swollen micelle to a small monomer-swollen
particle within a short time. As polymer is formed, the micelles
grow by addition of monomer from the aqueous phase, which diffuses
from the monomer droplets acting as reservoirs of monomer.

After about 2 to 3% polymerization, the polymer particles grow
much larger than the original micelles and absorb almost all the
emulsifier from the water phase. Any micelles not already activated disappear, and further polymerization takes place within the
polymer particles already formed.

The disappearance of the micelles divides the polymerization
reaction into two stages: stage 1, in which the particles are
initiated; and stage 2, in which the particles grow until the
supply of monomer or free radicals is depleted. Although the conversion is usually low at the end of stage 1, the course of the
reaction is already determined by the *number of particles*
initiated. This key parameter is a function of (1) the type and
concentration of emulsifier, (2) the concentration of electrolyte,
(3) the rate of radical formation, (4) temperature, (5) the type
and intensity of agitation, and (6) other less well-known and
poorly defined variables. Much of the variation in rates of emulsion polymerization can be attributed to variations in the number
of particles initiated. However, stage 2 is quite reproducible,
and once the number of particles has been determined the rate of

the remainder of the reaction can be predicted accurately.

In order to discuss stage 2, let it be assumed that 10^{16} particles/ml were formed in stage 1, and that radicals are being formed at a rate of 10^{14}/ml/sec. On the average, each particle would then receive a radical every 100 sec (neglecting radical termination outside the particles). A given radical, on entering the particle, initiates a polymeric radical, which grows for 100 sec, until another radical enters and causes termination. (Known termination constants indicate that two free radicals within the same polymer particle would mutually terminate within a few thousandths of a second.) The particle is then inactive until a third radical enters and initiates polymeric growth; a fourth radical then enters and causes termination, etc. On the average, each particle contains a growing radical half of the time and is inactive half of the time, and the overall rate of polymerization depends on the sum of the rates in all particles or on the number of particles. The molecular weight of the resulting polymer will depend on the length to which a polymer chain can grow in the time between successive entries of radicals, that is, on the number of particles relative to the number of free radicals formed per second.

If in the above example the number of particles was increased to 10^{17} particles/ml, the chain growth time on the average would be increased to 1000 sec, corresponding to a tenfold increase in polymer molecular weight. This result will only be obtained when the polymeric radicals are segregated so that they may grow close to one another (in neighboring particles) without the probability of imminent mutual termination.

In their work on the theory of emulsion polymerization, Smith and Ewart [2] derived the following steady-state equation in terms of the rate of radical entry into the particles, the rate of radical transfer out of the particles, and the rate of radical termination within the particles:

$$N_{n-1}(\rho/N) + N_{n+1}k_0 s\left[(n + 1)/v\right] + N_{n+2}k_t\left[(n + 2)(n + 1)/v\right]$$

$$= N_n\left\{(\rho/N) + k_0 s(n/v) + k_t n\left[(n - 1)/v\right]\right\} \qquad (8-1)$$

where N_n is the number of particles containing n radicals; v is the particle volume; s is the particle surface area; ρ is the rate of radical entry into the particles; k_0 is the rate constant for transfer of radicals out of the particles; and k_t is the rate constant for termination.

Smith and Ewart did not obtain a general solution of Eq. (8-1), but solved it for three limiting cases.

Case 1:   n < < 1
Case 2:   n ~ 0.5
Case 3:   n > > 1

Case 2 gave the best agreement with the experimental results for the polymerization of styrene. (It is also consistent with the concept of alternate growing chain and inactivated radical in each polymer particle.) Case 2 was used to develop the following expressions for the rate of polymerization R_p and the average degree of polymerization $\overline{DP}$:

$$R_p = k_p[M]N/2 \qquad (8-2)$$

$$\overline{DP} = k_p[M]N/\rho \qquad (8-3)$$

where k_p is the rate constant for propagation and [M] is the monomer concentration in the particles. An equation for the number of particles was also derived,

$$N = k(\rho/\mu)^{0.4}(a_s S)^{0.6} \qquad (8-4)$$

where μ is the rate of volume increase of the particles; a_s is the area occupied by one emulsifier molecule; S is the amount of emulsifier present; and k is a constant whose value is 0.37 (assuming that the micelles and polymer particles compete for the radicals in proportion to their respective total surface areas) and 0.53 (assuming that the primary radicals enter only micelles as long as micelles remain). The Smith-Ewart theory was experimentally tested and confirmed by Morton *et al.* [3], Bakker [4], Bartholome *et al.* [5], Gerrens [6], van der Hoff [7], and Vanderhoff *et al.* [8].

The literature on radiation-induced emulsion polymerization is surprisingly meager compared with bulk or solution polymerization. An early investigation by Ballantine [9] showed that the γ-induced emulsion polymerization of styrene is about 100 times faster and yields higher molecular weights (up to 2×10^6) than the γ-induced bulk polymerization. The high rate was attributed to the high radical yield (G_R value) of water, as compared with the G_R value of styrene. The temperature dependence of the reaction indicated an overall activation energy of 3.7 kcal/mole.

Okamura [10] investigated the γ-induced emulsion polymerization of styrene, methyl methacrylate, and vinyl acetate using anionic, nonionic, and cationic emulsifying agents, and dose rates of 270 to 2×10^4 rads/h. The rate increased with increasing emulsifier concentration, but was strongly dependent on the type of emulsifier. In the case of methyl methacrylate, the overall rate was proportional to the 0.25 power of the dose rate.

Chapiro and Maeda [11] polymerized styrene suspensions which contained polyvinyl alcohol as a protective colloid (but no emulsifier). The dose rates used were 8340, 2124, and 930 rads/h. A dose of 58,000 rads gave 60% conversion.

The following sections present details of some of the more recent studies of radiation-induced emulsion polymerization.

B. Comparison of Methyl Acrylate with
Other Monomers

Comparative studies on the radiation-induced emulsion polymerization of styrene (STY), methyl methacrylate (MMA), methyl
acrylate (MA), and ethyl acrylate (EA) were carried out by Hummel
et al. [12]. The major portion of the emphasis was placed on MA.
In each case the emulsion was contained in a dilatometer during
polymerization, and a linear relation was assumed between volume
contraction and percent conversion. A cylinder of cobalt-60 with
an activity of 6 to 7 Ci was employed as the source. The dose
rates in the dilatometer were 200, 50, and 22.2 rads/h, depending
on the distance from the source.

Temperature showed no effect on the rate of the radiation-
induced emulsion polymerization of MA in the range from 15° to 50°C.
The maximum overall rate appeared to depend on the 0.8 power of
monomer concentration. The maximum overall rate was always reached
after about 5 to 7 min. It was highest at the lowest concentration
of emulsifier (sodium dodecyl sulfate), and decreased continuously
as the emulsifier concentration increased. The number of particles
per milliliter of dispersion increased with increasing emulsifier
concentration, the diameter of the particles decreasing simultaneously. The molecular weight of the product increased with
increasing emulsifier concentration.

The overall rate (mg polymer/g/min) increased at the start,
due to the buildup of free radical concentration in the system.
Later the overall rate fell gradually (period I) and then fell more
rapidly (period II). Evidently there was greater complexity than
would be expected from the simple Smith-Ewart theory.

Hummel explained the results by postulating a strong gel
effect in the emulsion polymerization of MA. The gel effect may
have occurred earlier than in bulk systems, because the P/M ratio
in the monomer/polymer particles was already high in the early

stages of polymerization. This ratio determines the viscosity in
the particles and the speed of diffusion toward each other of the
two growing polymer chains. Especially in the case of MA, the
volume of the monomer/polymer particles was considered high in the
early stages of the polymerization, owing to the high capacity of
PMA particles for monomer absorption. It was postulated that the
high P/M ratio contributed to the strong and early gel effect. The
kinetics were too complex to be explained completely.

Since the results could not be interpreted in terms of the
simple Smith-Ewart theory, one would not expect (and did not ob-
serve) an overall rate independent of monomer/water ratio or
proportional to $[\text{emulsifier}]^{0.6}$ and to $[\text{dose rate}]^{0.4}$.

The overall polymerization rate for various monomers was found
to increase in the order: styrene, methyl methacrylate, ethyl
acrylate, and methyl acrylate. Using a dose rate of 200 rads/h,
the dosage needed to obtain 80 to 90% conversion was about 10^3 rads.
Such a dose was far less than that required to cause measurable
degradation or cross-linking of the polymers formed. The G value
for the formation of polymer molecules was orders of magnitude
higher than the G_R (H_2O) values given in the literature.

C. Emulsion Polymerization of Styrene

Vanderhoff *et al.* [13] developed a technique for emulsion
polymerization investigation known as "competitive particle growth."
This involves polymerizing monomer in a mixture of two monodisperse
latexes without initiating a new crop of particles. The different-
sized particles compete with each other for the available monomer
and free radicals. The particle sizes before and after γ-initiated
(^{60}Co) polymerization were determined by electron microscopy.

The mixed seed latexes were prepared by blending previously
prepared monodisperse polystyrene latexes. The experimental results
were expressed as the ratio of the final particle diameters (D_B/D_A)

plotted against the ratio of the final to the initial diameter of
the smaller particles (D_A/D_O). It was assumed that the rate of
particle volume increase with time (dV/dt) was proportional to the
nth power of particle diameter (D^n). Theoretical plots for values
of n = 1, n = 2, and n = 3 were computed and drawn for comparison.
Comparison of the experimental points for each system with the
curves for various values of n enable an estimate to be made of
the n value for each emulsion system. According to the Smith-Ewart
kinetic scheme, the rate of particle growth should be independent
of particle volume, corresponding to n = 0.

The results showed n = 3 for benzoyl peroxide initiation. The
results indicated identical n values for γ-ray and persulfate ion
initiation, with values ranging from 1.0 to 2.5 depending on conditions. All these values are larger than the n = 0 value predicted
by the Smith-Ewart theory.

Special series of experiments using γ-ray initiation were
carried out to determine the effect of emulsifier concentration
and polymerization temperature. The number of particles initiated a
and molecular weight of the polymer were correlated with the polymerization conditions. The monomer/polymer ratios ([M]/[P]) were
determined from the equilibrium swelling volumes of the final
latexes. The values of [M] were computed from the values of
[M]/[P].

It was found that the number of particles per cubic centimeter
of water increased with increasing emulsifier concentration. The
number of particles per cubic centimeter was proportional to the
square of emulsifier concentration for both persulfate initiation
and γ-ray initiation. This compares with the prediction of Smith
and Ewart [2] that the number of particles should vary as the 0.6
power of emulsifier concentration.

At constant emulsifier concentration, an increase in polymerization temperature caused the number of particles per cubic centimeter to decrease for γ-ray initiation; the opposite was found for

persulfate-ion initiation. The value of k_p was determined from the
Smith-Ewart equation,

$$\overline{DP}_n = k_p N[M]/\rho \qquad (8-5)$$

where ρ is the rate of radical entry into the particles, N is the
number of particles per cubic centimeter of water, and $\overline{DP}_n$ is the
number-average degree of polymerization. The calculated k_p values
increased as the number of particles decreased, that is, as the
particle size increased. Vanderhoff attributed this effect to an
apparent increase in k_p caused by an actual decrease in k_t (at
larger particle sizes). Such a decrease in k_t is sometimes called
the Trommsdorff effect or gel effect, and it takes place when the
mutual termination of two polymer radicals in a viscous medium is
diffusion controlled. At the same time the smaller monomer molecules can readily diffuse to the growing chain in the viscous
medium, causing a decrease in k_t without a significant change in
k_p.

Values of k_t calculated from γ-ray initiation at 50°C ranged
from 1.1×10^4 to 10×10^4 liters/mole sec at particle sizes
ranging from 860 to 1850 Å. These agreed reasonably well with
values of 3.1×10^4 to 29×10^4 liters/mole sec reported by van
der Hoff [14] for somewhat larger size particles. Such values are
so much smaller than the 1.84×10^7 value determined in dilute
solution [15] that they provide convincing support for the hypothesis of a decreased k_t in the viscous monomer-swollen polymer
particles.

D. Emulsion Polymerization of Vinyl Acetate

Stannett et al. [16] carried out studies of the γ-induced
polymerization of vinyl acetate in emulsion systems. The γ-
radiation was supplied by a 1500-Ci cobalt-60 source. The rate

of polymerization was followed dilatometrically, and the molecular weights of the polymers were determined by viscosity measurements in acetone solutions.

The investigation covered (1) an indepth study of the homopolymerization of vinyl acetate including possible posteffects, and (2) a study of the properties of vinyl acetate homopolymers and copolymers with several other monomers. The emulsion recipe included 60 g water, 15 ml vinyl acetate, and 0.75-1.6 g sodium lauryl sulfate. This recipe was irradiated in the dilatometer to give at complete conversion a latex of 20% solids. The conversion versus time plot was a typical S-shaped curve, with a low initial rate, linear midportion, and a gradual decrease in rate toward the end of the run. Complete conversion was obtained with a total dosage of less than 0.02 Mrad.

Several runs were carried out using an automatic recording dilatometer and the 20% solids recipe. In order to study the post-irradiation behavior, certain runs were interrupted by removing the radiation source at various conversions up to 50%. A 30-min waiting period did not cause any additional polymerization, and the rates were identical before and after the interruption. This contrasts with the behavior of other monomers, and Stannett postulated that the lack of posteffect during most of the polymerization of vinyl acetate may be due to the ease of chain transfer to monomer plus the high propagation rate. It has been computed [17] that in the emulsion polymerization system vinyl acetate transfers many times during the kinetic chain length as compared to styrene. The resulting radicals may easily diffuse out of the particles and eventually become terminated. However, at higher conversions, more and more macroradicals are produced which remain in the particles. Hence, posteffects are absent with vinyl acetate except at higher conversions.

The exponential dependence of rate on radiation intensity was found to be 0.71, 0.79, and 0.90 at 0°, 30°, and 50°C, respectively, as compared to 0.4 power dependence predicted by the Smith-Ewart theory. It was noted that the deviation from 0.4 power dependence has been observed by other investigators, and that it may be partly due to the solubility of vinyl acetate in water.

The dependence of the rate of polymerization and product molecular weight on emulsifier concentration was determined at $0°C$. It was found that rate was dependent on the first power of emulsifier concentration, while molecular weight was independent of this variable.

The rate of polymerization in a system containing a constant number of particles should be independent of the intensity, if the system follows the classic Smith-Ewart theory [2]. Two series of experiments were carried out at $50°C$, in the first of which the number of particles (amount of seed latex) was held constant and the intensity varied, while in the second series the intensity was held constant and the number of particles initially present was varied. A plot of log rate versus log intensity indicated rate dependence on the 0.26 power of intensity, which approaches the Smith-Ewart prediction of zero power dependence. Another log-log plot showed a rate dependence on the 0.7 power of the number of particles, which approaches the 1.0 power dependence predicted from Smith-Ewart kinetics (and differs from the 0.2 power dependence observed in separate runs employing persulfate initiation). Hence, the radiation-induced emulsion polymerization of vinyl acetate approaches "classic" behavior, even at $50°C$.

To summarize the comparison with the Smith-Ewart theory, the rate of the radiation-induced emulsion polymerization of vinyl acetate at $0°C$ showed a dependence on the 0.7-0.9 power of radiation intensity, the 0.9-1.0 power of emulsifier concentration, and the 0.7 power of the number of particles, as compared to the 0.4, 0.6, and 1.0 power, respectively, predicted by the theory. This contrasted with the persulfate-initiated reaction, which showed a rate dependent on the 1.0 power of initiator concentration, 0.2 power of emulsifier concentration, and 0.2 power of the number of particles. It was postulated that the closer approach to "classic" kinetics for the radiation-induced reaction may have been due to less polymerization in the aqueous phase in the absence of ionic end groups.

The yields of polymer per radiation dose were extremely high in the emulsion system. The product molecular weights were also extremely high, ranging from one to four million as compared to about 200,000 for most commercial polyvinyl acetate latices. Such high molecular weights would correlate with good scrub resistance, which is advantageous for certain film applications.

Several stable latices were prepared from copolymers of vinyl acetate with dibutyl maleate, 2-ethyl hexyl acrylate, n-butyl acrylate, and methyl acrylate. All the copolymerizations were carried out at 0°C, using 5% of a nonyl phenol polyethylene oxide-type emulsifier on a 50/50 monomer/water ratio by volume. The results clearly demonstrated that γ-radiation can be used to produce stable latices of high solids content and exceptionally high molecular weight which can be compounded into paints with improved scrub resistance.

E. Pilot Plant Emulsion Polymerization
Using Radiation Initiation

Greene *et al.* [18] have described the use of cobalt-60 radiation to produce a vinyl acetate/butyl acrylate copolymer latex by emulsion polymerization on a pilot-plant scale. The plant was similar to that which would be used for a commercial radiation process. The type of latex produced was similar to that now employed in the paint industry.

The reaction mixture was continuously circulated through a zone of high radiation intensity, where the initiating species were produced and most of the polymerization took place. Latex was produced by polymerization of the monomer, which was added in three ways: in batches, at a given rate until a specified quantity was delivered, and continuously. All materials used were of commercial grade, with no purification prior to reaction.

Whereas latex properties varied from batch to batch, the paint made from the latex had enamel holdout superior to that of paint produced from high-grade commercial latexes made by conventional methods. "Enamel holdout" can be defined as the suitability of a water-based flat paint to serve as a primer for enamel on unpainted woodwork. It was postulated that the improvement in this property resulted from the higher molecular weights of polymers produced by radiation initiation.

The operation of the pilot plant and evaluation of the product showed that radiation-initiated latex with acceptable properties can be produced in commercially feasible equipment. The work did not result in the development of a commercially feasible process, because of difficulties in residual monomer removal and polymer buildup on process lines. However, it was believed that such problems were soluble and did not represent an insurmountable difficulty in the use of radiation.

Comparison of radiation-initiation with a plant based on conventional initiation showed an estimated operating profit of 5 to 10% higher for the radiation plant with the difference in catalyst cost a principal contributing factor. It may also be possible to save money by using less emulsifier, since Lukhovitskii et al. [19] have found a stabilizing effect of ionizing radiation due to the formation of an additional negative charge on the latex particles.

In the work where this discovery was made, Lukhovitskii used γ-rays to polymerize a styrene emulsion containing 2% potassium laurate. Discontinuance of γ-radiation gave a decrease in polymerization rate, followed by a constant postpolymerization rate equal to about 10% of the rate during irradiation. When irradiation was stopped, the particles tended to coagulate, owing to the loss of the stabilizing effect of irradiation. Electron microscopy showed that particles formed during radiation initiation were two to three

times smaller than those formed by chemical initiation. An induction period was observed in chemically initiated polymerization,
but not in radiation-initiated polymerization. In the presence
of 2% emulsifier, the rate of postpolymerization does not increase
with dose rate at rates $\gtrsim$ 4000 rads/ml of emulsion.

II. POLYESTER SYSTEMS

A. Introduction

The curing of polyester systems (mainly coatings) by ionizing
radiation has been carried out with increasing success in the past
several years and has been adapted for commercial usage by a few
companies. The main ingredients of a typical coating would be a
long-chain, unsaturated polyester and a reactive monomer component
such as styrene. In conventional coating methods, the coated substrate is commonly passed through long heated ovens to accomplish
drying and/or curing. When radiation is used there is no solvent
to be removed, and the curing is caused by radiation in the absence
of heat. Furthermore, less plant space is used for the radiation-
curing method and there is a potentiality for lower cost in some
types of coating operations. Disadvantages of radiation include
the (usually) higher viscosity of the coatings and the need (at
present) for an inert atmosphere during curing.

The typical total dosage required for curing of a polyester
coating ranges from 1 to 10 Mrads. Computations [20] have shown
that such total dosages can be delivered more quickly by electron
accelerators than by x-ray machines or cobalt-60 sources. The
electron energy required to cure a coating is a function of the
thickness and density of the accelerator window, the air gap between
window and coating, and the composition of the coating. Most

coatings are between 0.4 and 8 mils in thickness and require electron energies of 25 to 150 keV for penetration of such thicknesses.
For typical accelerator windows and air gaps, accelerator ratings
of 150 to 600 kV are required.

As discussed previously, the absorption of radiation produces
free radicals, and these appear to be the active chemical species
that bring about the curing of coatings. The free radicals take
part in chain reactions (polymerization) which produce growing
chains of great length before they are terminated by reaction with
another free radical. The overall rate of polymerization or curing
is accelerated by deliberatley slowing down the termination reaction through the use of high viscosity formulations (gel effect).
The cured coating consists of a cross-linked, insoluble, and
infusible gel. Generally the percentage of gel present is used as
an indication of the degree of cure.

A higher frequency of unsaturation in the polymer and a higher
molecular weight generally increase the rate of gelation for any
particular monomer/polymer ratio. However, the molecular weight
cannot be too high, for such polymers are too highly viscous for
ease of application in a coating formulation. When polyunsaturated
monomers are employed, the frequency of unsaturation in the polymer
component can be decreased without sacrifice of gelation speed.

The effect of dose rate on curing is complex and not completely understood. The quantity generally evaluated is the time-
average dose rate. Special techniques have been used in applying
the beam to the sample, such as rotating over the sample a disk
containing a small slit. It appears that such variations are of
small importance, and the rate of cure depends basically on the
time-average dose rate. However, Hoffman and Smith [21] found that
fractionation of the total dose into several passes gave greater
gelation efficiency than if the total dose is absorbed in one pass.
Similar findings were reported by Pietsch [22] and by Blin and
Gaussens [23]. Too high a dose rate can be ineffective, if gelation occurs so rapidly that growing chains become trapped and im-
mobilized in the network. Hence, if the time-average dose rate is

so high that a rigid network is rapidly formed, it would be expected that further cure of the coating would be insensitive to dose
rate.

Whereas gelation can occur when unsaturated polymer molecules
are irradiated in the absence of monomer, gelation is more rapid
when the relatively immobile polymeric radicals are bridged by
mobile monomer molecules in a copolymerization chain reaction.
This effect is most noticeable when the monomeric component is
polyfunctional. However, as more and more monomer is added, viscosity drops and permits a higher rate of termination by recombination of macroradicals. The net result is a very low rate of
gelation at high monomer concentrations. Hence, there is a maximum
in the percent gel formation versus monomer concentration at any
dose [23].

The type of monomer used also has an effect on the rate of
cure (gelation). In work with γ-radiation, Ballentine and Manowitz
[24] found that styrene reacted more readily with fumarate or
maleate groups in polyesters than did vinyl acetate or methyl
methacrylate. Hoffman and Smith [21] observed that polyunsaturated
esters form gels more rapidly with styrene than with ethyl acrylate,
which, in turn, is more reactive than methyl methacrylate. Pietsch
[22] discovered that styrene is more efficient in cross-linking
polyesters than ring-substituted styrenes, acrylates, or dimethacrylates. Blin and Gaussens [23] demonstrated that vinyl acetate
is more efficient than methyl acrylate in the gelation of 80/20
polyester/monomer mixtures.

The following sections present more details concerning the
variables affecting the radiation curing of polyester/monomer
compositions.

B. Effect of Monomer Type

Pietsch [22] has evaluated a wide variety of monomer types in
the radiation curing of polyester/monomer coating compositions.
Two polyester prototypes were used throughout the study: (1) a
condensation product of maleic anhydride, phthalic anhydride, and
1,2-propylene glycol (PPM) of molecular weight 1700, and (2) a condensation product of 1,2-propylene glycol and maleic anyydride
(PM) of molecular weight 2400. Gelation characteristics of these
polyesters were evaluated when blended with the following monomers:
styrene, vinyl toluene, chlorostyrene, methyl methacrylate, acrylonitrile, divinylbenzene, ethylene glycol dimethacrylate, butylene
glycol dimethacrylate, tetraethylene glycol dimethacrylate, and
trimethylolpropane trimethacrylate.

The liquid resins were applied to small plywood panels and
irradiated with a 3-MeV van de Graaff generator or an 0.5-MeV electron accelerator. The dose from the 3-MeV van de Graaff generator
was administered in multiples of 1 Mrad/pass, using a dose rate of
0.7 Mrad/sec. The degree of cure was assessed by ATR infrared
spectroscopy and pencil hardness (P.H.) measurements. The liquid
resin surface was shielded from atmospheric oxygen by covering
with release paper during irradiation.

When using the PPM polyester, vinyl toluene showed no advantage in rate of cure over styrene as shown by the P.H. test, but
did lose unsaturation more rapidly as indicated by infrared.
Chlorostyrene was slower in curing under irradiation than either
styrene or vinyl toluene. For chlorostyrene alone, only 5% of the
total double bonds were converted at a dosage of 8 Mrads, increasing to 78% when half of the chlorostyrene was replaced by styrene.

Methyl methacrylate (with PPM) showed no improvement with
regard to P.H. over styrene. Actually, methyl methacrylate alone
had low conversion even at a total dose of 8 Mrads. A good cure
resulted at 9 Mrads if a 1/1 mixture of MMA/styrene was used.

Acrylonitrile also gave low conversion when used alone, and a higher degree of conversion for an AN/styrene mixture. With methacrylic acid alone, conversion reached a maximum of about 85% reacted double bonds at a total dose of 5 Mrads, and showed no further change when the dose was increased.

Several difunctional monomers were evaluated. No conversions over 90% were obtained unless a postcure was employed. Pietsch postulated that once the first vinyl group has reacted, the second vinyl group is confined to one position in the polymer network, thus hindering the obtaining of 100% cure. The best results for ethylene glycol dimethacrylate (EGDMA) were obtained by using a 1/1 mixture of styrene/EGDMA. The conversion of resin at 5 Mrads was 80%, which is about 10% less than that obtained by conventional curing methods. Butylene glycol dimethacrylate (BGDMA) performed similarly to EGDMA, as did tetraethylene glycol dimethacrylate (TEGDMA).

A parallel series of experiments was carried out using the PM polyester. One item of special interest was the comparison of formulations containing 28 and 38% styrene. For doses above 6 Mrads, there was hardly ant difference in P.H. for the two formulations. The relative efficiency of styrene and chlorostyrene was similar to what it was in the PPM polyesters. Similar comments apply to EGDMA and BGDMA.

Pietsch also carried out some experiments designed to determine whether the level of cure is different when a given total dose is administered in one pass or is divided among several passes. It was found that at least two exposures were required to obtain a good cure, and increasing the number of exposures beyond that had no beneficial effect. Administering the dose in two passes increased the P.H. considerably without any significant change in the percent conversion of unsaturation for the PPM polyester. The beneficial effect of two passes was much less pronounced for the PM polyester.

A postcure effect was observed, consisting of an increase in

the PH of the polyester resin coating on aging. Although the increase in hardness was smaller, it was observed at various total
doses and dose rates. This was attributed to a slow reaction of
trapped radicals with styrene molecules which migrated into their
vicinity.

C. Effect of Polyester Backbone Reactivity

While Pietsch investigated the effect of using different reactive monomers, Hoffman *et al.* [25] concentrated on differences in
the polyester backbone and used styrene as the only monomer. One
of the important variables was found to be the molecular weight of
the polyester backbone. An electron accelerator was used as the
radiation source in this work.

The molar ratio composiitons of the polyesters evaluated were
as follows:

    PA/MA/PG = 1/1/2
    IP/FA/DEG = 2/1/3
    IP/MA/PG = 3/1/4
    IP/MA/PG = 2/1/3
    IP/MA/PG = 1/1/2

where the abbreviations had the following meanings: IP, isophthalic
acid; PA, phthalic anhydride; MA, maleic anhydride; FA, fumaric
acid; PG, propylene glycol; DEG, diethylene glycol.

In each case the polyester was shaken with the proper amount
of styrene to form a smooth blend. A small sample was placed in
an aluminum foil cup and passed under the electron accelerator
beam without exclusion of air. The doses were delivered in multiple
passes of 2 Mrads/pass using 500-KeV electrons. The conveyor moved
beneath the accelerator window at such a rate that the sample
received 0.4 Mrad/sec.

The samples were weighed after irradiation and allowed to stand 96 h to remove volatile monomer. They were then placed in an excess of methyl ethyl ketone, drained, and weighed again to estimate swollen gel weight. After drying in a vacuum they were weighed to obtain percent gel. Volume fraction of cross-linked polymer in swollen gel, V_2, was then computed.

$$V_2 = \frac{w_2/\rho_2}{w_2/\rho_2 + w_1/\rho_1} \tag{8-5}$$

where w_1 and w_2 are weight fractions of solvent and polymer, respectively, in swollen gel, and ρ_1 and ρ_2 are densities of solvent and polymer, assumed to be 0.8 and 1.2 g/cm^2, respectively.

Results were compiled in the form of percent gel versus dose curves, percent gel versus initial percent styrene after 9 Mrads, volume fraction of polymer versus percent gel curves, and volume fraction of polymer versus percent initial styrene for all polyesters studied.

The following conclusions were drawn concerning the effects of polyester composition and molecular weight on curing efficiency:

1. At low styrene concentrations, 0 to 10%, the mode of cross-linking was mainly through the reaction of adjacent free radicals on the polyester backbone. Gelation efficiency increased for both greater molecular weight and higher frequency of polyester unsaturation (more double bonds). The molecular weight appeared to be more important than unsaturation frequency, other things being equal. There was also a higher rate of gelation for chains of equal molecular weight and double-bond frequency when DEG was substituted for PG. That may have been due to the greater radiation sensitivity of DEG compared to PG or to the greater flexibility of polyester chains containing DEG.

2. At intermediate styrene concentrations, 10 to 30%, the maxima in the curves of gel content versus styrene concentration were attributed to the predominance of copolymerization of styrene

with backbone unsaturation as a mechanism of cross-linking and gelation. Polyester-polyester interactions perhaps contributed to cross-linking, but they became less important as styrene concentration increased. In these formulations the frequency of unsaturation along the backbone was probably more important in determining gelation rate than was the molecular weight of the polyester.

The high viscosity of the formulations probably slowed the termination step relative to the propagation step. The importance of the gel effect was shown by adding 27% of an acrylic polymer containing no unsaturation to the solution of polyester/styrene. The increased viscosity increased the initial gelation rate by about 50%, evidently by enhancing propagation at the expense of termination.

3. At styrene concentrations above 30%, the rapid decrease in gel formation rate was probably caused by the decrease in viscosity and hence in the gel effect. Also, as styrene concentration increased, the more highly unsaturated and higher molecular weight polyesters were found to cross-link less rapidly than the less unsaturated and lower molecular weight polyesters. Hoffman *et al.* [25] called this a "crossover" effect and postulated that it was caused by different compatibilities of the polyester chains with styrene, so that the shorter chains or the least unsaturated chains (with a greater aromatic content where IP was substituted for MA) would be more compatible with styrene and would therefore react more efficiently with it at higher styrene concentrations.

Conclusions were also drawn [25] as to the effect of double-bond type and location in the polyester.

1. For chain and or pendant unsaturation, the expected order of decreasing activity is

$$-CH\!=\!CH_2 \ > \ \underset{\displaystyle -\overset{\textstyle CH_3}{\underset{|}{C}}\!=\!CH_2}{} \ > \ -CH_2-CH\!=\!CH_2$$

2. Internal double bonds, —CH=CH—, would not be as reactive as chain-end double bonds, while the latter should be sterically more accessible than pendant double bonds. (The highly reactive maleate/styrene or fumarate/styrene pairs would be exceptions to the rule.)

In the course of this research, Hoffman also obtained some interesting information on the effect of dose rate. Using the IP/MA/PG resin at a 1/1/2 ratio and employing a polyester/styrene 70/30 blend, the rate of gel formation was studied over a range of dose rates from 0.12 to 165 Mrads/min. A plot of log initial gelation rate versus log dose rate was drawn. From 0.12 to 10 Mrads/min, the gelation rate depended on the 0.95 power of the dose rate. Above 10 Mrads, the gelation rate depended on the 0.35 power of the dose rate. In the region from 0.12 to 10 Mrads/min, Hoffman postulates a unimolecular termination mechanism as polymer radicals become occluded or buried in the very viscous gel matrix. Whereas the mechanism above 10 Mrads/min has not been resolved, the efficiency of gel formation must decrease because of the recombination of primary radicals with each other or with short growing copolymer chains.

The ideal coating formulation would have low viscosity for ease of application to the substrate, but would quickly increase in viscosity after a small radiation dose. This is in conflict with the need for a high molecular weight backbone polyester, and a compromise is required to obtain a formulation having optimum handling and curing properties.

D. Kinetics of Styrene/Polyester Reaction

The most systematic analysis of the γ-activated styrene/ polyester reaction was made by Burlant and Hinsch [26]. The polyester used contained 1.5 double bonds per molecule of number-average molecular weight of 890 and was prepared from a 1/2/3 mole ratio of maleac acid/phthalic acid/propylene glycol. Solutions of ester in styrene were transferred under nitrogen to fixed-thickness liquid infrared cells. Irradiations were with γ-rays from ^{60}Co using dose rates from 8.33 x 10^3 to 7.10 x 10^3 rads/min at 33°C. Infrared analyses of styrene and ester unsaturation were performed within a few minutes after irradiation. Extraction with hot benzene left behind the insoluble gel fraction, which was weighed after drying to constant weight in a vacuum.

Earlier kinetic analyses by Gordon and McMillan [44] of vinyl/divinyl copolymerization showed that the conventional radical scheme holds up to the gel point, which takes place at less than 4% conversion. Beyond that point there was observed an increase in rate, attributed to the gel effect on the termination step. At still higher degrees of cross-linking the propagation step slowed down and the overall rate decreased.

Burlant employed ester concentrations ranging from 65 to 2.4%, corresponding to ester/styrene unsaturation ratios from 0.327 to 0.004. Analysis of the system was aided by the following factors: (1) disappearance of ester and styrene unsaturation could be followed separately by infrared analysis (baseline technique); (2) the isolated gel could be hydrolyzed to give a tractable styrene/maleic acid copolymer, so that the average degree of polymerization of the growing chains could be estimated; (3) initiation by γ-rays at 30°C eliminated certain difficulties caused by the use of chemical initiators at higher temperatures; and (4) the observed absence of dark reactions simplified the kinetic analysis.

Gelation took place at about 1% unsaturation conversion. At that point the system was composed of a gel, highly swollen with the mixture of monomers, in which were incorporated pendant vinyl groups. Assuming bimolecular chain termination, the existence of a single effective propagation constant, and single termination constant, then at the steady state the concentration of radicals separately in the sol and gel was given by:

$$\frac{d[S\cdot]}{dt} = k_1(1-g) - k_t[S\cdot]\left(2[S\cdot] + [G\cdot]\right) - k_p[S\cdot]M_2 = 0 \qquad (8\text{-}6)$$

and

$$\frac{d[G\cdot]}{dt} = k_1(g) - k_t[S\cdot][G\cdot] + k_p[S\cdot]M_2 = 0 \qquad (8\text{-}7)$$

where $[S\cdot]$ is concentration in moles per liter of soluble radicals at time t; k_1 is initiation rate constant in mole per liter minute; g is gel fraction; (1-g) is sol fraction = S; k_t is termination rate constant in liters per mole minute; k_p is propagation rate constant in liters per mole minute; and M_2 is overall concentration of ester double bonds at time t in moles per liter. The quantity $k_1(1-g)$ describes the rate of radical formation in the sol phase. The last two terms in Eq. (8-6) indicate the various rates at which radicals are removed from the sol phase (by combination with another radical or by incorporation into the gel).

The overall rate of unsaturation disappearance must be

$$\frac{d\alpha}{dt} = k_p(1-\alpha)\left([S\cdot] + [G\cdot]\right) \qquad (8\text{-}8)$$

Solution of (8-6) and (8-7) followed by substitution of the results into (8-8) yields the following expression for the overall rate of unsaturation disappearance.

$$\frac{d\alpha}{dt} = \frac{k_p{}^2}{2k_t}\left(\frac{M_2(1-\alpha)}{(1-2g)}\right)$$

$$+ \frac{k_p{}^2}{2k_t}\left[\sqrt{\left(\frac{M_2(1-\alpha)}{(1-2g)}\right)^2 + \frac{2k_1k_t(1-d)^2}{k_p{}^2(1-2g)}}\right] \qquad (8\text{-}9)$$

Equation (8-9) can be applied to high, low, and intermediate concentrations of ester in styrene.

Burlant considered solutions containing over 50% ester in styrene as "high" concentrations. Plots of conversion versus time were found to be linear for 50 and 65% ester in styrene. Also, each component entered the gel at a constant molar ratio of styrene to ester double bond of 2/1. Such behavior would not be predicted from the known copolymer reactivity ratios. However, the experimental data established the following relations between M_2 and α:

$$M_2 = M_2^{\;0} - \alpha/3 \; (M_1^{\;0} + M_2^{\;0}) \tag{8-10}$$

$$M_1 = M_1^{\;0} - 2\alpha/3 \; (M_1^{\;0} + M_2^{\;0}) \tag{8-11}$$

where $M_1^{\;0}$ and $M_2^{\;0}$ are the initial molar concentrations of styrene and ester, respectively. Using Eqs. (8-10) and (8-11) and making certain simplifications based on gel fraction/conversion plots, Eq. (8-9) can be reduced to

$$\frac{d\alpha}{dt} = 2A_3 \left(\frac{(1-A_6\alpha)(1-\alpha)}{(1-2g)} \right) \tag{8-12}$$

where

$$A_3 = (k_p^{\;2}/2k_t)M_2^{\;0} \tag{8-13}$$

and

$$A_6 = (M_1^{\;0} + M_2^{\;0})/3M_2^{\;0} \tag{8-14}$$

Furthermore, curve fitting of gel fraction/conversion data showed that

$$(1-2g) = 1 - 2A_1\alpha + 2A_2\alpha^2 \tag{8-15}$$

where A_1 = 1.48 and 2.10 and A_2 = 0.85 and 1.85 for the 65 and 50% ester solutions, respectively.

Substitution of Eq. (8-15) into Eq. (8-12), followed by integration, showed that up to 30% conversion, approximately,

$$\alpha = 2 A_3 t \tag{8-16}$$

This was confirmed experimentally by a linear plot of α versus $2A_3 t$ having a slope of 1, so that

$$d\alpha/dt = 2 A_3 \tag{8-17}$$

Combination of Eq. (8-10) and (8-11) with (8-17) indicated the rate of styrene disappearance to be

$$\frac{d\alpha_1}{dt} = 4/3 \, A_3 [(M_1{}^0 + M_2{}^0)]/M_1{}^0 \tag{8-18}$$

For dilute ester solutions of 2.4 and 4.6% ester in styrene, a linear plot of styrene conversion versus time was obtained up to about 40% conversion. For such cases with M_2 less than 0.008 mole/liter, Eq. (8-9) reduced to

$$d\alpha/dt = k_p (k_1/2k_t)^{1/2} [(1-\alpha)/(1-2g)^{1/2}] \tag{8-19}$$

From infrared it was not possible to estimate the low rate of ester disappearance, α_2. However, simplification was possible since the dilute solution copolymerizations appeared to follow the conventional copolymer composition equation [27], with reactivity ratios r_1 and r_2 for styrene and ester, respectively. Hence,

$$r_1 = k_{M^0 M_1}/k_{M_1^0 M_2} \tag{8-20}$$

$$r_2 = k_{M_2^0 M_2}/k_{M_2^0 M_1} \tag{8-21}$$

and it was possible to rewrite Eq. (8-19) as

$$\frac{d\alpha_1}{dt} = kp \left(\frac{k_1}{2k_t}\right)^{\frac{1}{2}} \left(\frac{(1-\alpha_1)}{(1-2g)^{\frac{1}{2}}}\right) \left(\frac{(1+M\theta)(r_1+M\theta)}{(r_1+2M\theta)}\right) \tag{8-22}$$

where

$$M = M_2^{\ 0}/M_1^{\ 0} \tag{8-23}$$

and

$$\theta = (1-\alpha_2)/(1-\alpha_1) \tag{8-24}$$

For small values of M_1 and with r_1 between 0.1 and 0.9, Eq. (8-22) reduces to

$$\frac{d\alpha_1}{dt} = k_p \left(\frac{k_1}{2k_t}\right)^{\frac{1}{2}} \frac{(1-\alpha_1)}{(1-2g)^{\frac{1}{2}}} \tag{8-25}$$

Experimentally, it was found that $(1-\alpha_1)$ versus $(1-2g)^{\frac{1}{2}}$ gave a linear plot with slope equal to 1, so that Eq. (8-25) becomes, up to 40% conversion,

$$\frac{d\alpha_1}{dt} = k_p \left(\frac{k_1}{2k_t}\right)^{\frac{1}{2}} \tag{8-26}$$

Simplification of equations was not so convenient for intermediate ester concentrations, namely, 9.8, 20.0, and 30.0%. In fact, the simple kinetic model predicted rates up to threefold slower than observed. The discrepancy was attributed to processes neglected in the model, such as the disappearance into the gel phase of branched sol copolymer containing several radical sites, the increase with conversion of unreacted double bonds in the gel, etc. The following empirical expression in agreement with the data was developed

$$\frac{d\alpha_1}{dt} = 2A_3 \left(\frac{(1+M)(r_1+M)}{(r_1+2M)} \right) \qquad\qquad (8\text{-}27)$$

in accordance with the linearity of $d\alpha_1/dt$ versus $M_2{}^0$. The value of $k_p{}^2/k_t$ was calculated from the slope of the line. Equation (8-27) is independent of r_1 for $0 < r_1 < 0.9$, and indicates that α_1 is linear with time for a particular initial composition.

Calculated values for $k_p{}^2/k_t$ from Eqs. (8-17) and (8-27) were as follows.

Initial (ester)/(styrene)	$k_p{}^2/k_t$, liter mole^{-1} min^{-1} x 10^2
0.327	2.52
0.175	0.79
0.019	0.43

The value of $k_p{}^2/k_t$ of 0.43 x 10^{-2} may be compared with the reported one of 0.27 x 10^{-2} liters mole^{-1} min^{-1} for irradiation of bulk styrene under similar dose rate and temperature conditions [28].

The value of k_1 computed from the above value of $k_p{}^2/k_t$ substituted in Eq. (8-25) was 3.6 x 10^{-5} mole liter^{-1} min^{-1}. The G(radical) computed from this k_1 was 4.2. The reported G(radical) for bulk styrene is 0.9 [28].

E. Coating Techniques Used with Radiation Curing

In one reported investigation [25] of the radiation curing of polyester/styrene coating compositions, a source of 500 keV electrons at beam currents up to 20 mA was employed. It was possible for the electron beam to be spread out or "scanned" up to 72 in.

in width. The 500 keV electrons were found to penetrate up to 50
mils in an absorber of density equal to 1. That particular
electron processing system consisted of three major components:
power supply, electron accelerator, and control consoles. Electrical power was transmitted to the accelerator by a flexible cable
leading from the power supply. It was possible to accommodate up to
three electron accelerator assemblies from any one power supply.
That made it feasible to irradiate several production lines simultaneously or to fractionate the total dosage into two or three
doses in series from one production line. Local radiation shielding was required for the accelerator assembly only; the power
supply did not require shielding.

Whereas gelation kinetics may depend on a local dose rate in
a small element of volume, the time-average dose rate is usually
measured in practice. The time-average dose rate is a function of
electron energy, window thickness, area scanned, distance between
window and substrate, and thickness and composition of coating and
substrate. Hoffman [30] has expressed the time-average dose rate
through the thickness of the coating as

$$\text{Dose rate} = \frac{1}{\rho X_0} \int_0^{X_0} \frac{i}{WL} \left(\frac{dE}{dX}\right)_i \quad dX \qquad (8\text{-}28)$$

where

ρ = coating density, g/cm^3

X_0 = coating thickness, cm

i = time-average number of electrons arriving at coating
surface, each having an average energy E, electrons/sec

W = scan width (transverse direction)

L = scan length (machine direction)

$\left(\frac{dE}{dX}\right)_i$= linear rate of energy transfer in the coating for each
electron of average energy E, ergs/cm electron.

The dose delivered to a substrate moving at a velocity v is then

$$\text{Dose} = \frac{L}{v} \times (\text{dose rate}) = \frac{1}{v\rho X_0 W} \int_0^{X_0} i \left(\frac{dE}{dX}\right)_i \ dX \qquad (8\text{-}29)$$

so that

$$\text{Dose} \cong \frac{\text{constant}}{v(\rho X_0)} \qquad \begin{array}{l}\text{(for a particular set of} \\ \text{irradiation conditions)}\end{array} \qquad (8\text{-}30)$$

or

$$\text{Dose} \cong \frac{\text{constant}'}{v} \qquad \begin{array}{l}\text{(for a particular set of} \\ \text{irradiation and coating} \\ \text{conditions)}\end{array} \qquad (8\text{-}31)$$

Curing difficulties are sometimes caused by inhibitors or dissolved air that may be present in the coating. Any type of free radical scavengers in the coating will retard the copolymerization reactions until such compounds are used up. For most inhibitors including dissolved air, this may require up to 1 Mrad or more before gelation proceeds efficiently. Air is relatively easy to remove by degassing the monomer before dissolving the polyester, following which the solution may be stored under nitrogen or some other inert gas. However, if oxygen is present in the atmosphere above a coating during irradiation, the surface cure will be inhibited to some extent. The oxygen molecules will not diffuse much below the outer surface layer, but this can produce a very thin top layer that is tacky and only partially cross-linked.

For these reasons, radiation curing of polyester coatings is generally carried out in a nonoxidizing atmosphere. One method that has been used is to maintain a reducing atmosphere by incomplete combustion of desulfurized natural gas [29]. Cost will be important in the choice of a suitable inert gas atmosphere. It is also possible that at high intensities in air, oxygen, or ozone will form surface peroxides that will decompose to form radicals and initiate curing, thus limiting the inhibiting effect of oxygen.

Another problem that sometimes arises is the evaporation loss
of volatile monomer because of the large coating surface area exposed before the monomer is consumed by the curing reaction. This
may lead, in turn, to two other problems: (1) Control of the
gelation reaction (cure) may then become difficult because the
mechanism is very sensitive to monomer content, and (2) a monomer
recovery system may then become required, which is a costly item.
Hence, monomer loss should be minimized by employing low-volatility
monomers and by covering the coating or enclosing the coating system
until it is cured.

For reasons cited above, most coating compositions developed
for radiation cure are very viscous, and when they are applied at
high speed there may be problems in attaining thin, smooth, bubble-
free coatings. Vanderbie et al. [29] commented that curtain
coating is more feasible at high speeds, whereas roll or spray
coating may be used at lower speeds or lower viscosities.

Each type of coating technique has certain advantages and
disadvantages. Curtain coating gives the best appearance, particularly at speeds over 300 ft/min, but requires a large space and
may entrain bubbles. Roll coating requires the least space, but
a ribbed pattern may appear at higher speeds or higher coating
viscosities, and foreign particles may cause streaks. Spray
coating is difficult to use at high speeds or with high viscosities, and an orange peel pattern may result.

Various pigments may be added to the coating formulation to
increase the viscosity without having much effect on the dosage
needed for curing. If greater build is required, successive
applications and irradiation cycles may be used, although adequate
adhesion between the separate coats may be a problem. The goal
would be to apply a low viscosity coating in one coat, expose it
first to a low dose rate in order to raise the viscosity rapidly
in the dose-rate-sensitive cure region, then irradiate with a
higher dose rate beam to complete the cure of the highly viscous
gel which is more insensitive to dose rate.

The resistance of a coating to undesirable changes during
aging is dependent on a number of factors. Acrylic or other non-
aromatic monomers and polymers generally have better aging properties than systems based on styrene or other aromatic materials.
Unreacted double bonds in the coating may serve as groups readily
susceptible to oxidation. Good adhesion of the coating to the
substrate surface is correlated with better aging resistance. Adhesion to metal surfaces can be enhanced by the inclusion of carboxyl or epoxide groups in the coating. Better weathering resistance
has been claimed for radiation-cured styrene/polyester coatings than
for chemically cured coatings [31].

III. THERMOSETTING SYSTEMS IN GENERAL

A. Introduction

For most thermosetting plastics, not much research has been
done on radiation curing and few articles have been published
dealing with this subject. The work that has been done shows
several potential advantages over thermal curing including better
control of curing rate by control of radiation dose rate; longer
pot life, because of the elimination of chemical catalysts; reduced shrinkage because of room temperature curing; synthesis of
new polymers where chemical catalysts are not available; and improved aging stability because of the elimination of catalyst
residues that may promote degradation. Some authors have claimed
that radiation promotes a higher degree of cure than can be obtained with chemical catalysts [32].

One general principle has also been discovered in regard to
the radiation curing of ureas, melamines, and phenolics, namely,
that curing is promoted by the inclusion of chlorine-containing

additives. This finding and other details are discussed in the
following sections.

B.   Polyurethanes and Polythioethers

Klotz et al. [33] produced polyurethane castings by γ-irradiation of liquid unsaturated polyurethanes. The best elastomers
were made from tolylene diisocyanate, polypropylene glycol, and
glyceryl monolinoleate. (The products were "thermosetting" in
the sense that an insoluble, three-dimensional network was produced
by the gelation reactionl.) Free radical initiators gave erratic
behavior, and ultraviolet irradiation could not be used because of
opacity, but γ-radiation from a 10,000-Ci cobalt-60 source gave
excellent results. Samples were irradiated in 1-in. diam test
tubes, except for mechanical property specimens, which were irradiated in dumbbell-shaped molds.

Preliminary experiments involved the screening of compositions
prepared from 15 unsaturated diols, 2 diisocyanates, and 2 polymeric saturated diols. It was found that dose rates of (7 to 9) x 10^5
rads/h produced a promising elastomer from tolylene diisocyanate,
polypropylene glycol, and glyceryl monolinoleate. All further work
was done on various ratios of these principal ingredients.

A cross-linked elastomer of good appearance was obtained from
these ingredients after 16 h of irradiation (total dosage of
1.3×10^7 rads). A liquid prepolymer of polypropylene glycol (PPG)
and tolylene diisocyanate (TDI) alone gave only a slight increase
in viscosity after 19 h of irradiation, showing that the curing
takes place mainly through the unsaturated groups of the glyceryl
monolinoleate (GML).

Nearly unlimited pot life was shown by one sample of polymer
which remained castable after 2 weeks at room temperature, then
cured normally when irradiated. The advantage of radiation curing
for this composition was shown by the failure of thermal cures of

the same formulation plus catalyst. Samples containing from 0.05 to 0.5% methyl ethyl ketone peroxide remained liquid after 48 h at 43°C and after 24 h at 82°C.

Irradiation of a typical GML/TDI/PPG formulation produced insolubility after 3 h (3×10^6 rads), but hardness continued to increase during 21 additional hours of irradiation. The mechanical properties could be altered greatly by changing the saturated to unsaturated diol ratio or by changing the NCO/OH ratio. Typical tensile strength values ranged from 83 to 103 psi, while elongation at break ranged from 80 to 130%.

Cross-link density determinations from stress relaxation data indicated an order of magnitude of 1.00 to 1.67×10^{-4} moles of cross-links/cm^3. The G values of cross-linking ranged from 74 to 145 and were observed to increase as the concentration of GML decreased. This was attributed to possibly less cross-linker consumption in ring formation (intramolecular) reactions as its concentration decreased. Similar work by Charlesby et al. [34] on maleic acid polyesters indicated a G (cross-linking) value of 200 and led to the conclusion that the mechanism of cross-linking resembled a free radical addition polymerization of short chain length.

Klotz et al. [33] also investigated the polymerization reactions of dithols and trithiols with diolefins and triolefins, using the same cobalt-60 source mentioned above. The addition reaction of a thiol to a double bond has been known since the work of Posner [35]. The most extensive work on polymers made by this reaction was done by Marvel and has been summarized by Davis and Fettes [36].

Earlier work on such systems used free radical initiators or ultraviolet irradiation. Klotz found poor control with free radical initiators, either giving little curing or else the reaction was so rapid that the pot life was short. On the other hand, a pot life of several days was noted in the absence of catalyst. Ultraviolet radiation was not useful because many of the systems were opaque. With γ-radiation from a cobalt-60 source, the curing was rapid and the products were strong and elastomeric.

Other advantages obtained through the use of radiation included: (1) low shrinkage after gelation, because gelation took
place at a high extent of reaction: The mechanism of the reaction
resembles a conventional polycondensation reaction; (2) low polymerization exotherms: about 14 kcal per equivalent compared to 52
kcal for urethane formation; (3) high efficiency of radiation usage
(G values), owing to the chain mechanism; (4) good thermal stability owing to the absence of catalyst; and (5) in contrast to an
ionic polycondensation, it was possible to obtain an increased pot
life by adding a free radical inhibitor.

Bis- and tris(mercaptoacetates) were prepared for this study
by the acid-catalyzed esterification of diols or triols with
mercaptoacetic acid. The bis- and tris(allyl carbonates) were prepared by the base-catalyzed reaction of diols or triols with ally
carbonates with polypropylene glycol (PPG) or polytetramethylene
ether diol (PRME) were prepared prior to incorporation in the final
urethane resin formulation. In one experiment, for example, a
formulation was prepared from 54.17 parts by weight of PTME bis-
(mercaptoacetate), 43.63 parts PTME bis(allyl carbonate), and 2.20
parts acetyl triallyl citrate. This formulation was irradiated at
a rate of 10^6 rads/h for 0.7, 3, and 22 h. After 0.7 h, there was
some cross-linking but the sample was not completely cured. After
3 h, the sample exhibited good elongation and tensile strength.
Not much difference was noted between the 3 and 22-h samples.

A similar formulation (containing triallyl cyanurate in place
of the citrate) showed a decrease in swelling ratio and sol fraction up to a dose of about 2 Mrads at 80°C. Further irradiation up
to 16.4 Mrads caused no further change in these properties. The
same formulation had a tensile strength of 134 psi after a dose of
1.1 Mrads, compared to 297 psi after 4.5 Mrads. The overall
shrinkage during cure of this formulation was about 1.6%, which is
about half the shrinkage normally observed for conventionally
(thermally) cured elastomeric polyurethanes. Since gelation takes

place at high extents of reaction, the shrinkage leading to internal strains after gelation will generally be low in radiation cures.

A better understanding of the curing reaction was obtained by
investigating the kinetics of the γ-induced reaction between thiols
and olefins. An equimolar mixture of PTME bis(mercaptoacetate) and
PTME bis(allyl carbonate) was irradiated at a dose rate of
1.9×10^5 rads/h. The reaction was followed by potentiometric
titration of unreacted thiol with alcoholic silver nitrate. Second-
order kinetics was observed between 40 and 75% of reaction, corresponding to

$$\frac{-d\,(RSH)}{dt} = k\,(RSH)\,(CH_2{=}CHR') \tag{8-32}$$

which can be integrated when olefin concentration equals thiol
concentration to give

$$\frac{1}{(RSH)} - \frac{1}{(RSH)_0} = kt \tag{8-33}$$

In accordance with Eq. (8-33), a plot of the reciprocal of thiol
molarity versus seconds gave a straight line plot, indicating k
equal to 7.38×10^{-4} liter mole^{-1} sec^{-1}.

Since the overall rate decreased as monomer disappeared at
constant dose rate, the quantum yield decreased with time. Still
it was possible to compute an instantaneous quantum yield Q from

$$Q = -\frac{d\,(RSH)}{rdt} \tag{8-34}$$

where r is the dose rate. Substitution from Eq. (8-32) gives

$$Q = \frac{k\,(RSH)^2}{r} \tag{8-35}$$

Substitution of proper values for k and r gave an initial quantum
yield for thioether linkage formation of 5.7×10^7. Noting that

one quantum equals 1.25 MeV in this case, the yield based on energy
absorbed is G(thioether bonds) = 4560 per 100 eV. When the reaction is 95% complete, G(thioether bonds) is 11.4. At this point
-G (C—H) is of the same order of magnitude [37], so that decomposition becomes a competing reaction.

C. Thermosetting Acrylic Resins

    Nordstrom and Hinsch [38] reacted acrylic copolymers contain-
ing pendant epoxy groups with unsaturated acids to obtain polymers
for coating applications which cross-link when irradiated with a
beam of electrons. All the primary events in polymers such as
radical formation, cross=link formation, and scission have G values
of the order of 1 to 10 [39]. Because of the long kinetic chain
lengths, G values for polymerization of vinyl monomers range from
10^2 to 10^4. These facts led Nordstrom to formulate radiation-
curable coating compositions by dissolving polymers bearing pendant
vinyl unsaturation in readily polymerizable vinyl monomers.

    These resins resemble thermosetting resins in that a three-
dimensional network is formed during cure (gelation). Nordstrom
investigated the effects of double bond concentration and changes
in the chemical structure of the acrylic copolymer backbone on
curing characteristics and coating properties.

    The type of polymer evaluated can be indicated by

$$\{A\!-\!B\!-\!C\}_X$$
$$|$$
$$D$$

where A is methyl methacrylate, B is ethyl acrylate, C is glycidyl
methacrylate, and D is acrylic acid. The backbone consists of a
copolymer of nonfunctional acrylic monomers, A and B, containing
a certain amount of functional monomer C (glycidyl methacrylate),
which bears the glycidyl group as a short side chain containing

reactive epoxy.   The backbone was prepared by the usual free raci-
cal-initiated solution polymerization method.

The desired amount of difunctional monomer, D, generally
acrylic or methacrylic acid, was then added to react with the pendant epoxy side groups and yielding the final structure

$$
\begin{array}{l}
\quad\quad CH_3 \\
\quad\quad | \\
-(CH_2-C-)-copolymer- \\
\quad\quad | \\
\quad\quad C=O \\
\quad\quad | \\
\quad\quad O \\
\quad\quad | \\
\quad\quad CH_2 \\
\quad\quad | \\
\quad\quad CHOH \\
\quad\quad | \\
\quad\quad CH_2O-C-CH=CH_2 \\
\quad\quad\quad\quad || \\
\quad\quad\quad\quad O
\end{array}
$$

I

thus leaving the double bonds of the D monomer to take part in
cross-linking reactions between the backbone copolymer chains.   In
producing structure I, the acid-epoxy reaction was followed by
measuring the residual acid and epoxy content.   Under suitable conditions the reaction attained 90% conversion of epoxy to unsaturated hydroxy ester in several hours.   After adding inhibitor, the
inert solvents were distilled off and the product resin dissolved
in the desired monomer(s) for application to the substrate prior
to curing.

The resin solutions were drawn down on phosphated steel panels
to give a final cured coating thickness of about 1 mil.   The
coatings were cured by irradiation in nitrogen with a 275-kV electron beam at a distance of 3 in., to a total dose of about 15 Mrads.
Film properties were observed to change at low doses and then reach
a plateau region of no further change with additional dosage, in
line with Charlesby and Wycherly's results [40].

The properties of a series of coatings were studied, keeping

the ratio of methyl methacrylate constant while varying unsaturation by changing the amount of glycidyl methacrylate in the backbone. The resins were dissolved at 65% solids in methyl methacrylate, applied to steel panels, and cured with a 15 Mrad dose.
Comparative assessments of film hardness were obtained by measuring
pencil hardness, solvent resistance (a measure of cross-link
density), mandrel bending, and impact strength. Hardness and solvent resistance of cured films were found to increase with increasing cross-link density (unsaturation concentration), whereas the
bending and impact properties decreased.

The effects of varying the "hardness" of the nonreactive backbone were studied. As a measure of backbone hardness, the glass
transition temperature of the uncured polymer, T_g, was computed
from the equation [41]

$$\frac{1}{T_g} = \sum \frac{W_n}{T_{gn}} \tag{8-36}$$

where W_n is the weight fraction of the individual monomers and T_{gn}
is the glass transition of the homopolymer of the nth monomer in
absolute degrees. (The T_g values used were MMA = 105°C, EA = -22°C,
and glycidyl methacrylate = 55°C.) For different MMA/EA ratios,
the computed T_g values ranged from -11° to 28°C. The cross-link
density of all films in this series should have been equal if
saturated portions of the polymers had no effect on cross-linking.
Actually, the resistance to the solvent rub test using methyl
ethyl ketone was nearly the same for all resins in the series. The
pencil hardness of the films in the series increased as the transition temperatures of the backbones increased. Large variations
in flexibility (mandrel bend) were not observed in this series.

The most pronounced difference in this series was in the viscosity of the resin solution at 65% nonvolatiles in methyl methacrylate. Molecular weight determinations showed that these differences were mainly due to changes in molecular weight. The
resin viscosity was known to be an important variable, because it

controlled the level of monomer addition required to reduce the
viscosity adequately for coating application. The final monomer/
polymer ratio, in turn, had a strong influence on rate of cure and
on cured coating properties.

It was found possible to prepare paints from such unsaturated
acrylic resins with most conventional pigments in conventional
dispersing equipment. It was possible to apply such paints with a
variety of commercial application methods, including external mix
air spray, electrostatic spray, roller coat, curtain coat, or silk
screen.

The properties of radiation-cured acrylic paint on plywood
were compared with those of an established heat-cured alkyl paint.
A standard test of 25 cycles in and out of boiling water, 25
cycles in and out of room temperature water, and 10 freeze-thaw cycles was passed by both the radiation-cured acrylic and the commercial alkyd. Two thousand hours exposure to the carbon arc in a
weatherometer produced no change in the radiation-cured acrylic,
but caused severe yellowing of the commercial alkyd.

D. Melamines, Ureas, Phenolics, and Epoxies

Kaetsu *et al.* [42] investigated the radiation curing of urea-
formaldehyde, melamine-formaldehyde, and phenol-formaldehyde prepolymer systems. It was found that such systems could be cured
by the action of ionizing radiation in the presence of chlorine-
containing additives such as methylene dichloride, chloroform,
carbon tetrachloride, trichloroethylene, aluminum trichloride,
sodium chloride, potassium chloride, chloral hydrate, and polyvinyl
chloride.

For examole, 100 parts of formaldehyde and 30 parts of urea
were mixed at pH 7-8 and heated 1 h at 90°C to yield a methylol
urea compound. Formic acid was added to reach pH 4.8, followed by

heating 3 h at 90°C and neutralization with sodium hydroxide to give the prepolymer. Time of cure was taken as the length of time required to produce a hard, brittle resin (gelation).

Such a urea-formaldehyde prepolymer containing 2.4 wt% chloral hydrate cured at 55 h using a dose rate of 3×10^6 R/h, but cured after only 20 h with 9.0% chloral hydrate at the same dose rate. At a dose rate of 10×10^6 R/h, a similar prepolymer containing 17% $AlCl_3$ cured in about 30 h, whereas one containing 17% carbon tetrachloride cured in about 5 h. Curing time was controlled by selection of additive type, concentration, and dose rate. Curing times for phenol-formaldehyde and melamine-formaldehyde systems were not given. The observed cures were attributed to reactive species formed by the radiolysis of the additives.

Cordischi *et al.* [43] made a systematic study of the radiation-induced polymerization of aliphatic and alicyclic epoxy compounds the liquid and solid state. Some alicyclic epoxides and tetrafluoroethylene oxide gave high molecular polymers, whereas other epoxides gave low molecular weight products or no polymer at all. Tetrafluoroethylene oxide polymerized only if irradiated in the solid state. Aliphatic epoxides had no tendency to polymerize. Cyclohexene, cyclopentene, and cycloheptene oxides were polymerized by irradiation or by ionic catalysis.

REFERENCES

1. Harkins, W. D., *The Physical Chemistry of Surface Films*, Chap. 5, Reinhold, New York, 1952.
2. Smith, W. V. and Ewart, R. H., *J. Chem. Phys.*, 16, 592 (1948).
3. Morton, M., Salatiello, P. P., and Landfield, H., *J. Polymer Sci.*, 8, 111 (1952).

4. Bakker, J., thesis, *Kinetics of the Emulsion Polymerization of Styrene*, University of Utrecht, The Netherlands, 1952.

5. Bartholome, E., Gerrens, H., Herbeck, R., and Weitz, H. M., *Z. Elektrochem.*, 60, 334 (1956).

6. Gerrens, H., *Z. Elektrochem.*, 60, 400 (1956).

7. van der Hoff, B. M., *J. Phys. Chem.*, 60, 1250 (1956).

8. Vanderhoff, J. W., Vitkuske, J. F., Bradford, E. B., and Alfrey, T., Jr., *J. Polymer Sci.*, 20, 225 (1956).

9. Ballantine, D. S., Rept. Brookhaven Natl. Lab. T-50 No. 294, p. 18; T-53 No. 317, 7 (1954).

10. Okamura, S., International Conference on Application of Large Radiation Sources in Industry and Especially to Chemical Processes, Warsaw, Sept. 8, 1959.

11. Chapiro, A. and Maeda, N., *J. Chem. Phys.*, 56(2), 230 (1957).

12. Hummel, D., Ley, G., and Schneider, C., *Ado. Chem. Ser.*, 34, 60 (1962).

13. Vanderhoff, J. W., Bradford, E. B., Tarkowski, H. L., and Wilkinson, B. W., *J. Polym. Sci.*, 50, 265 (1961).

14. van der Hoff, B. M. E., *J. Polymer. Sci.*, 33, 487 (1958).

15. Matheson, M. S., Aurer, E. E., Bevilacqua, E. B., and Hart, E. J., *J. Amer. Chem. Soc.*, 73, 1700 (1951).

16. Stannett, V., Gervasi, J. A., Kearney, J. J., and Araki, K., *J. Appl. Polymer Sci.*, 13, 1175 (1969).

17. Patsiga, R. A., Litt, M., and Stannett, V., *J. Phys. Chem.*, 64, 801 (1960).

18. Greene, R. E., Allen, R. S., Ransohoff, J. A., and Woodard, D. G., *Isotopes Radiat. Tech.*, 9, 92 (1971).

19. Lukhovitskii, V. I., Polikarpov, V. V., Lebedeva, A. M., Lagucheva, R. M., and Karpov, V. L., *Vysokomol. Soedin., Ser. A*, 1968, 10(6), 1297.

20. British Pat. 949,191, February, 1964, to Tube Investments (Group Services Limited).

21. Hoffman, A. S. and Smith, D. E., *Mod. Plast.*, 43(10), 111 (1966).

22. Pietsch, G. J., *Ind. Eng. Chem. Prod. Res. Develop.*, 9, 149 (1970).

23. Blin, M. F. and Gaussens, G., in *Large Radiation Sources for Industrial Processes*, Intern. At. At. Energ. Agency, 1969 (STI/PUB/236), p. 499.

24. Ballentine, D. S. and Manowitz, B., USAEC Report BNL-389, Brookhaven National Laboratory, May, 1956.

25. Hoffman, A. S., Jameson, J. T., Salmon, W. A., Smith, D. E., and Trageser, D. A., *Ind. Eng. Chem. Prod. Res. Develop.*, 9 (2), 158 (1970).

26. Burlant, W. and Hinsch, J., *J. Polymer Sci.*, A2, 2135 (1964).

27. Alfrey, T., Bohrer, J., and Mark, H., *Copolymerization*, Interscience, New York, 1952, p. 10.

28. Chapiro, A., *J. Chim. Phys.*, 47, 764 (1950).

29. Vanderbie, W., Spencer, P., and Swanholm, C., *Paint Varn. Prod.*, 59(3), 39 (1969).

30. Hoffman, A. S., *Isotopes Radiat. Tech.*, 9(1), 84 (1971).

31. Miettinen, J. K. and Autio, T., in *Large Radiation Sources for Industrial Processes*, Symposium Proceedings, Munich, 1969, p. 513, Intern. At. Energ. Agency, Vienna, 1969 (STI/PUB/236).

32. Omelchenko, S. I., Videnina, N. G., Matjushova, V. G., Chervetsova, I. N., and Pyankov, G. N., *Ind. Eng. Chem. Prod. Res. Develop.*, 9(2), 143 (1970).

33. Klotz, M. A., Carleton, L. T., Saltonstall, C. W., Kispersky, J. P., Phillips, R. E., Grakauskas, V., and Mishude, E., *Ind. Eng. Chem. Prod. Res. Develop.*, 7(3), 165 (1968).

34. Charlesby, A., Wycherly, V., and Greenwood, T. T., *Proc. Roy. Soc.*, A244, 54 (1958).

35. Posner, T., *Chem. Ber.*, 38, 646 (1905).

36. Davis, F. O. and Fettes, E. M., in *Polyethers* (N. G. Gaylord, ed.), Pt. 3, Interscience, New York, 1962, p. 11.

37. Bovey, F. A., *The Effects of Ionizing Radiation on Natural and Synthetic High Polymers*, Interscience, New York, 1958, p. 43.

38. Nordstrom, J. D. and Hinsch, J. E., *Ind. Eng. Chem. Prod. Res. Develop.*, 9(2), 155 (1970).

39. Charlesby, A., *Atomic Radiation and Polymers*, Chap. 3, Pergamon Press, New York, 1960.

40. Charlesby, A. and Wycherly, J., *J. Appl. Radiat. Isotopes*, 2, 26 (1957).

41. Fox, T. G., *Bull. Amer. Phys. Soc.*, 1, 123 (1956).

42. Kaetsu, I., Hayashi, K., and Okamura, S., *J. Macromol. Sci. Chem.*, A2(6), 1271 (1968).

43. Cordischi, D., Mele, A., and Somogyi, A., *Proc. Tihany Symp. Radiat. Chem., 2nd, Tihany, Hung., 1966*, p. 83-91 (1967).

44. Gordon, M. and McMillan, I., *Makromol. Chem.*, 23, 188 (1957).

Chapter 9

RADIOLYTIC GRAFTING OF MONOMERS
ON POLYMERIC FILMS

I.   INTRODUCTION

While a homopolymer is made up of chains containing monomer
units that are all alike, a *graft* copolymer is composed of chains
containing two or more chemically different types of monomer units.
A graft copolymer, in the more generally accepted sense of the
term, is made up of a backbone chain consisting entirely of monomer
X, attached to one or more side chains composed of monomer Y, as
shown diagrammatically in the following example.

$$\begin{array}{l} \quad\quad\quad\overset{\displaystyle\rceil\!-Y\!-\!Y\!-\!Y\!-\!Y\!-\!Y}{\big|} \\ X\!-\!X\!-\!X\!-\!X\!-\!X\!-\!X\!-\!X\!-\!X\!-\!X\!-\!X \\ \quad\quad\underset{\displaystyle\lfloor\!-Y\!-\!Y\!-\!Y\!-\!Y\!-\!Y}{\big|} \end{array}$$

One of the reasons such graft copolymers are of interest to the
polymer chemist is that a film or fiber of a chain of X's can be
surface grafted with branches of Y's to increase surface water
repellency, dyeability, solvent resistance, light resistance,
mildew resistance, etc. Often the grafting of monomer Y on the
surface of a polymer, or within the polymer to a certain depth,
produces significant changes in physical properties of the polymer.

The synthesis of graft copolymers is based on the production of one or more *active sites* on the backbone of the original polymer. Monomer Y units then add successively at each active site, producing long chains of Y extending out from the backbone as shown in the above drawing. Although not always the case, the active site is most commonly the locus where an unpaired electron exists on a free radical, and to this spot monomer Y is added by an addition mechanism to produce a side chain of Y units.

Radiochemical methods of producing graft copolymers are often easier to employ and control than conventional chemical methods. The claim has been made that owing to the unselective absorption of radiation in matter, it is possible (in principle) to combine any monomer/polymer pair by this technique. The grafting reaction occurs not only at the polymer surface, but in cases where the monomer penetrates the polymer it may occur deep within the interior of the solid polymer.

A variation on the above method is to produce free radicals in the polymer by irradiation with or without the presence of oxygen, to store the polymer at low temperature to preserve the entrapped radicals, and then to react the stored polymer with monomer under different conditions and possibly at a different location. Grafting can also be caused through reactive groupings which can be produced in the polymer by irradiation in the presence of properly selected reagents. For example, Cl or Br groups can be introduced into a polymer by prior irradiation in the presence of chlorine or bromine gas. Various other reactive grouping can be incorporated in a polymer by irradiation in the presence of suitable organic additives.

The direct method of irradiating the polymer in the presence of monomer is perhaps the most efficient method of grafting, because the free radicals are utilized to initiate chains of Y as fast as they are produced. This is the simplest technique and involves a single step. A disadvantage is the fact that some

homopolymerization of Y takes place, initiated possibly by free
radicals formed in the irradiation of monomer Y. Homopolymerization is promoted in cases where monomer Y is present in excess,
which is generally true when this technique is employed.

In the more indirect method, the polymer may be irradiated in
oxygen to produce a variety of free radicals through the oxidation
process. Such free radicals can be preserved within the polymer
by storage under appropriate conditions and then employed later to
initiate the grafting of monomeric side chains. One defect of
this approach is that the prior irradiation in oxygen may cause
some undesirable oxidative changes in the properties of the original polymer. This method is technically more difficult and time
consuming because it takes place in two steps. A possible advantage is that the peroxidized polymer may be sent to a different
geographic location for the final grafting step if desired.

The main problem in the indirect method is that only a portion
of the radicals formed remain trapped after the irradiation. The
efficiency of radical retention can be improved by using the radicals immediately after irradiation or by storing the irradiated
sample at low temperature prior to use. A key advantage of the indirect method is that very little homopolymerization of the added
monomer takes place.

The kinetics of monomer addition on the polymer backbone are
assumed to be similar to those of conventional free radical addition polymerization, subject to certain effects peculiar to the
conditions of grafting, which will be illustrated below. In most
experimental investigations the *grafting ratio* P/P_0 is measured,
that is, the weight ratio of the grafted polymer to the original
polymer. The *percentage of grafting*, $100 (P-P_0)/P_0$, is commonly
reported and evaluated as a function of several controllable
variables. The *grafting efficiency* is defined as the ratio of the
weight of grafted monomer to the total weight of monomer converted
in both grafting and homopolymerization.

II. THEORETICAL CONSIDERATIONS IN GRAFTING

Experimental observations on radiolytic grafting have attracted the interest of a large number of investigators. Some phenomena particularly worth noting include the generation and preserva-
tion of free radicals in the polymer substrate, the effect of
oxygen on radical generation, the dependence of grafting rate on
dose rate, the importance of chain transfer, the promotion of
grafting when termination is hindered, and the limitation of
grafting sometimes caused by the slow diffusion of monomer into
the polymer substrate. These will be discussed in turn.

A. Entrapment of Free Radicals

The state of knowledge is meager concerning the mechanism of
formation of trapped radicals and their subsequent chemistry. The
efficiency of such radicals in the initiation of grafting depends
on the yield of trapped radicals formed in the various polymers
under different irradiation conditions. In most cases the concentration of radicals in the polymer increases as the dose increases,
but reaches a maximum value after a time and no longer increases
with increasing dose. Hence, there is an optimum dose for each
system, and this dose can be determined only by experimentation.

In general, the lifetime of the trapped radicals is much
longer at low temperatures. Oxygen is known to destroy the radicals, and higher radical yields may be obtained if the irradiation
is carried out in the absence of oxygen. The presence and nature
of such free radicals has been studied through the use of ESR
spectroscopy.

B. Effect of Oxygen on Radical Generation

Irradiation of a polymer in oxygen generally results in the production of peroxidized polymer.

$$R\cdot + O_2 \rightarrow RO_2\cdot \qquad\qquad (9\text{-}1)$$

Such peroxidic radicals may remain trapped in the polymer. The
length of time they remain uncombined depends on the physical state
of the polymer, the temperature, the dose rate, and the total dose
in a somewhat complicated manner. A peroxidic radical may disappear through hydrogen abstraction,

$$RO_2\cdot + R'H \rightarrow RO_2H + R'\cdot \qquad\qquad (9\text{-}2)$$

or the two radicals may combine with loss of oxygen

$$RO_2\cdot + RO_2\cdot \rightarrow ROOR + O_2 \qquad\qquad (9\text{-}3)$$

Hydrogen abstraction (9-2) and recombination (9-3) are in competition with each other. When hydrogen abstraction predominates, the
main product is hydroperoxides. When recombination is the main
reaction, the principal product is peroxides [1].

Reaction (9-2) requires a certain energy of activation and
will hence tend to occur at high temperature and in polymers
containing tertiary (labile) hydrogens. Reaction (9-3) would be
expected to predominate at low temperature and in linear polymers
which do not contain tertiary hydrogen.

Important differences between polymers containing peroxides
or hydroperoxides are said to become evident when the polymers are
used to initiate graft copolymerization. The thermal breakdown of

a polymeric peroxide, ROOR, produces two polymeric radicals, RO·, which in the presence of a monomer leads to the production of two grafted branches. Under similar mild heating, a polymeric hydroperoxide, ROOH, will form one grafted branch and one molecule of homopolymer initiated by the HO· radical. The difference can be shown by comparing the grafting reactions initiated by radiation-peroxidized polyethylene (PE) and polypropylene (PP). The PE contains peroxides which are stable at room temperature and retain the ability to initiate graft copolymerization for periods of 1 year or more. Also, when used to initiate graft copolymerization, the PE produces only small amounts of homopolymer. On the other hand, PP which has been irradiated in air initiates grafting at room temperature accompanied by the production of a relatively large amount of homopolymer. It should further be noted that PP is more easily peroxidized by low radiation doses than PE, suggesting that hydroperoxides are formed in PP by means of a chain reaction involving abstraction of tertiary hydrogens [Reaction (9-2)].

During irradiation at higher temperature (60°C), PE may also be peroxidized by a chain reaction resulting in the formation of hydroperoxides [2].

C. Effect of Dose Rate

This discussion can best be divided into two parts: (1) effect of dose and dose rate in the direct grafting method and (2) effect of dose and dose rate in grafting on peroxidized polymer.

In the direct grafting method, the number of branches formed and their length are influenced by the dose and dose rate, respectively. The number of radicals formed (number of branches) is determined directly by the size of the dose and the G_R value, which is the number of radicals formed per 100 eV absorbed. The dose rate, which evidently determines the rate of initiation of polymerization, will therefore affect the kinetic chain length and consequently the length of the grafted branches. The length of the branches also depends on the concentration of monomer, the reaction temperature, and the viscosity of the reaction medium.

In preparing peroxidized polymer for the initiation of grafting, the concentration of polymeric peroxides will increase with radiation dose. At low radiation doses, the concentration of peroxides increases linearly with dose. At very high doses, an equilibrium state may be attained in which the rate of formation of peroxides equals their rate of disappearance.

In view of the fact that the number of peroxidic sites in the polymer determines the number of branches, the number of branches can be regulated by controlling the amount of the preirradiation dose. Furthermore, if the grafting is conducted at a certain temperature, the concentration of peroxides determines the rate of initiation of polymerization and therefore influences the length of the grafted branches. Hence, in the peroxidation method of grafting, the preirradiation dose affects both the number and length of the grafted branches. Furthermore, branch length is influenced by monomer concentration and reaction temperature.

The effect of dose rate on peroxidation yield cannot be foreseen without a detailed determination of reaction mechanism. For example, if the supply of oxygen is limited by its slow rate of diffusion into the polymer, high dose rates may cause exhaustion of oxygen within the polymer and therefore favor other reactions such as cross-linking and degradation.

D. Effect of Chain Transfer

If the monomer or solvent has a high transfer constant, additional homopolymer will be produced through transfer to monomer or solvent as follows:

$$X{-}{\top}{-}X\ +\ AB\ \longrightarrow\ X{-}{\top}{-}X\ +\ A\cdot \qquad (9\text{-}4)$$
$$\underset{Y\cdot}{} \qquad\qquad \underset{YB}{}$$

$$A\cdot\ +\ nY\ \longrightarrow\ AY_n \qquad\qquad (9\text{-}5)$$

Here AB is the monomer or solvent or any added substance which
transfers readily. Reactions (9-4) and (9-5) evidently tend to
decrease the grafting efficiency.

However, if chain transfer occurs mainly toward the polymer,
each polymeric radical will form several grafted branches. This
reaction decreases the length of the grafted branches, but does
not change the grafting yield. In more general terms, when any
monomer is polymerized in the presence of a polymer, chain transfer
to polymer will take place to some extent and this leads to the
formation of some graft copolymer as follows:

$$X{-}{\top}{-}X\ +\ Y_n\cdot\ \longrightarrow\ X{-}{\top}{-}X\ +\ Y_n B$$
$$\underset{B}{} \qquad\qquad\qquad \underset{\cdot}{} \qquad\qquad (9\text{-}6)$$

$$X{-}{\top}{-}X\ +\ nY\ \longrightarrow\ X{-}{\top}{-}X$$
$$\underset{\cdot}{} \qquad\qquad\qquad \underset{Y_n}{} \qquad (9\text{-}7)$$

Such a process can increase the overall grafting efficiency
considerably, especially if the polymer has a high transfer constant with respect to monomer Y. This transfer to polymer can be
favored in some systems by increasing the temperature.

E.   The Termination Step

Many graft copolymerizations take place in a reaction medium
of high viscosity. Because of this high viscosity, the termination

step (radical combination) may become difficult or essentially impossible. Hence many growing chains remain unterminated and remain
"trapped" at the end of the grafting process. Similar behavior is
characteristic of any monomer exhibiting the "gel effect" in its
polymerization.

The action of the termination step in causing possible cross-
linking of the original polymeric substrate should be mentioned.
For example, if the termination of the polymerization of monomer Y
occurs through the combination of two growing chains, the final
graft copolymer will contain molecules in which two chains of
polymer X_n are linked by a chain of polymer Y_m.

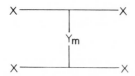

Such a process could lead to a three-dimensional network at
high conversion in which all chains of X are cross-linked by chains
of Y. This would be especially true if Y happens to be a polyfunctional monomer. Such radiation-activated cross-linking of
polymers by polyfunctional monomers has been observed by several
investigators [3-6].

In such reactions the formation of cross-links can be avoided
by the use of a suitable transfer agent. The length of the grafted
branches will depend on the concentration of the transfer agent.
Such a technique will lead to an increased amount of homopolymer
through reactions (9-4) and (9-5).

F. The Diffusion Effect

One very important factor mentioned by most writers in this
field is the possible slowing of the grafting reaction owing to the
slowness of monomer diffusion into the solid polymer. For example,

Hoffman *et al.* [7] carried out radiolytic grafting of styrene on polyethylene film and postulated that monomer concentration within the film may decrease during grafting owing to the inability of the diffusion of monomer into the film to keep pace with its rate of reaction in the film. In line with this postulate, he observed that grafting was faster for thin films than for thick films on the basis of percentage of total film weight. He introduced the idea of a critical film thickness, above which the rate of grafting on a weight percentage basis should fall off rapidly.

Chapiro and Matsumoto [8] found that the rate of grafting of styrene on polyvinyl chloride film at 20°C was entirely controlled by the rate of monomer diffusion into the interior of the film. When the reaction temperature was increased to 60°C, the monomer diffused into the film more rapidly, reaction rate was no longer limited by diffusion rate, and the reaction kinetics became "normal."

The situation that probably exists inside the polymer film can be illustrated by reference to Fig. 9-1, which compares a thin film (A) with a much thicker film (B). Considering film (A) first, the solid curve shows the monomer concentration plotted versus film thickness after a constant rate of grafting is attained (linear portion of grafting versus time curve). The concentration of monomer has not quite reached zero at the center of the film, but it is much lower than at the film surface. The dotted line indicates the equilibrium concentration, C_o, which would be the uniform concentration attained if the film were simply immersed in monomer and allowed to become saturated with monomer in the absence of the grafting reaction.

In the thicker film (B) the monomer reacts and is consumed before it can diffuse to the center of the film. The supply of monomer in a considerable portion of the center of the film is completely exhausted. Under these circumstances, the monomer penetrates to a limited depth, d, and all the grafting takes place within a distance d of the surface of the film.

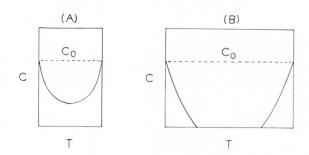

FIG. 9-1. Comparison of monomer concentration curves in thick film (B) and thin film (A), resulting from diffusion effects.

Factors that would cause diffusion to slow down or limit grafting rate are the following.

1. Low intrinsic diffusion constant in the solid polymeric film

2. Low temperature (causing decreased diffusion constant)

3. High intrinsic grafting rate

4. High dose rate (causing high grafting rate)

Factors 3 and 4 cause rapid consumption of monomer within the film, whereas factors 1 and 2 limit the speed with which fresh monomer can get into the film.

On the other hand, the following conditions would maximize the rate of monomer diffusion into the film and minimize the rate of monomer consumption within the film and would, therefore, tend to eliminate the effect of diffusion on grafting rate.

1. High intrinsic diffusion constant in the solid polymeric film

2. High temperature (causing increased diffusion constant)

3. Low intrinsic grafting rate

4. Low dose rate (causing low grafting rate)

III. EXPERIMENTAL TECHNIQUES

Although numerous variations exist in the details of the technique, the two main procedures for grafting monomer on polymer can
be described as (a) irradiation of polymer before or during immersion in liquid monomer and (b) irradiation of polymer before or
during immersion in monomer vapor.

A. Liquid-Phase Monomer

A certain amount of homopolymerization will occur whenever
solid polymer is irradiated while immersed in liquid monomer. Such
homopolymerization takes place both inside of and on the surface of
the polymer. The distribution of homopolymer within the film is
often inhomogeneous [9].

Excessive homopolymerization can be prevented by first irradiating the film (in air or vacuum) followed by immersion of the film
in liquid monomer. When the polymer is irradiated at low temperature in the presence of oxygen, peroxidic derivatives are formed
which later can be decomposed at high temperature when the polymer
is immersed in liquid monomer, thus causing the grafting reaction
to take place. The grafting reaction will also take place when the
polymer is irradiated at low temperature in the presence or absence
of oxygen, and the monomer is added later at the same low temperature. The rate of grafting is influenced by the temperature at
which irradiation is done and also by the temperature at which
monomer is added.

It is often found that there is a fundamental difference in
behavior of films irradiated in air and in vacuum. For example,
when polyethylene is irradiated in air it will form peroxides,
whereas the radicals produced by irradiation in a vacuum will either
cause cross-linking or become trapped in the unreacted state. Even

in the presence of air some trapped free radicals are formed.

The experimental procedure for immersion of film in liquid monomer followed by irradiation is relatively simple. After the purification steps, the monomer is outgassed in a vacuum system to remove oxygen which acts as an inhibitor for most polymerization reactions. The outgassing is accomplished by several successive cycles of freezing the monomer, pumping down, and thawing the monomer again. In the last step, the monomer is distilled into a small capsule or tube containing the polymeric film, and the capsule is hermetically sealed off from the vacuum system and then subjected to radiation from a suitable source.

When the film is irradiated prior to immersion in monomer, a more elaborate experimental procedure is required. Using the apparatus shown in Fig. 9-2, distilled monomer is first added to tube I which is connected to a high vacuum line at A. The monomer is outgassed by repeated freezing and thawing, then hermetically

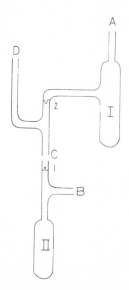

FIG. 9-2. Apparatus for monomer distillation and purification.

sealed off at A while freezing down in tube I under vacuum. The
tube containing the outgassed monomer is then resealed to the
vacuum line at D. In the meantime, tube II containing the film has
been pumped down, sealed off at B, and irradiated. The tube containing the irradiated film, tube II, is then sealed onto the above
apparatus at C after inserting a small piece of soft iron bar
between the break-seal points 1 and 2. The entire system is again
evacuated to about 10^{-5} mm Hg and removed from the vacuum line.
The two break seals 1 and 2 are then broken by manipulating the
soft iron bar with a magnet, and the monomer is admitted to tube II
containing the irradiated film. Tube II is then placed in a suitable temperature control device while carrying out the grafting
reaction.

B. Vapor-Phase Monomer

Much work had been done on liquid-phase grafting before the
monomer vapor grafting technique was developed. It has been observed that a higher degree of grafting is sometimes obtainable by
grafting with monomer vapor rather than liquid monomer [10,19]. The
almost complete lack of homopolymer formation minimizes the after-
treatment for purification of the graft polymer.

One method of employing monomer vapor is the direct grafting
technique, whereby polymer film or fiber is irradiated while surrounded by monomer vapor. Takamatsu and Shinohara [11] have used
this method to study grafting on polyethylene, polyvinyl chloride,
and polytetrafluoroethylene.

The indirect method involving preirradiation of polypropylene
fiber has been employed by Hayakawa and Kawase [10]. Polypropylene
fibers were irradiated at dry ice temperature in air, under which
conditions it has been shown that a considerable concentration of
peroxidic radicals is produced and trapped within the polymer.
These radicals can be employed later for efficient initiation of

graft copolymerization at room temperature. Such radicals are
known to be quite stable at dry ice temperature.

IV. GRAFTING RESULTS FOR TYPICAL MONOMER/POLYMER SYSTEMS

A. Styrene on Polyvinyl Chloride

Takamatsu and Shinohara [11] investigated the direct vapor-
phase grafting of styrene on polyvinyl chloride film. A certain
amount of styrene was placed in the lower part of a glass ampule
and PVC film in its upper part. The ampule was connected to a
vacuum line and outgassed by a freeze-thaw cycle and then hermetically sealed. The polymer film in the upper part of the ampue was
then irradiated by γ-rays from a cobalt-60 source, while the lower
part of the ampule containing styrene was shielded form the rays by
lead. Grafting experiments with liquid monomer (direct method)
were also carried out.

The degree of grafting was found to increase linearly with
irradiation time for both monomer vapor and liquid monomer. The
rate of grafting for a given dose rate was always larger for monomer vapor than for liquid monomer.

For all runs, the rate of grafting defined as increase in
degree of grafting per unit time was calculated. Logarithm of
grafting rate cersus dosage rate gave linear plots for both monomer
vapor and liquid. Also, grafting rate was proportional to the
square root of dose rate over a considerable range of dose rate
values for both monomer vapor and liquid.

Studies of the effect of monomer concentration gave results of
considerable interest. It should first be noted that when PVC is
placed in styrene vapor or liquid styrene the monomer diffuses into
the interior of the polymer. When PVC is placed in liquid styrene,

only about an hour is required for the monomer to reach its equilibrium concentration in the polymer of 5.2 moles/liter at 20°C.
When PVC is placed in styrene vapor, the internal concentration of
styrene rises very slowly and after 100 hours has still not reached
its equilibrium value.

Polyvinyl chloride films containing styrene at various concentrations were prepared by keeping PVC films in styrene vapor at
room temperature for various periods of time. Irradiation of the
films was then carried out for periods of 1 to 3 h at a dose rate
of 4.8×10^4 R/h. Separate curves of degree of grafting versus
styrene concentration were plotted for irradiation times of 1.0,
1.5, 2.0, and 3.0 h. Each curve passed through a maximum at some
styrene concentration between about 3 and 4 moles/liter of PVC.
Hence the rate of grafting is small at low styrene concentration,
becomes larger as the concentration increases, and again becomes
smaller at higher concentrations.

In the grafting of liquid styrene on PVC, the concentration of
monomer in polymer reaches its equilibrium value of 5.2 moles/liter
during the time the sample is being prepared for irradiation.
While the grafting reaction is taking place, monomer is being consumed within the film, but fresh monomer is rapidly diffusing into
the film from the outside. The monomer diffuses into the film at a
rate of 1.4×10^{-4} moles/liter/sec, as shown in separate experiments involving immersion of film samples in liquid styrene. The
rate of monomer consumption within the film owing to grafting is
about 2.5×10^{-5} moles/liter/sec at a dose rate of 5×10^4 R/h.
Hence the monomer diffuses into the film much more rapidly than it
is consumed, resulting in a high and constant concentration of
monomer in the film during the grafting reaction.

Under the conditions employed by Takamatsu, the irradiation
usually started about 20 h after the film had been positioned in
the monomer vapor. Hence, some question arises about the concentration of monomer in polymer at the start of the irradiation. In
any case, at a low dose rate of $(5-10) \times 10^2$ R/h the monomer is

used up at a rate of (3-4) x 10^{-5} moles/liter/sec, while fresh
monomer diffuses into the film at a rate of 2 x 10^{-5} moles/liter/
sec. Thus at low dose rate, the monomer is supplied almost as fast
as it is used up and hence should remain at an almost constant concentration within the film. At a high dose rate exceeding 10^3 R/h,
the monomer is consumed within the film more rapidly than it can
diffuse in, and the monomer concentration in the film will drop in
some undetermined way. (It was not established whether the monomer
concentration continued to drop throughout the grafting reaction or
became relatively constant at some low level.)

When the inital rate of grafting is plotted against styrene
concentration, using data from both vapor-phase and liquid-phase
grafting, it is found that the maximum rate of grafting occurs at
a styrene concentration of 3.2 moles/liter and is smaller for both
lower and higher concentrations. In order to explain the decrease
in rate above 3.2 moles/liter, the author assumes that the rate
constant for the termination reaction increases with styrene concentration. Termination involves the recombination of polymeric
free radicals, and there is some plausibility to the theory that
the styrene plasticizes the PVC and increases the molecular motion
of the polymer so that radical recombination is expedited. The
plasticizing action of styrene in PVC was demonstrated by measuring
the elastic modulus of PVC containing various concentrations of
styrene.

B. Butadiene on Polyethylene

The use of the indirect grafting method is illustrated by the
work of Furuhashi et al. [12] on the grafting of butadiene on polyethylene film. Two types of experimental technique were used in
this work: (1) the grafting of butadiene on polyethylene (PE) film
preirradiated in air and (2) the grafting of butadiene on PE preirradiated in vacuo. Several workers had already reported on the
grafting of various monomer on preirradiated polyethylene [13-15].

In a previous investigation, Ohnishi *et al*. [16] had studied
the behavior of free radicals in PE irradiated *in vacuo* which was
later exposed to air.   Ohnishi found that an allyl radical was
formed during irradiation of PE *in vacuo*, and that a peroxy radical
was formed immediately on the introduction of air from a fraction
of the allyl radicals.   The objective of the work of Furuhashi *et
al*. [12] was to discover which radical (trapped nonoxidized radical
or trapped peroxy radical) initiates low temperature grafting on
polyethylene preirradiated in air.   This was to be done by comparing the grafting on PE preirradiated in air with that on PE preirradiated *in vacuo*.

In carrying out the investigation, Furuhashi studied both high
and low density PE films of several different thicknesses.   Irradiations of film (in air or *in vacuo*) were carried out using γ-rays
from a cobalt-60 source at dose rates ranging from 1.05×10^4 to
1.4×10^5 rads/h.   Butadiene was introduced to the film ampules
prior to irradiation, and the grafted film products were purified
by the usual extraction techniques.

Plots of percent grafting versus time for high density polyethylene (PEH) preirradiated in air showed a high initial rate which
gradually decreased almost to zero after a day or so.   Similar results and similar grafting rates were obtained for PEH preirradiated
in vacuo.   Arrhenius plots for a temperature range from 0° to 50°C
indicated an activation energy of 9.7 kcal/mole for grafting on PEH
whether preirradiation was in air or *in vacuo*.   Similar results were
obtained for low density polyethylene (PEL), and an energy of activation of 8.4 kcal/mole was obtained.   However, with PEL the time
to reach maximum grafting and the saturation value of the percent
grafting were small compared to PEH.

For preirradiation in air and *in vacuo*, an increase in temperature of irradiation produced a decrease in percent of grafting.
The effect is more pronounced for preirradiation in air.   Increased
temperature may remove free radicals through recombination.   In air,
increased temperature will favor combination of graft-active

radicals with oxygen to form a peroxy radical which decays quickly
(20 min), was shown by Ohnishi *et al.* [16].

A plot of percent grafting versus PE density gave a straight
line for film irradiated both in air and *in vacuo*. The density in
the experiments ranged from 0.916 to 0.957. The author also found
it impossible to graft butadiene on amorphous ethylene/propylene
copolymer under the same experimental conditions. From this the
author concluded that crystallinity of the PE is one of the most
important factors affecting grafting of butadiene by the preirradiation technique. However, it was pointed out that the ratio
of concentrations of trapped radicals in irradiated PEH and PEL is
5/3 [17], and this value is approximately equal to the ratio of
their crystallinities (90/55). This does not explain the ratio
of grafting on high density PE to that on low density PE, namely,
7/1, observed by Furuhashi. Hence, Furuhashi concluded that the
effect of crystallinity on percent grafting is due not only to the
trapping of free radicals but also to other factors, such as the
gel effect.

Percent grafting after a fixed time (24 h) at a fixed temperature (50°C) was measured for a large number of films after
irradiation and storage in order to determine the effects of irradiation conditions, storage conditions, and film type. In these
studies, the measured percent grafting was used as a sensitive
indicator of the number of graft-active radicals which remained
in each film. For example, several PEH films of various thicknesses were preirradiated in air at 15°C, then stored in air at
50°C prior to grafting. It was observed that the percent grafting
decayed more rapidly with storage time for thin films than for
thick films (Fig. 9-3). This indicates that graft-active radicals
are being destroyed by diffusion of oxygen into the film, and
that oxygen diffuses more rapidly into thin films as expected.

Another group of PEH films was preirradiated *in vacuo*, and then
stored at 50°C for various periods of time *in vacuo* and in air prior

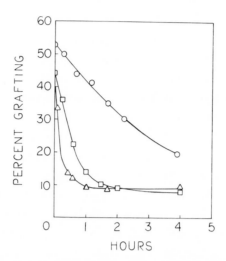

FIG. 9-3. Decay curves of per cent grafting of butadiene, polymerized at 50°C for 24 h, onto high density polyethylene films of various thicknesses preirradiated in air at 15°C at 1.4×10^5 rad/h with storage time at 50°C in air: (O) 0.5 mm; (□) 0.1 mm; (Δ) 0.05 mm. [From A. Furuhashi, *J. Appl. Polym. Sci.*, <u>12</u>, 2209 (1968); John Wiley and Sons, New York.)

to grafting. The decay of percent grafting with storage time was much smaller for storage *in vacuo* than in air, regardless of film thickness. Again for storage in air, the decay was more rapid for thin than for thick film, again implying faster diffusion of oxygen into the thin film. An important point is that when films are stored in air at 50°C, there is little difference in rate of decay for films preirradiated in air or *in vacuo*.

These results indicate that the graft-active radical is in-activated essentially by oxygen diffusing into the film and that the presence of oxygen during preirradiation in air has little effect on the grafting reaction.

When films of different densities preirradiated *in vacuo* at 15°C are stored in air at 50°C, the relative amount of grafting produced decays most rapidly for film of 0.940 density, with intermediate speed for film of 0.949 density, and slowest of all for film of 0.957 density. Where similar films are stored *in vacuo* at 50°C, the rate of decay is hardly affected by film density, and the rate of decay is much less at all densities than where the films are stored in air at 50°C. These data correlate well with the fact that oxygen diffuses more rapidly in low density film than in high density film. Similar results are obtained for films of different densities preirradiated in air at 15°C and stored in air at 50°C.

From the above data showing extensive similarities in behavior of film preirradiated in air and *in vacuo*, the author concludes that the grafting of butadiene on PE film preirradiated in air (or *in vacuo*) is initiated by trapped nonoxidized radical (allyl radical) in the polyethylene.

The author also drew some conclusions about the region of PE in which the graft-active radicals are trapped. It was found that the percent grafting produced on high density PE preirradiated in air at 15°C hardly decreases at all after 24 h of storage in air below 15°C. Hence, the graft-active radical is hardly oxygenated when stored in air for 24 h below 15°C. Furthermore, as mentioned above, the percent grafting is proportional to the density of the polyethylene. Also, in amorphous ethylene/propylene copolymer the grafting reaction does not take place under the same experimental conditions. All these facts constitute evidence that the trapping of radicals takes place and the grafting reaction occurs in the crystalline portion of the polyethylene.

This conclusion correlates well with the electron-spin resonance data reported by Ohnishi *et al.* [16]. Ohnishi found evidence that the oxidation of the allyl radical takes place in several steps. In the first step, immediately after the introduction of air, part of the allyl radical reacts to form a peroxide

radical, which decays rapidly to about one-half its original concentration in about 20 min. In the second step (from 3 to 130 h),
some oxygenated radicals of another type appeared. In the third
step (over 130 h), the allyl radical decayed almost entirely to
form oxidation products, and conjugated polyenyl radicals remained.

Furuhashi has postulated that the portion of allyl radicals
which decays in the first step is produced in the amorphous region
(where oxygen can diffuse in most rapidly). Other allyl radicals,
which decay in the third step, are assumed to be produced in the
crystalline part of the PE which is less accessible to oxygen and
are able to initiate the grafting reaction after 24 h of storage
in air below 15°C.

Furuhashi notes that percent grafting decreases substantially
after storage in air at 30° or 50°C, following irradiation in air
at 15°C. Also, it decreases substantially even after storage at
50°C *in vacuo*. Hence, the graft-active radical is deactivated by
oxygen when storage takes place in air, but also apparently disappears when stored *in vacuo*, even at a temperature below the
melting point of the polyethylene crystallites. For preirradiation
in vacuo, the decline in percent grafting with increased irradiation
temperature suggests that the graft-active radicals recombine more
readily as the temperature increases. Since a radical cannot
migrate along a chain, the polymer chain must have a certain amount
of mobility in order to enable the allyl radicals to recombine.

However, independent evidence indicates that the crystallite
of PE is not destroyed by brafting. This would mean that the
grafting may take place on the surface of the crystallite, that is,
in the semicrystalline region. Evidence is lacking on the behavior
of radicals trapped in the inner part of the crystallite.
Furuhashi suggests that alkyl radicals may possibly migrate from
the inner part of the crystalite to the region of grafting and
become converted to allyl radicals.

C. Styrene on Polymethylpentene

Useful information on the grafting mechanism can often be obtained by comparing the grafting behavior of structurally or chemically similar polymer types. For example, in recent research by Wilson [69], grafting rates of styrene on polymethylpentene, polypropylene, and polyethylene were measured at room temperature and at the same dosage rate and found to be in the order polyethylene > polypropylene > polymethylpentene. All three of these are hydrocarbon polymers, and such comparative information helps to determine how much effect degree of chain branching has on rate of grafting. Polypropylene contains one branch (and one tertiary hydrogen) for every three carbon atoms, and the same ratio exists for polymethylpentene if both main and side chains are included.

Polyethylene
$$\left[-CH_2-CH_2- \right]_n$$

Polypropylene
$$\left[\begin{array}{c} -CH_2-CH- \\ | \\ CH_3 \end{array} \right]_n$$

Polymethylpentene
$$\left[\begin{array}{c} -CH-CH_2- \\ | \\ CH_2 \\ | \\ HC(CH_3)_2 \end{array} \right]_n$$

Polyethylene contains very minor amounts of chain branching as revealed by infrared studies.

With regard to the hypothesis that branching promotes grafting the comparatively high rate for polyethylene effectively negates this hypothesis. While branching may play some part in determining the rate of grafting, it must not be the dominant factor, at least for this series of polymers.

Such rate comparisons between different polymers also lead to
other conclusions. The generally accepted expression for the rate
of radiation-induced polymerization by a free radical mechanism
can be written [11]

$$\text{Rate} = k_p k_t^{-\frac{1}{2}} R_i^{\frac{1}{2}}[M] \qquad\qquad (9\text{-}8)$$

where k_p and k_t are rate constants for propagation and termination,
respectively, and [M] equals monomer concentration. This equation
was derived from liquid-phase polymerization and can only be applied
with several precautions to the grafting reaction in a solid
(polymeric) phase.

Equation (9-8) suggests that rate of grafting should increase
as monomer concentration [M] increases inside the polymer film.
However, saturation styrene absorption measurements showed absorbed
styrene concentration to be in the following order for the three
polymers [69]: polymethylpentene > polypropylene > polyethylene,
which is just the reverse of the order of the grafting rates given
above. This finding suggests that styrene may have a plasticizing
action, increasing the value of k_t and thus decreasing the overall
rate of grafting.

In support of this theory, it will be recalled that Takamutsu
and Shinohara [11] showed that an increase in styrene concentration
caused an increase in k_t for the styrene/polyvinyl chloride
grafting system.

A comparative study was made of the grafting behavior of
styrene, methyl methacrylate, and acrylonitrile on polymethylpentene. Styrene grafted readily, and the amount grafted versus
time gave a linear plot. Methyl methacrylate started off slowly
and rapidly increased in grafting rate, probably owing to the gel
effect often noted in methyl methacrylate polymerization. Acrylonitrile did not graft at all, and it was observed in independent
experiments that polymethypentene films did not swell where immersed in acrylonitrile. Hence, comparison of monomer behavior

gave a clue to the dependence of grafting on swelling for this particular polymer. (A similar relation has been observed for most monomer/polymer grafting systems.)

In the grafting of styrene on polymethylpentene, the swelling of the film increased progressively as the grafting proceeded, and a graphical plot of percent grafting versus uncorrected swelling was found to be linear (grafting proportional to swelling). A similar plot for polyethylene was also linear, but at all values for percent grafting the percent swelling for polymethylpentene was greater than for polyethylene, as would be expected in view of the greater styrene absorption for polymethylpentene than for polyethylene.

Polymethylpentene showed uncorrected swelling of about 200% after 45 h of irradiation in styrene. This is not surprising in view of the changed character of the grafted polymer. The increased percentage of grafted styrene branches in a polymer makes it capable of absorbing a much higher percentage of liquid styrene. A similar effect was observed by Kawase and Hayakawa [18], who studied the saturation absorption of several alkyl methacrylates by polypropylene. They noted that polypropylene absorbed several times as much of a particular alkyl methacrylate after radiolytic grafting of the alkyl methacrylate on the polypropylene as it did in the ungrafted state.

The 200% swelling of polymethylpentene in styrene would correspond to a large increase in styrene concentration [M] in the polymethylpentene film. According to Eq. (9-8), this should cause a great increase in grafting rate, but the grafting rate remained constant (linear plot of grafting versus time). In order to account for the constancy in rate, some other factor in Eq. (9-8) must also be changing to balance the effect of the increase in [M]. It is reasonable to postulate that the plasticizing action of increased monomer concentration produces an increase in the rate of polymeric radical recombination, k_t, thus balancing the effect of increased [M] and preserving rate constancy.

Grafting rates in the styrene/polymethylpentene system were measured at several different temperatures, and essentially no change in rate was found over a range from 23° to 85°C. The lack of rate variation indicates a zero energy of activation for the grafting of styrene on polymethylpentene in this temperature range.

Ballantine et al. [9] studied the grafting of styrene on polyethylene in the range from 0° to 53°C, using polyethylene films that had been irradiated at 30°C in vacuo. The initial grafting rate in that range was essentially independent of temperature, but the longer grafting runs showed greater amounts of styrene grafted at elevated temperatures. Ballantine stated that a change in temperature influences grafting rate by its effect on three parameters: (1) rate of styrene diffusion into the film, (2) propagation rate, and (3) termination rate involving radical recombination, which is a diffusion-controlled process. A major factor preventing higher grafting rates as temperature increases may be the greater free radical mobility and the resulting increase in termination rate.

In the case of styrene/polymethylpentene [69], the possibility of some cationic polymerization cannot be ruled out. The monomer was dried by storage over anhydrous calcium sulfate, which is not generally considered the most rigorous method of drying. However, the monomer may have been dried more thoroughly than expected, and Ueno et al. [74] have shown that radiolytic styrene polymerization is predominantly cationic under very dry conditions. The bulk polymerization of styrene dried with calcium hydride was studied from -20° to 70°C by Ueno, and the rate showed a maximum at about 30°C, the height of the maximum varying with the batch of styrene used. Earlier results [75] on styrene dried with sodium alloy showed a negligible activation energy in the range from -20° to 80°C.

Metz [76] also reported that the radiolytic bulk polymerization of very dry styrene takes place by predominantly ionic rather than free radical mechanisms. Metz resorted to extensive baking of the

glass apparatus and rigorous drying treatments to produce
"extremely dry" styrene.

A mathematical analysis has been made [118] of the interaction
of diffusion and reaction kinetics during the grafting of styrene
on polymethylpentene film. While polymerization rate is conventionally taken as proportional to monomer concentration [Eq. (9-8)],
it was noted above that this effect may be balanced by the plasticizing action of increased monomer concentration which increases
k_t and reduces the overall polymerization rate. If it is assumed
that grafting rate is approximately independent of [M], the relation between styrene diffusion rate and grafting rate can be
expressed [118] as follows:

$$\frac{\partial C}{\partial t} = D \frac{\partial^2 C}{\partial X^2} - k \qquad (9-9)$$

where C is the concentration of styrene in the film, k is the rate
of styrene consumption in the film, X is the coordinate normal to
the film surface, and t is time. Equation (9-9) has been solved
[119] to give a rather lengthy expression, which reduces to the
following simple form at large values of t representing the
attainment of a steady state:

$$C = C_0 + \frac{k(X^2 - X1)}{2D} \qquad (9-10)$$

where C_0 equals the saturation concentration of styrene in the film
and 1 equals film thickness.

In Ref. [118], a "critical film thickness" was defined as that
thickness above which the grafting rate per unit surface area becomes constant (or decreases) as film thickness continues to increase. A more quantitative definition is that thickness for which
the steady-state monomer concentration becomes zero in the center
of the film, but only in the center of the film. Hence, the
critical film thickness can be computed by setting C = 0 and

$X/1 = 0.5$ in Eq. (9-10), giving

$$1^2 = 8DC_0/k \tag{9-11}$$

Experimental measurements at 25°C indicated

$$D = 3.5 \times 10^{-8} \text{ cm}^2 \text{ sec}^{-1} \tag{9-12}$$

$$C_0 = 1.88 \text{ moles/kg} \tag{9-13}$$

$$k = 7.53 \times 10^{-4} \text{ moles/kg/sec} \tag{9-14}$$

Substitution of these values in Eq. (9-11) yields

$$1^2 = \frac{(8)(3.5 \times 10^{-8})(1.88)}{(7.53 \times 10^{-4})} \tag{9-15}$$

$$1 = \text{critical thickness} = 0.0265 \text{ cm} \tag{9-16}$$

This may be compared with the experimentally observed critical thickness of 0.020 cm [118]. The agreement is good, especially in view of the fact that k, C_0, and D were determined in separate and independent experiments.

For very thick films, there is no monomer in the center portion of the film after the steady state is attained, because the monomer is consumed before it can reach the center of the film. Only the surface layer of the film would contain monomer to a certain depth (penetration depth). A plot of concentration versus film thickness, as in Fig. 9-1, would show C = 0 at a certain depth of penetration, p, into the film. Solving for p by setting C = 0 in Eq. (9-10) shows that

$$p = \frac{k1 - (k^2 1^2 - 8kC_0 D)^{\frac{1}{2}}}{2k} = \frac{1 - (1^2 - h^2)^{\frac{1}{2}}}{2} \tag{9-17}$$

where the symbol h is used to indicate the critical thicknes..

Since the grafting rate is proportional to the penetration depth, Eq. (9-17) indicates that the grafting rate per square centimeter decreases as the film thickness increases above critical thickness. This means that the grafting rate per square centimeter should go through a maximum at the critical thickness. The experimental results for the styrene/PMP system showed that the grafting rate increased until the critical thickness was reached, but then it leveled out and did not decline above the critical thickness [118]. Some of the reasons that the postulated diffusion model does not agree completely with the experimental facts include: (1) assumption of a constant value for D, whereas D is known to vary as a function of monomer concentration, (2) the large amount of film swelling caused by monomer absorption and not taken into account by the theory, and (3) the fact that the assumption of a constant grafting rate k may be an oversimplification. Since the correctness of these approximations will vary from one monomer/film system to another, the agreement between theory and observation will probably vary from system to system.

D. Grafting on Miscellaneous Materials

Attempts have been made to carry out radiolytic grafting of vinyl monomers on a wide variety of miscellaneous materials, including metals, vulcanized rubber, glass fibers, clays, ionic salts, inorganic oxides, and graphite. It has been postulated that when an inorganic material and a polymer come into intimate contact, their chemical structures and surface geometrics determine what sort of bonds (if any) will be produced between the two.

Any weak bonding that results may be less strong than actual grafting, unless the surfaces bear atomic groups that can react with each other. In that case, organic bonds will be produced between the surfaces. Reactions of this sort may involve the oxide surface of metals or the sulfur present in vulcanized rubber.

Treatments capable of producing reactive groups such as surface oxidation of metals or oxidative production of carbonyl groups may promote chemical bonding. The known improved adhesion between metal and a polymer caused by irradiation may result from the production of metal-oxygen-polymer bonds betwen the metal oxide surface and the polymer.

Pretreatment of surface is sometimes employed to introduce mutually reactive chemical groupings. This seems to occur when silanes are used to treat glass fibers used in reinforced plastic laminates. The resulting polymer insolubilization and data from infrared provide evidence for the formation of covalent bonding between the polymer and the glass fiber. Solubility data indicated a more definite presence of grafting when the treated glass fiber was added during polymerization than when mixed with the already polymerized resin. Under some conditions, however, the glass fiber can promote polymerization without the occurrence of grafting.

An investigation of some interest involves the use of $TiCl_4$ for glass fiber pretreatment. When no $TiCl_4$ was present, polymerization of vinyl monomers containing glass resulted in the formation of soluble polymers. However, when glass fibers carrying adsorbed $TiCl_4$ were present in the monomer, production of an insoluble grafted polymer was observed. The ESR spectra of $TiCl_4$/ glass/monomer systems indicated the formation of paramagnetic complexes on the glass fiber surface. Such complexes acted as centers for the grafting of the monomer. Photomicrographs revealed the grafts as isolated buds on the fibers.

In the investigation of polymer/filler systems, all degrees of interaction may be found ranging from physical adsorption to covalent grafting. The degree of interaction depends on the presence of potentially reactive groups as well as surface morphology and geometry and is strongly affected by the presence of surface discontinuities and irregularities. Such interaction may cause reduced mobility of the adsorbed layer, similar to the effect of cross-linking.

Changes in surface chemistry or morphology may be induced
deliberately to promote grafting. For example, freshly formed
surfaces have been prepared on inorganic materials by mechanical
attrition methods, including subjection to ultrasonic vibration.
Another method is to irradiate the material with ionizing radiation.

Electron-spin-resonance spectra can be used to follow surface
changes resulting from such treatments. Inorganic surfaces may
form centers of free radical character or electron-donor capability
that are able to initiate polymerization. Grafting on the inorganic substance may occur if attrition takes place in the presence of a vinyl monomer. With metals it is postulated that polymerization may result from direct transfer of an electron from the
metal to the monomer when the escape energy of the electron is
exceeded.

Mechanical attrition may break covalent bonds in SiO_2, TiO_2,
graphite, and other inorganics and in oxide films on metal surfaces,
resulting in the production of radicals that can initiate polymerization. Vinyl polymer chains may become attached to carbon
black without any prior surface treatment. Mechanical dispersion
of a polyethylene/carbon black mixture resulted in formation of a
graft copolymer as demonstrated by electron microscope inspection
and x-ray analysis. Polymer may be produced on the freshly formed
surfaces on sliding metals, but may not involve grafting to the
surface.

The other method of changing surface morphology is irradiation
with ionizing radiation. This technique may be accompanied by a
chemical surface pretreatment and/or mechanical grinding. Numerous
studies have shown that surface defects result from the irradiation
of metals and inorganic materials. It has been postulated that the
increased rate of radiation-induced polymerization in the presence
of inorganic additives is due to the fact that the energetics of
radiolysis of the monomer favor the adsorbed state rather than the
liquid state. The radicals formed by radiolysis then initiate

polymerization. There are two possible results: if the bond with the carbon is sufficiently strong, grafting to the inorganic substrate results; and if the bond is weak, a polymer molecule detaches from the surface to form homopolymer. These postulates are in agreement with the experimental observations.

Seitz [70] studied the effect of irradiation on metals and observed atomic displacements and the production of interstitial atoms and vacancies. Similar effects were produced in graphite. Kazanskii and co-workers [71-73] investigated the ESR spectra of organic radicals adsorbed on silica gel. In one instance, ethane was adsorbed on silica gel at room temperature, cooled to -196°C, and irradiated at the latter temperature with γ-rays from cobalt-60. Electron-spin-resonance spectral studies showed ethyl free radicals on the surface of the gel that were quite stable, quite loosely bonded to the surface, and probably capable of taking part in heterogeneous catalysis.

Vlasov [77] studied the production of a semiconducting polymer on a glass fiber base. Gas-phase polymerization was caused by irradiating phenylacetylene or acrylonitrile on glass fiber. The apparatus was made of two connecting bulbs, joined at one end to a vacuum pump [78]. Liquid monomer was introduced into one of the bulbs, which was partly immersed in a cooling bath. The glass fibers were put in the other bulb which was held at a controlled temperature and fitted with an aluminum-foil window for irradiation. Polymerization was carried out after evacuating the system by heating each bulb to a selected temperature. Polymer up to 20% of fiber weight was found deposited on the fiber in a process referred to as grafting, although no proof of grafting was presented.

The same apparatus was employed in a study of Tsetlin [79],who stated that grafting of methyl methacrylate, acrylonitrile, and vinyl chloride was obtained on carbon black, silica gel, and magnesium oxide by irradiation with β-rays. Methyl methacrylate was found to polymerize on the surface of BeO but not on ZnO. It was

postulated that the polymerization took place by a free radical
mechanism and was initiated by oxygen-ion radicals formed at the
inorganic surface.

Tsetlin [80] also irradiated MgO and SiO_2 with x-rays in the
presence of gaseous monomers and produced polymers insoluble in
selected solvents. A soluble polymer was produced by dissolving
the inorganic substrate in hydrogen fluoride, providing strong
evidence that actual grafting had taken place.

Polystyrene grafted on silica gel and sulfonated produced
materials having characteristics of both ion-exchange resins and
mineral solvents. Such materials undergo both ionic and molecular
sorption. Related products were made by grafting acrylic acid and
styrene on silica gel followed by sulfonation [81].

Morosov et al. [82] studied reaction rates of grafting of
acrylonitrile on glass fibers, using x-ray initiation. Graphical
plots of initial polymerization rate versus monomer vapor pressure
showed an upper limiting rate at a vapor pressure at which the
amount of adsorbed monomer reached an upper limit. Hence, the rate
of polymerization is strongly dependent on the amount of adsorbed
monomer. With porous and nonporous fibers the rate was proportional
to the first power and square root of the dose rate, respectively.
Evidence was presented to show that the polymerization takes place
by a free radical mechanism.

Berezkin [83] studied the kinetics of the gas-phase polymerization of acrylonitrile on a mineral powder, employing cobalt-60
as a γ-ray source. The mineral powder was a brick material which
was irradiated in an apparatus of the type employed by Vlasov [78].
The amount of graft polymer formed was measured for various irradiation times, while holding dose rate and monomer vapor pressure
constant.

Other investigators have considered it more efficient to use
liquid-phase monomers in grafting on inorganic fillers [84]. For
example, inorganic fillers such as kaolin, glass fiber, silica gel,

and asbestos were immersed in styrene or methyl methacrylate under nitrogen atmospheres and subjected to γ-radiation from cobalt-60. The refractive index of the monomer solution was used to estimate the amount of homopolymer formed. The amount of grafted polymer was determined from the loss of weight on calcination of the dried filler after thorough extraction with a suitable solvent. Silica was more effective than kaolin or asbestos in increasing the grafting rate. It was postulated that polystyrene was held to the filler surfaces by chemical bonding (grafting) while polymethyl methacrylate was held partly by hydrogen bonding as well as by grafting. The yields of homopolymer and grafted polymer increased with temperature, corresponding to a free radical polymerization mechanism in both cases. The apparent acceleration of polymerization in the presence of added fillers has been noted by other investigators [85].

It should be noted that radicals adsorbed on a solid surface can take part in chain-termination reactions. Hence, the polymer molecular weight and the amount of grafted polymer per unit area are lower for materials such as silica gel having high specific surface, and higher for glass fibers and other materials having smooth surfaces.

Yasuda and Refojo [86] studied the grafting of vinyl pyrrolidone on sheets of polydimethyl siloxane using high energy electrons. The grafting reaction was promoted by the presence of water, but water decreased the depth of monomer penetration into the sheet. To determine the depth of penetration, the grafted sheet was stained with dyes (such as eosin) which are known to be suitable for staining proteins. Water content of the grafting solution and radiation dose determined whether the grafting took place on the surface only or homogeneously throughout the sheet. The hydrophilicity of the grafted product increased as the amount of vinyl pyrrolidone grafted increased.

Warrick [87] suspended small particles of silica, silicates,

metal halides, and oxides of magnesium, calcium, strontium, and
barium in a mixture of an organopolysiloxane and a vinyl monomer
and produced graft copolymer by irradiating with ionizing radiation
from a Van de Graaff generator. It was stated that the process
increased the thermal stability, improved stability toward hydrolysis, and changed the solubility properties of the inorganic
materials.

In a similar type of process, Ingersoll [88] irradiated
silica, zirconia, alumina, or aluminum silicate with ionizing radiation and then treated the irradiated materials with tetrafluoroethylene or methyl methacrylate. Reduced solubility indicated that
a part of the polymer formed was bound to the inorganic material.
The process is an example of the indirect technique of grafting,
in which the monomer is not irradiated. It was found that the
irradiated inorganic materials could be held for several days at
low temperature in the absence of oxygen and still retain their
ability to initiate polymerization of the monomers. Grinding of
the irradiated inorganics increased their ability to initiate polymerization of the monomers.

Nahin [89] claims to have prepared a new class of materials,
polyorganosilicate graft polymers, having unusually high thermal
resistance and solvent resistance. They were prepared by first
reacting the surface of clays with hexamethylene diamine or polyvinyl alcohol. The surfaces were then cross-linked to polyethylene
by irradiation with γ-rays. The product is said to contain organic
chains chemically bonded to large silicate sheets and is claimed
to have high strength and resistance to organic solvents.

In one process [90], the filler employed was the adduct of a
clay with an organic base such as oleylamine or 4-vinylpyridine.
The filler was incorporated into a thermoplastic polymer such as
polyethylene or polystyrene, heat fused, and then subjected to
intense irradiation with β- or γ-rays to induce cross-linking. It
was claimed that the presence of the filler did not destroy the
tensile strength or other desirable properties of the polymers.

REFERENCES

1. Chapiro, A., *J. Chim Phys.* , 52, 246 (1955).
2. Sack, H., *J. Pol. Sci.*, 34, 434 (1959).
3. Pinner, S. H., *Nature*, 183, 1108 (1959).
4. Lyons, B. J., *Nature*, 185, 604 (1960).
5. Pinner, S. H. and Wycherly, V., *J. Appl. Pol. Sci.*, 3, 338 (1960).
6. Pinner, S. H., Greenwood, T. T., and Lloyd, D. G., *Nature*, 184, 1303 (1957).
7. Hoffman, A. S., Gilliland, E. R., Merrell, E. W., and Stockmayer, W. H., *J. Pol. Sci.*, 34, 461 (1959).
8. Chapiro, A. and Matsumoto, A., *J. Pol. Sci.*, 57, 743 (1962).
9. Ballantine, D., Glines, A., Adler, G., and Metz, D. J., *J. Pol. Sci.*, 34, 419 (1959).
10. Hayakawa, K. and Kawase, K., *J. Pol. Sci., Pt. Al*, 5, 439 (1967).
11. Takamatsu, T. and Shinohara, K., *J. Pol. Sci., Pt. Al*, 4, 197 (1966).
12. Furuhashi, A., Mukozaka, H., and Matsuo, H., *J. Appl. Pol. Sci.*, 12, 2201 (1968).
13. Tamura, H., Tachibana, H., Takamatsu, T., and Shinohara, K., *Rept. Progr. Pol. Phys. Japan*, 6, 269 (1963).
14. Tamura, H., Tachibana, H., Takamatsu, T., and Shinohara, K., *Rept. Progr. Pol. Phys. Japan*, 6, 273 (1963).
15. Ueda, K., *Kobunshi Kagaku*, 23, 222 (1966).
16. Ohnishi, S., Sugimoto, S., and Nitta, I., *J. Pol. Sci. Pt. A.*, 1, 605 (1963).
17. Kashiwabara, H. and Shinohara, K., *J. Phys. Soc. Japan*, 15, 1129 (1960).
18. Kawase, K. and Hayakawa, K., *Radiat. Res.*, 30, 116 (1967).
19. Takamatsu, *Rept. Inst. Phys. Chem. Res.*, 40, 203 (1964).
20. Sakurada, I., Okada, T., and Kugo, E., *Doitai to Hoshasen*, 2, 316 (1959).

21. Nakagawa, T. and Manaka, K., *Genshiryoku Kogyo*, <u>11</u>(U), 47 (1965).

22. Tanner, D., U.S. Pat. 3,290,415 to DuPont (1966).

23. Scardaville, P., Wetherell, T., and Sears, L,, Fr. 1,470,131 (1967).

24. Henglein, A. and Schnabel, W., *Makromol. Chem.*, <u>25</u>, 119 (1957).

25. Chapiro, A., *J. Pol. Sci.*, <u>29</u>, 321 (1958).

26. Henglein, A., Schnabel, W., and Heine, K., *Angew. Chem.*, <u>70</u>, 461 (1958).

27. Dobo, J., *Magyar. Chem. Lapja*, <u>13</u>, 76 (1958).

28. Houillers, Fr. 1,160,106 (1958).

29. Hayden, P. and Roberts, R., *Intern. J. Appl. Radiat. Isotopes*, <u>7</u>, 317 (1960).

30. Sakurada, I., Okada., T., and Eiko, K., *Large Radiation Sources in Industry, Proc. Conf. Warsaw*, <u>1</u>, 447 (1960).

31. Smith, R. R., Mann, D. C., and Salmon, J. F., Brit. 871,572 (1961).

32. Shinohara, K. and Takamatsu, T., *Rika Gaku Kenkyusho Hokoku*, <u>36</u>, 652 (1960).

33. Durup, J. and Magat, M., *Fr. addn. 74,130* (1961).

34. Sobue, H., Shimokawa, Y., and Tajima, Y., *Kogyo Kagaku Zosshi*, <u>64</u>, 1505 (1961).

35. Usmanov, K., Tillaev, R. S., Musaev, U., and Tursunov, D., *Tr. Tashkentsk, Konf. po Mirnomu Ispolz, At. Energii, Akad, Nauk Uz. SSR*, <u>1</u>, 298 (1961).

36. Usmanov, K., Tillaev, R. S., Musaev, U., and Yuldaskava, K., *Khim. i Fiz.-Khim. Prirodn i Sintetich Polimerov, Akad. Nauk Uz. SSR, Inst. Khim. Polimerov*, <u>No. 2</u>, 175 (1964).

37. Umezawa, M. and Hirota, K., *Kobunshi Kagaku*, <u>21</u>(4), 222 (1964).

38. Chapiro, A., Jendrychowska-Bonamour, A. M., and Roussel, P., *C. R., Ser. C*, <u>262</u>(14), 1132 (1966).

39. Tsurugi, T., *Genshiryoku Kogu*, <u>12</u>(9), 38 (1966).

40. Kawai, Y., Fuse, M., Kambe, H., and Meta, I., *Kogyo Kagaku Zasshi*, <u>69</u>(6), 1237 (1966).

41. Chapiro, A. and Jendrychowska-Bonamour, A. M., Fr. 1,479,453 (1967).

42. Jendrychowska-Bonamour, A. M., *Eur. Pol. J.*, <u>4</u>(5), 627 (1968).

43. Bernstein, B. S. and Odian, G., U.S. At. Energ. Comm., RAI-329 (1963).

44. Fr. 1,441,518 to Kurashiki Rayon Co. (1966).

45. Furuhashi, A., Mukozaka, H., and Matsuo, H., *J. Appl. Pol. Sci.*, <u>12</u>(10), 220. (1968).

46. Furuhashi, A., *Denki Tsushin Kenkyusho Kenku Jitsuyoka Khokou*, <u>17</u>(8), 1673 (1968).

47. Nobel-Bozel, Fr. 1,210,101 (1960).

48. Hayden, P. and Roberts, R., *Intern. J. Appl. Radiat. Isotopes*, <u>7</u>, 317 (1960).

49. Odian, G., Acker, I., Ratchik, E., Sobel, M., and Klein, R., U.S. At. Energ. Comm., RAI-301 (1962).

50. Odian, G., Acker, I., and Sobel, M., *J. Appl. Pol. Sci.*, <u>7</u>, 245 (1963).

51. Sagane, N. and Kaetsu, I., Japan 6,809,070 (1968).

52. Takamatsu, T., *Rika Gaky Kenkyusho Hokoku*, <u>37</u>, 1-5 (1961).

53. Nobel-Bozel, Brit. 883,473 (1961).

54. Angier, D. J. and Turner, D. T., *J. Pol. Sci.*, <u>28</u>, 265 (1958).

55. Cooper, W., Vaughan, G., Miller, S., and Fielden, M., *J. Pol. Sci.*, <u>34</u>, 651 (1959).

56. Cooper, W. and Vaughan, G.,*J. Pol. Sci.*, <u>37</u>, 241 (1959).

57. Cooper, W., Vaughan, G., and Madden, R. W., *J. Appl. Pol. Sci.*, <u>1</u>, 329 (1959).

58. Cockbain, Pindle, T. D. and Turner, D. T., *J. Pol. Sci.*, <u>39</u>, 419 (1959).

59. Graham, R. K., Gluckman, M. S., and Kampf, M. J., *J. Pol. Sci.*, <u>38</u>, 417 (1959).

60. Sakurada, T., Okada, T., and Kugo, E., *Daitai to Hoshasen*, <u>2</u>, 306 (1959).

61. Dobo, J. and Somogyi, A., *J. Chim. Phys.*, <u>56</u>, 863 (1959).

62. Graham, K. and Gluckman, S., Ger. 1,024,240 (1958).

63. Houilleres, Fr. 1,160,108 (1958).

64. Dobo, J., Somogyi, M., and Kiss, L., *Large Radiation Sources in Industry Proc. Conf. Warsaw*, 1, 423 (1960).

65. Sakurada, I., Okada, T., and Eiko, K., *Large Radiation Sources in Industry, Proc. Conf. Warsaw*, 1, 447 (1960).

66. Somogyi, A., Geczy, I., and Dobo, J., *Magyar Indomanyos Akad. Kem. Tudomanyok Osztalyanak Kozlimenyii*, 15, 17 (1961).

67. Sakurada, I., Matsuzawa, S., and Kubota, Y., *Makromol. Chem.*, 69, 115 (1963).

68. Koshijima, T. and Musaki, E., *Nippon Mokuzai Gahkaishi*, 10, 110 (1964).

69   Wilson, J. E., *J. Macromol. Sci. Chem.*, A5(4), 777 (1971).

70. Seitz, F., *Science*, 138, 563 (1962).

71. Kazanskii, V. B. and Paniskii, G. B., *Kinetics Analysis (USSR)*, 2, 460 (1961).

72. Kazanskii, V. B., Paniskii, G. B., Aleksandrov, I. V., and Zhidomirov, G. M., *Fiz. Tverd. Tela.*, 5, 649 (1963).

73. Kazanskii, V. B. and Paniskii, G. B., *3rd Intern. Congr. Catasysts, Amsterdam, 1964*, 1, 367 (1965).

74. Ueno, K., Williams, F., Hayashi, K., and Okamura, S., *Trans. Faraday Soc.*, 63, 1478 (1967).

75. Ueno, K., Hayashi, K., and Okamura, S., *Polymer*, 7, 431 (1966).

76. Metz, D. J., *Advances in Chemistry Series, No. 66*, American Chemical Society, 1967, p. 170.

77. Vlasov, A. V., *Dokl. Akad. Nauk SSSR*, 158(1), 141 (1964). (Chem. Abstr. 61:15495b.)

78. Vlasov, A. V., *Chim. Volokna*, No. 6, 24 (1963).   (Chem. Abstr. 60:12156g.)

79. Tsetlin, B. L., *Tr. 2-go Vses. Soveshch po Radiats. Khim. Akad. Nauk, Moscow, 1960*, p. 497 (1962).

80. Tsetlin, B. L., *Ind. Uses Large Radiat. Sources, Conf. Proc., Salzburg, 1963*, Vol. I, Intern. At. Energ. Agency, Vienna, 1963, (STI/PUB/75), p. 289.

81. Egorov, E. V., Novikov, P. D., Razgon, D. R., and Tsetlin, B. L., *Dokl. Akad. Nauk SSSR*, 146, 1360 (1962).

82. Morosov, Y. L., Vitushkin, N. I., Glazunov, P. Y., Rafikov, S. R., Khomutov, A. I., and Tsetlin, B. L, *Radiats. Khim. Polim. Mater. Simp., Moscow, 1964*, p. 160 (1966).

83. Berezkin, U. G., *Kinetics Catalysis (USSE)*, 7, 170 (1966).

84. Uskov, I. A. and Tertykh, L. I., *Pol. Sci. (USSR)*, 8(1), 24 (1966).

85. Worrall, R. and Pinner, S. H., *J. Pol. Sci.*, 34, 229 (1959).

86. Yasuda, H. and Refojo, M. F., *J. Pol. Sci.*, Pt. A, 2, 5093 (1964).

87. Warrick, E. L., U.S. Pat. 2,959,569 (1960).

88. Ingersoll, H. G., Brit. Pat. 866,768 (1961).

89. Nahin, P. G., *Chem. Eng. News*, 42(43), 70 (1964).

90. Nahin, P. G. and Backlund, P. S., U.S. Pat. 3,084,177 (1963).

91. Chen, W. K. W., Mesrobian, R. B., Ballantine, B. S., Metz, G. J., and Glines, A., *J. Pol. Sci.*, 23, 903 (1957).

92. Sebban-Danon, J., *J. Pol. Sci.*, 29, 367 (1958).

93. Sebban-Danon, J., *J. Pol. Sci.*, 48, 121 (1960).

94. Turner, D. T., *J. Pol. Sci.*, 35, 17 (1959).

95. Minnema, L., Hazenberg, J. F. A., Callaghan, L., and Pinner, S. H., *J. Appl. Pol. Sci.*, 4, 246 (1960).

96. Sakurada, I., Okada, T., and Kugo, E., *Doitai to Hoshasen*, 2, 296 (1959).

97. Mesrobian, R. B., *Proc. UN Intern. Conf. Peaceful Uses At. Energ., 2nd, Geneva*, 29, 196 (1959).

98. Houilleres, Fr. 1,160,107 (1958).

99. Odian, G. G., Sobel, M., Rossi, A., and Klein, R., *J. Pol. Sci.*, 55, 663 (1961).

100. Chapiro, A. and Magat, M., Fr. addn. 72,899 (1960).

101. Odian, G. , Rossi, A., and Trachtenberg, S. N., *Amer. Chem. Soc., Div. Pol. Chem., Preprints*, 1(1), 5-9 (1960).

102. Takamatsu, T. and Shinohara, K., *Rika Gaku Kenkyusho Hokoku*, 38, 1 (1962).

103. Matsuo, H., Iino, K., and Kondo, M., *J. Appl. Pol. Sci.*, 7(5), 1833 (1963).

104. Jones, T. T., *Preprints Papers Intern. Symp. Free Radicals, 5th, Uppsala, 1961*, 27-1-27-16.

105. Dobo, J., Somogyi, A., and Czvikovsky, K., *J. Pol. Sci., Pt. C*, No. 4, 1173 (1964).

106. Ethylene Plastiques, Fr. 1,334,305 (1963).

107. Lamb, J. A. and Weak, R. E., *J. Pol. Sci., Pt. A*, 2, (6), 2875 (1964).

108. Koshijima, T., *Nippon Mokuzai Gakkaishi*, 12(3), 144 (1966).

109. Munari, S., Vigo, F., Tealdo, G., and Rossi, C., *J. Appl. Sci.*, 11(8), 1563 (1967).

110. Wagner, J. and Marchal, J., *C. R. Acad. Sci., Ser. C*, 264(15), 1263 (1967).

111. Steinberg, N., Henley, E. J., and Dougherty, T. J., *Chem. Engr. Progr., Symp. Ser.*, 62(68), 28 (1966).

112. Sakurada, I., *Bull. Inst. Chem. Res., Kyoto Univ.*, 46(1), 13 (1968).

113. Doumas, A. C., to Dow Chemical, U.S. Pat. 3,342,899 (1967).

114. Ueda, K., *Kobunshi Kagaku*, 23(252), 221 (1966).

115. Maratova, N. K., *Vyskomol. Soedin., Ser. B*, 10(8), 566 (1968).

116. Usmanov, K. U., *Khim. Volokna*, No. 6, 32 (1967).

117. Usmanov, K. U., *Usb. Khim. Zh.*, 12(4), 56 (1968).

118. Wilson, J. E., *J. Macromol. Sci. Chem.*, A6(2), 391 (1972).

119. Wilson, J. E., *J. Chem. Phys.*, 22, 333 (1954).

APPENDIX

Table 9-A lists references to radiolytic grafting of various monomers on polymeric film substrates for the period 1957 to 1969, inclusive. The monomers are listed alphabetically, and under each monomer the films grafted with that monomer are given. Several references to grafting on miscellaneous materials are also included, such as glass, magnesium oxide, aluminum oxide, and carbon black. A glance at the table shows that γ-rays are most commonly used to initiate grafting, with cobalt-60 being the usual source of such rays.

TABLE 9-A

Grafting on Film Substrate and Miscellaneous Substrate Forms

Grafted monomer	Substrate	Substrate form	Source	Investigator and reference
Acrylamide	Polyvinyl alcohol	Film	Cobalt-60	Sakurada [20]
	Polytetrafluoro-ethylene	-	Cobalt-60	Nakagawa [21]
Acrylic acid	Polyvinyl alcohol	Film	Cobalt-60	Sakurada [20]
	Polyethylene	Film	-	Tanner [22]
	Polytetrafluoro-ethylene	-	Cobalt-60	Nakagawa [21]
	Polyethylene	Film	Electron beam	Scardaville [23]
	Silica gel	Powder	-	Egorov [81]
Acrylonitrile	Polyvinyl-pyrrolidone	Film	Cobalt-60	Henglein [24]
	Carbon black	Powder	-	Tsetlin [79]
	Polyethylene	Film	Cobalt-60	Chapiro [25]

Material	Form	Radiation	Reference
Polymethacrylic ester	Film	Cobalt-60	Henglein [26]
Silica gel	Powder	-	Tsetlin [79]
Polyethylene	Sheet	Electron beam	Dobo [27]
MgO	Powder	-	Tsetlin [79]
Polyvinyl alcohol	Film	Cobalt-60	Sakurada [20]
Polyethylene	-	Cobalt-60	Houillers [28]
Polyvinylacetate	Emulsion	Cobalt-60	Hayden [29]
Glass	Fiber	-	Vlasov [77]
Polymethylmethacrylate	Emulsion	Cobalt-60	Hayden [29]
Glass	Fiber	x-Rays	Morosov [82]
Polyvinyl alcohol	Film	Cobalt-60	Sakurada [30]
Polyethylene	Film	Cobalt-60	Smith [31]
Polyethylene	-	-	Shinohara [32]
Polyvinylchloride	-	-	Shinohara [32]
Polytetrafluoroethylene	-	-	Shinohara [32]

Monomer	Substrate	Form	Source	Reference
Acrylonitrile (continued)	Brick	Powder	Cobalt-60	Berezkin [83]
	Polystyrene	Solution	Cobalt-60	Durup [33]
	Polypropylene	Film	-	Sobue [34]
	Polystyrene	-	Cobalt-60	Usmanov [35]
	Polystyrene	-	Cobalt-60	Usmanov [36]
	Polyvinyl chloride	-	Cobalt-60	Usmanov [36]
	Chlorinated poly-vinyl chloride	-	Cobalt-60	Usmanov [36]
	Polypropylene	-	α-Rays	Umezawa [37]
	Polyvinyl chloride	Film	Cobalt-60	Chapiro [38]
	Polybutadiene	-	-	Tsurugi [39]
	Polydichloro-acetaldehyde	-	-	Kawai [40]
	Polyvinyl chloride	Film	Cobalt-60	Chapiro [41]
	Polyvinyl chloride	-	-	Jendrychowska-Bonamour [42]
Allyl acrylate	Polyethylene	-	-	Bernstein [43]
	Polyisobutylene	-	-	Bernstein [43]

Monomer	Polymer	Form	Source	Reference
	Polypropylene	–	–	Bernstein [43]
	Cellulose acetate	–	–	Bernstein [43]
	Polystyrene	–	–	Bernstein [43]
	Polyvinyl alcohol	–	–	Bernstein [43]
Allyl methacrylate	Cellulose acetate	–	–	Bernstein [43]
	Polyethylene	–	–	Bernstein [43]
	Polyisobutylene	–	–	Bernstein [43]
	Polypropylene	–	–	Bernstein [43]
	Polystyrene	–	–	Bernstein [43]
	Polyvinyl alcohol	–	–	Bernstein [43]
Butadiene	Polyvinyl chloride	–	–	[44]
	Polyethylene	Film	Cobalt-60	Furuhashi [45]
	Polyethylene	Film	Cobalt-60	Furuhashi [46]
Butyl acrylate	Cellulose acetate	Powder	Cobalt-60	[47]
Butyl methacrylate	Polytetrafluoro-ethylene	Emulsion	Cobalt-60	Hayden [48]
t-Butylaminoethyl methacrylate	Polyethylene	–	Cobalt-60	Odian [49]

Monomer	Polymer		Radiation source	Reference
t-Butylaminoethyl (continued)	Polypropylene	-	Cobalt-60	Odian [49]
	Polyvinyl chloride	-	Cobalt-60	Odian [49]
	Polytetrafluoro ethylene	-	Cobalt-60	Odian [49]
	Polyamide	-	Cobalt-60	Odian [49]
	Polyethylene	-	Cobalt-60	Odian [50]
	Polypropylene	-	Cobalt-60	Odian [50]
	Polyvinyl chloride	-	Cobalt-60	Odian [50]
	Polyamide	-	Cobalt-60	Odian [50]
Diallyl maleate	Polyethylene	-	-	Bernstein [43]
	Polypropylene	-	-	Bernstein [43]
	Polyisobutylene	-	-	Bernstein [43]
	Cellulose acetate	-	-	Bernstein [43]
	Polyvinyl alcohol	-	-	Bernstein [43]
	Polystyrene	-	-	Bernstein [43]
Diallyl phosphate	Polyethylene	-	-	Odian [49]
Diallyl phthalate	Polyacrylamide	-	Cobalt-60	Sagane [51]

Monomer	Polymer	State	Source	Reference
Ethyl acrylate	Polymeth-acrylamide	-	Cobalt-60	Sagane [51]
	Polymethyl acrylate	Emulsion	Cobalt-60	Hayden [48]
	Polytetrafluoro-ethylene	Emulsion	Cobalt-60	Hayden [48]
	Polyethylene	Film	-	Takamatsu [52]
	Polyvinyl chlordie	Film	-	Takamatsu [52]
	Polyvinyl acetate	Emulsion	-	[53]
Isobutylene	Polyvinyl chloride	:	-	Jendrychowska-Bonamour [42]
Isoprene	Polyvinyl chloride	-	-	[44]
Methyl acrylate	Polyethylene	Film	-	Takamatsu [52]
	Polyvinyl chloride	Film	-	Takamatsu [52]
	Polyethylene	-	Cobalt-60	Odian [50]
	Polypropylene	-	Cobalt-60	Odian [50]
	Polyvinyl chloride	-	Cobalt-60	Odian [50]
	Polyamide	-	Cobalt-60	Odian [50]
	Polytetrafluoro-ethylene	-	Cobalt-60	Odian [50]

Monomer	Polymer	Form	Radiation source	Reference
Methyl methacrylate	Natural rubber	—	—	Angier [54]
	Natural rubber	—	—	Cooper [55]
	Natural rubber	—	—	Cooper [56]
	Natural rubber	—	—	Cooper [57]
	Natural rubber	—	—	Cockbain [58]
	Polymethyl methacrylate	—	—	Graham [59]
	Polyethylene	Film	Cobalt-60	Chapiro [25]
	Polyethylene	Sheet	Electron beam	Dobo [27]
	Polyvinyl alcohol	Film	Cobalt-60	Sakurada [60]
	Polyethylene	—	x-Rays	Dobo [61]
	Polybutyl acrylate	—	Cobalt-60	Graham [62]
	Polymethyl methacrylate	—	Cobalt-60	Houilleres [63]
	Polyethyl acrylate	—	Cobalt-60	Hayden [29]
	Polytetrafluoroethylene	—	Cobalt-60	Hayden [29]
	Polyethylene	—	—	Dobo [64]
	Polyvinyl alcohol	Film	Cobalt-60	Sakurada [65]

Polyethylene	Film	–	Takamatsu [52]
Polyvinyl chloride	Film	–	Takamatsu [52]
Polyethylene	–	–	Somogyi [66]
Polystyrene	Solution	Cobalt-60	Durup [33]
Polyvinyl acetate	Emulsion		[53]
Polyvinyl alcohol	Film	Cobalt-60	Sakurada [67]
Polystyrene	–	Cobalt-60	Chapiro [38]
Lignin	–	–	Koshijima [68]
Polyvinyl chloride	–	–	Jendrychowska-Bonamour [42]
Carbon black	Powder	β-Rays	Tsetlin [80]
Silica gel	Powder	β-Rays	Tsetlin [80]
MgO	Powder	β-Rays	Tsetlin [80]
BeO	Powder	β-Rays	Tsetlin [80]
Silica gel	Powder	Cobalt-60	Uskov [84]
Kaolin	Powder	Cobalt-60	Uskov [84]
Asbestos	Fiber	Cobalt-60	Uskov [84]
Glass	Fiber	Cobalt-60	Uskov [84]

Monomer	Substrate	Form	Radiation	Reference
Methyl methacrylate (continued)	Silica	Powder	–	Ingersoll [88]
	Alumina	Powder	–	Ingersoll [88]
	Zirconia	Powder	–	Ingersoll [88]
	Aluminum silicate	Powder	–	Ingersoll [88]
α-Methyl styrene	Polyvinyl chloride	–	–	[44]
Phenylacetylene	Glass	Fiber	–	Vlasov [77]
Styrene	Polyethylene	Film	–	Chen [91]
	Polyethylene	Film	–	Hoffman [7]
	Polyethylene	Film	–	Ballantine [9]
	Polyisobutylene	–	–	Sebban-Danon [92]
	Polyisobutylene	–	–	Sebban-Danon [93]
	Natural rubber	–	–	Turner [94]
	Polyethylene	Film	Cobalt-60	Minnema [95]
	Polyisobutylene	Film	γ-Rays	Henglein [26]
	Polyethylene	Sheet	Electron beam	Dobo [27]
	Polyvinyl alcohol	Film	Cobalt-60	Sakurada [96]
	Polyethylene	–	x-Rays	Dobo [61]

Polyethylene	Film	γ-Rays	Mesrobian [97]
Polyethylene	–	γ-Rays	Houilleres [28]
Polyethylene	–	γ-Rays	Houilleres [98]
Polystyrene	–	γ-Rays	Houilleres [63]
Polytetrafluoro-ethylene	Emulsion	γ-Rays	Hayden [48]
Polyethylene	–	–	Dobo [64]
Polyvinyl alcohol	Film	Cobalt-60	Sakurada [65]
Polyethylene	–	–	Shinohara [32]
Polyvinyl chloride	–	–	Shinohara [32]
Polytetrafluoro-ethylene	–	–	Shinohara [32]
Polyethylene	–	x-Rays	Somogyi [66]
Polyvinyl chloride	–	x-Rays	Somogyi [66]
Polyethylene	–	–	Odian [99]
Polyvinyl acetate	Emulsion	–	[53]
Polytetrafluoro ethylene	Sheets	Cobalt-60	Chapiro [100]
Polyethylene	–	–	Odian [101]

Styrene
(continued)

Substrate	Form	Method	Reference
Polyethylene	-	-	Odian [49]
Polypropylene	-	-	Odian [49]
Polyvinyl chloride	-	-	Odian [49]
Polytetrafluoroethylene	-	-	Odian [49]
Polyamide	-	-	Odian [49]
Polyethylene	Film	Electron beam	Takamatsu [102]
Polyvinyl chloride	Film	Electron beam	Takamatsu [102]
Polypropylene	Film	Cobalt-60	Matsuo [103]
Polyethylene	-	-	Jones [104]
Polytetrafluoroethylene	Film	Cobalt-60	Dobo [105]
Polyester	-	Cobalt-60	[106]
Polyethylene	-	-	Lamb [107]
Polypropylene	-	α-Rays	Umezawa [37]
Polymethyl methacrylate	-	γ-Rays	Chapiro [38]
Polyvinyl chloride	Film	γ-Rays	Chapiro [38]
Polydichloroacetaldehyde	-	-	Kawai [40]
Polyvinyl chloride	-	-	[44]

Monomer	Substrate	Form	Radiation	Reference
	Lignin	–	–	Koshijima [108]
	Polytetrafluoro-ethylene	Film	–	Munari [109]
	Polyoxyethylene	–	–	Wagner [110]
	Polyethylene	–	Cobalt-60	Steinberg [111]
	Polyacrylamide	–	γ-Rays	Sagane [51]
	Polymethacrylamide	–	γ-Rays	Sagane [51]
	Polyvinyl alcohol	–	γ-Rays	Sakurada [112]
	Polyvinyl chloride	–	–	Jendrychowska-Bonamour [42]
	Silica gel	Powder	–	Egorov [81]
	Silica gel	Powder	Cobalt-60	Uskov [84]
	Kaolin	Powder	Cobalt-60	Uskov [84]
	Asbestos	Powder	Cobalt-60	Uskov [84]
	Glass	Fiber	Cobalt-60	Uskov [84]
Tetrafluoroethylene	Polytetrafluoro-ethylene	–	γ-Rays	Houilleres [63]
	Polytetrafluoro-ethylene	Suspension	Cobalt-60	Doumas [113]
	Silica	Powder	–	Ingersoll [88]
	Alumina	Powder	–	Ingersoll [88]

Tetrafluoroethylene (continued)	Zirconia	Powder	-	Ingersoll [88]
	Aluminum silicate	Powder	-	Ingersoll [88]
Triallyl phosphate	Polyethylene	-	-	Odian [49]
Vinyl acelate	Polyvinyl alcohol	Film	Cobalt-60	Sakurada [20]
	Polytetrafluoroethylene	Emulsion	γ-Rays	Hayden [29]
	Polyvinyl alcohol	Film	Cobalt-60	Sakurada [30]
	Polyethylene	┆	x-Rays	Somogyi [66]
	Polyvinyl chloride	-	x-Rays	Somogyi [66]
	Cellulose acetate	Powder	Cobalt-60	[47]
	Polyethylene	Solution	Cobalt-60	[106]
	Polystyrene	-	-	Lamb [107]
	Polydichloro-acetaldehyde	-	-	Kawai [40]
Vinyl carbazole	Polyethylene	Film	Cobalt-60	Chapiro [25]
Vinyl chloride	Polyethylene	-	γ-Rays	Houilleres [28]
	Natural rubber	-	Cobalt-60	Usmanov [35]
	Polyethylene	-	-	Ueda [114]
	Carbon black	Powder	β-Rays	Tsetlin [79]
	Silica gel	Powder	β-Rays	Tsetlin [79]
	Magnesium oxide	Powder	β-Rays	Tsetlin [79]

Monomer	Polymer	Radiation	Form	Reference
Vinyl chloroacetate	Polyethylene	Electron beam	Sheet	Dobo [27]
Vinylene carbonate	Polymethyl methacrylate	β-Rays	–	Maratova [115]
Vinylidene chloride	Polyethylene	γ-Rays	–	Houilleres [98]
Vinyl fluoride	Polyamide	γ-Rays	–	Usmanov [116]
	Chlorinated poly-vinyl chloride	γ-Rays	—	Usmanov [116]
	Polyvinyl chloride	Cobalt-60	–	Usmanov [117]
2-Vinyl pyridine	Polyvinyl alcohol	Cobalt-60	Film	Sakurada [20]
4-Vinyl pyridine	Polyethylene	–	Film	Takamatsu [52]
	Polyvinyl chloride	–	Film	Takamatsu [52]
Vinyl pyrrolidone	Polymethacrylic ester	γ-Rays	Film	Henglein [26]
Vinyl stearate	Polydimethyl siloxane	Electron beam	Sheet	Yasuda [86]
	Polyethylene	–	–	Odian [49]

Chapter 10

RADIOLYTIC GRAFTING ON FIBERS

I.   INTRODUCTION

During the past 12 years, interest in radiation grafting of
monomers on fibers has grown extensively. Radiolytic grafting of
vinyl monomers has been carried out on fibers of polyethylene,
polyvinyl alcohol, polyethylene terephthalate, kapron, lavsan,
polycaprylamide, polypropylene, polytetrafluoroethylene, wool,
polyesters, cellulose esters, polyvinyl chloride, and cotton
(cellulose).

The reason for such widespread research in fiber grafting is
the realization that it may well be possible to produce desirable
physical and chemical modifications in the properties of polymeric
fibers. In this connection, it should be noted that graft polymerization usually leaves the main chain of the fiber polymer unchanged. For this reason, grafting will probably affect certain
properties of the side chains without altering many of the fundamental properties of the fiber. This phenomenon is especially
desirable in the case of textile fibers, which must have certain
structural symmetry and regularity in order to retain acceptable
elastic moduli and softening points. Fairly extensive grafting
does not destroy the basically necessary fiber properties. On the
other hand, this same feature of the grafting reaction implies

that it probably cannot be employed to produce large increases in mechanical strength. Some of the fiber properties which can easily be modified by radiolytic grafting are dirt-release properties, antistatic character, and dyeability.

Certain general principles regarding radiolytic grafting on fibers are worth mentioning at this point. The molecular weights of the added side chains are usually quite high, because the termination (radical recombination) reaction is often severely retarded by the high viscosity of the polymeric reaction medium. Hence, the molecular weight of a grafted side chain is generally about one million. Since the molecular weight of the substrate fiber is often much lower, it follows that only a small fraction of the fiber molecules become grafted.

This reasoning has been confirmed by studies of monomer/fiber grafting [12, 13]. It might be more desirable, property-wise, to have a large number of short side chains rather than a small number of long chains. But such a result would correspond to a lower G value, implying that a larger irradiation dose would be required to produce the same percentage of grafting. Such a large dose might cause unwanted side reactions. An alternative way of obtaining more frequent side chains would be the employment of a chain-transfer agent. The chain-transfer agent, however, would probably result in a large amount of undesired homopolymer. Actually, the possible effect of such homopolymers on the properties of the grafted fiber is not clear and also perhaps not important in many cases.

The three techniques of radiolytic fiber grafting are the mutual, preirradiation, and peroxide techniques. In the *mutual* method, the monomer is padded onto the fibers which are then subjected to irradiation. In the *preirradiation* method, the fiber is first irradiated and then brought into contact with liquid or gaseous monomer. In the *peroxide technique*, the fiber is irradiated in air to form polymeric peroxides, which are later employed

to decompose and initiate grafting when the fiber is immersed in
gaseous or liquid monomer.

Many investigators have studied the role of diffusion in the
grafting reaction. For grafting to occur at a significant rate, an
adequate quantity of monomer must come in contact with the active
free radicals inside the fiber to initiate polymer chains at a
faster rate than they become terminated. In some cases, as for dry
cotton, the monomer does not diffuse to the active centers fast
enough to produce a significant rate of grafting. The introduction
of a small percentage of water into the cotton increases the monomer diffusion rate and enables grafting at a relatively high rate.
Diffusion of monomer to the active radical site is a universal requirement, regardless of conditions or whether the grafting is done
in liquid or vapor monomer. In the case of hydrophobic polymer
fibers, the monomer may cause enough swelling of the fiber to
facilitate diffusion without the addition of small amounts of water
or methanol.

A knowledge of the effect of dose rate, and the interaction of
dose rate with monomer diffusion, allows a degree of control of the
grafting process. In the mutual process, the dose rate has a
strong influence on grafting rate. As dose rate is increased, a
point is reached where the monomer is consumed (by grafting) before
it can reach the center of the fiber, causing a marked reduction in
grafting yield per weight of fiber. In using the preirradiation
method, the dose rate has less of an effect on grafting yield, and
considerable grafting efficiency can be obtained even at very large
dose rates.

Because of the slowness of monomer diffusion, the grafting
tends to concentrate at or near the surface of the fiber. By a
planned choice of swelling solvent, dose rate, and monomer contact
time, it has been found possible to exercise some control over the
depth to which the grafting penetrates in the fiber.

The following sections discuss in more detail some of the
better-known monomer/fiber grafting systems.

II. GRAFTING ON COTTON

More radiation-grafting research has been done on cotton than
on any other fiber. Perhaps the most extensive work in this field
has been carried out by Arthur *et al.*[4]. Arthur's technique
usually involved two steps:

1. Radiolytic grafting of selected vinyl monomers on
cellulose

2. Later treatment of the grafted product with cross-linking
agents to give durable-press materials

Both ionic and free-radical mechanisms for vinyl monomer grafting
on cellulose have been reported [1-3]. In the grafting procedure,
the vinyl monomer becomes distributed within the structure of the
cotton fiber.

Use of radiolytic methods for grafting on cellulose have
several advantages over other techniques of grafting.

1. The rates of the primary reactions caused by irradiation
of cellulose are temperature independent. This allows selection of
a temperature of irradiation which may be desirable for other
reasons, and does not necessitate the use of elevated temperature
as in certain other grafting techniques.

2. Irradiation with ionizing radiation does not require the
use of catalysts or other additives, which may be difficult to
remove subsequently from the cellulose.

3. Irradiation of cellulose produces long-lived free radical
sites in the cellulose structure, and leads logically to a two-step
process for grafting [4, 5]. First, cellulose is irradiated to
produce free radical sites on the molecule. Second, irradiated
cellulose is treated with a vinyl monomer, causing the monomer to
graft onto the cellulose and form a copolymer. An advantage of
such a two-step technique is that it minimizes the amount of homopolymer formed within the cellulose fiber structure.

4. Vinyl monomer-solvent systems can be selected that tend to produce the desired dimensional changes in the cellulose structure, influence the rate of diffusion of monomer into the cellulose structure, and distribute the grafted copolymer within the fiber structure.

Gamma-radiation from cobalt-60 has often been used to graft various vinyl monomers on cellulose. Such radiation is very penetrating in character, leading to uniform dosage rate and hence uniform reaction rate throughout the bulk of fiber, yarn, or fabric being irradiated.

The primary effect of γ-radiation on cellulose has been described as chemically nonspecific, involving the ejection of an energetic electron from an atomic orbital of one of the elements of the cellulose molecule [6]. Interaction of such energetic electrons with cellulose produces physical and chemical changes in the molecule.

One of the chemical reactions is dehydrogenation, which causes the formation of a free radical site at the point where a hydrogen atom is lost. Such radical sites may be produced in the amorphous regions, partially ordered regions, or highly ordered (crystalline) regions of the cellulose fiber. The most probable site for the formation of a stabilized free radical on the anhydroglucose ring of the cellulose molecule is C_5.

Electron-spin-resonance spectroscopy indicates that natural cotton cellulose may have a stabilized free radical site on C_5, whereas mercerized cotton cellulose may have stabilized radical sites on C_5 and C_6 [4].

It has been noted that as the fraction of amorphous regions in cellulose increases, the rate of vinyl monomer grafting on the cellulose decreases. It has been postulated that free radical sites in the amorphous regions react quickly with water or solvent and become terminated.

In the ordered regions of the cellulose, the monomer must first diffuse to the free radical sites before graft copolymerization can be initiated. The fraction of free radical sites which actually initiate polymerization has been found to be low. The free radical sites located within highly ordered or crystalline regions of the fiber have less tendency to be terminated when treated with solvents that cause considerable change in the dimensions of the fibers. Because of this inaccessibility of radical sites in the highly ordered regions, some investigators have postulated that the free radical sites in the amorphous and partially ordered regions initiate graft copolymerization reactions [5].

In preparing the cotton for irradiation, it is advisable to dry it to less than 2% moisture content. Electron-spin-resonance spectroscopy showed that irradiated cotton with less than 2% moisture had three times the concentration of free radicals as in irradiated cotton containing more than 6% moisture.

Similar ESR studies have demonstrated that about 30% of the "stable" free radical sites in irradiated cotton are not accessible to the monomer-solvent system. Irradiation of the dry cotton may be conducted in air, but is preferably carried out under nitrogen or in a vacuum. Oxygen must be excluded during the graft copolymerization reaction to obtain acceptable rates.

The formation of a cellulose-polyvinyl copolymer is predicated on the diffusion of vinyl monomer to the free radical sites in

sufficient concentration to permit a graft copolymerization reaction. With regard to technique, the stabilizer usually present
in the vinyl monomer should be removed by distillation or some
other effective procedure. The monomer can best be utilized if it
is first dissolved in a solvent which is known to produce dimensional changes in the cellulose fiber.

Monomer solvent systems that have been used effectively include acrylonitrile-dimethyl sulfoxide; styrene-methanol; acrylo-
nitrile-aqueous zinc chloride; acrylonitrile-dimethylformamide;
methyl methacrylate-methanol; and vinyl acetate-methanol [7, 8].

Since cellulose is depolymerized by ionizing radiation, the
minimum irradiation dosage should be employed in producing free
radical sites prior to grafting. A dosage of 1 Mrad forms about
one free radical site per molecule of cellulose. Such a dosage
produces about 5 cleavages/molecule, resulting in a decrease in
breaking strength of only 5 to 7% [9].

If the cotton is irradiated while in contact with monomer
solution (one-step process), a ratio of one molecule of grafted
polymer to one molecule of cellulose can be attained. The one-step
process also initiates grafting at short-lived free radical sites.
The main disadvantage of the one-step process is that more extractable homopolymer is formed and less efficient use is made of the
vinyl monomer employed.

It has been found permissible to allow a lapse of several days
between the irradiation of the cotton and the addition of monomer
for the copolymerization reaction. In one experiment, cotton was
irradiated with γ-rays under nitrogen to a total dosage of 1 Mrad.
The irradiated cotton was immersed in a styrene/methanol solution
for 1 h, and a weight increase due to grafting of 38% was obtained.
When the immersion in styrene solution was delayed to 48 h after
irradiation, the grafting weight increase was decreased to 32% for
equivalent reaction time. If a lapse of 240 h was allowed after
irradiation, the weight of styrene grafted was only 25%. Comparable

results were obtained in a similar sequence of runs with methyl
methacrylate, except that a much larger weight increase owing to
grafting was observed in each case.

The location of the grafted copolymer inside the fiber structure can be regulated to a certain extent by the choice of solvent
and selection of technique of irradiation. For example, in a two-
step process, when solvents are used which cause dimensional
changes in the fiber by swelling it and rounding out its cross-
section, the grafted copolymer can be placed mainly in the outer
layers of the fiber.

When a one-step process is used, the fiber cross section
generally becomes swollen and rounded. Depending on the type of
monomer and solvent used, the copolymer may be lcated in the surface of the fiber, uniformly throughout the fiber, or in circular
layers around the center of the fiber cross section.

"Sensitizers" capable of promoting grafting to cotton include
acetic acid, methanol, water, and concentrated zinc chloride
solution [14-16]. The main function of these ingredients is to
swelll the fiber, thus promoting diffusion of monomer into the
fiber and bringing more monomer in contact with the active radical
sites. They may also possibly function by producing additonal
free radicals through radiolysis.

According to Arthur and Daigle [7], the molecular weight of
grafted styrene branches ranges between 300,000 and 600,000, in
agreement with the results of other researchers. Using a postirradiation method, it was found possible to increase the molecular
weight of such branches to three million [17].

Razikov *et al.* [18] presented electron/microscope evidence
that methacrylic acid diffuses into the unordered and densely
packed parts of the cellulose structure and produces a compaction
of the supramolecular structure. The width of the fibrillar forma-
tions in ungrafted cellulose, the methacrylamide graft, and the
methyl acrylate graft were 100, 130, and 150 Å, respectively. It

was observed that methacrylate diffuses uniformly through all
sections of the cellulose, thus producing uniform grafting across
the diameter of the fiber.

Rutherford and co-workers [19] found that the rate of grafting
on cotton was increased by partial acetylation, hydroxyethylation,
and cyanoethylation. However, the rate was unchanged by partial
methylation or carboxymethylation. Arthur et al. [15] observed
that acrylonitrile grafted on cyanoethylated cotton produced
thermoplastic products.

Armstrong and Rutherford [14] observed little modification in
cotton mechanical properties produced by the grafting of butadiene,
ethyl acrylate, acrylonitrile, or vinyl acetate. Grafting of 3.5%
acrylonitrile on cotton gave a product highly resistant to attract
by microorganisms [19]. Arthur et al. [15] noted that cellulose
grafted with styrene or acrylonitrile exhibited a decrease in
resiliency index with increasing temperature. Grafting of styrene
on cotton resulted in lower absorption of water vapor, reduced
heats of wetting by water, and better resistance to mineral acid
[20].

Blouin et al. [21] used postirradiation grafting in zinc
chloride solution of methyl methacrylate, acrylonitrile, and vinyl
acetate to avoid significant radiation damage while attaining a
wide range in percent monomer grafted. Grafted yarn prepared in
this way showed thermoplastic properties, increased elongation
at break, and decreased stiffness. Cotton grafted with acrylonitrile had improved abrasion and rot resistance. Fabrics of improved abrasion resistance were also obtained by grafting with
vinyl acetate.

Usmanov and Azizov [23] grafted cotton fiber with styrene,
methyl methacrylate, methacrylic acid, acrylonitrile, and methacrylamide. The strength of the fibers was essentially unaffected,
but increases were observed in resistance to microorganisms,
thermal stability, dye affinity, and adhesiveness. The type of

grafted monomer and conditions of grafting determine whether water vapor absorption will be increased or decreased [22].

Armstrong *et al.* [19] were able to improve dry- and wet-crease recovery of cotton cloth by grafting with allyl acrylate. Grafting with acrylic acid improved wet- but not dry-crease recovery. Dry-crease recovery but not wet-crease recovery was improved by grafting methylol acrylamide on the cotton cloth.

III.   RADIOLYTIC GRAFTING ON WOOL

An extensive investigation of the radilytic grafting of vinyl monomers on wool was made by Campbell *et al.* [10]. Styrene dissolved in a swelling agent such as methanol was employed as the monomer solution. Methanol concentrations from 7 to 18% were employed, and percent grafting was found to increase with percent methanol at constant irradiation dosage.

It was postulated that addition of water or methanol was necessary to swell the wool fibers and permit the diffusion of styrene to the free radical sites in the wool produced by irradiation.

The irradiations were carried out in a 1500-Ci cobalt-60 source. Dosage rate was changed by varying the distance from the source.

After grafting, the residual styrene homopolymer was removed from the wool by benzene extraction. The grafted polystyrene was then separated from the wool by two separate 24-h treatments with a two-phase toluene/sodium hypochlorite solution. For the ESR measurements, samples were sealed in 3-mm-o.d. quartz tubes after degassing both wool and monomer solution. They were then irradiated for selected periods of time, and ESR measurements were

carried out. Results were presented either as first or second
derivatives of the resonance absorption curve. The number of free
radicals was estimated by integration of the first derivative curve
and comparison with known radical concentrations supplied by DPPH.

Several parallel runs were made to evaluate postirradiation
effects. In one series, styrene at 3.06 M in dioxane containing
10% methanol was used, whereas in the other, styrene was at 3.06 M
in dioxane containing 18% methanol. Two dosage rates, 0.05 Mrad/h
and 0.20 Mrad/h, were employed. Two procedures were used. In one
case, tubes were frozen in liquid nitrogen immediately after irradiation and then cut open and poured into methanol. In the
other, the tubes were left overnight at room temperature and then
opened.

The results showed that at the higher dose rate a great deal
of postpolymerization took place at both levels of swelling. In
all cases, the tubes left 24 h prior to opening showed more grafting than those opened immediately. All subsequent experiments
were carried out so as to minimize the posteffect by opening each
tube directly after irradiation.

The effect of dose rate variation was studied in a range of
dose rates from 7.5 x 10^{-4} to 0.2 Mrad/h, using both 10 and 18%
methanol in the monomer solution. Grafting rate increased with
dose rate in all cases, but the grafting yield per megarad increased as the dose rate decreased, for both methanol concentrations. Such behavior implies that the monomer does not diffuse
into the fiber fast enough to be utilized efficiently at high dose
rates, and hence the reaction is diffusion controlled at all dose
rates employed.

Increasing the methanol content of the monomer solution continued to increase the grafting yield per megarad up to 40%
methanol. Above that concentration, polymer started to precipitate
as it was formed and results became erratic. An independent
sequence of experiments showed that the yield of homopolymer in

the free solution did not change when the methanol concentration
was increased from 0 to 18%. This indicates that the effect of
increasing rate with increasing methanol concentration is not
caused by changes in the radical yield through radiolysis of the
solvent system.

Grafting in the presence of air using cobalt-60 radiation was
also studied. Considerable grafting took place in air, and in some
cases there was no difference between the air and vacuum grafting
rate. The author postulated that the styrene competes successfully with the oxygen in the diffusion to and reaction with the
free radicals in the wool formed by irradiation.

In the previous work, the wool had been hydrolyzed away from
the grafted material in order to isolate the polystyrene side
chains. Sodium hydroxide had been used, but the hydrolysis of the
side chains was incomplete. A new technique was developed, involving two separate treatments of the grafted wool with a two-phase
toluene/5% aqueous sodium hypochlorite solution, each for 24 h.
This was found to completely remove the wool and render the polystyrene completely soluble in benzene and toluene. Samples of
grafted wool prepared at 10 and 18% methanol and over a wide range
of dose rates were treated this way, and the intrinsic viscosity of
the resulting isolated polystyrene side chains was measured. Little
difference in the intrinsic viscosities was found for the 10 and 18%
methanol contents. The viscosity average molecular weights of the
side chains were found proportional to the square root of the dose
rate up to about 0.02 Mrad/h. On the other hand, change in dose
rate apparently caused little change in molecular weight of the
grafted polymer. At low dose rates the molecular weights of the
grafted polymer were less than those of the side-chain polystyrene.
The molecular weights of the grafted polymer ranged from 25,000 to
100,000. Under similar grafting conditions, cellulose and cellulose acetate are known to form grafted polymers with molecular
weights of approximately one million [11]. This indicates that
wool grafting is unique in haveing a large number of very short
grafted side chains.

The short side chains are probably caused by a strong chain-transfer reaction that is more marked in the case of wool than for the cellulosics. Additional evidence is found in the high G (graft initiation) values of 20 to 60 for wool, which are remarkably high when compared to a G(radical) value for dry wool of less than 1. If the chains were terminated mainly by chain transfer, the molecular weight would be

$$\text{M.W.} = \frac{k_p (R \cdot) (M)}{k_{tr} (R \cdot) (\text{wool})} = \frac{k_p (M)}{k_{tr} (\text{wool})}$$

where k_p and k_{tr} are the propagation and chain-transfer rate constants, respectively. Wool contains many functional groups which have high chain-transfer constants, including disulfide and thiol groups. Hence, it is probable that the low molecular weights result from chain transfer by the wool.

The first-derivative ESR spectrum for γ-irradiated dry wool has been investigated by Kenny and Nicholls [38]. This spectrum is based on several component spectra and is asymmetric. From the change in the spectru, the change in free radical concentration with dose has been estimated as demonstrated in Fig. 10-1 [10]. The G(R·) value computed from the slope of the linear curve through the origin was 0.8. This value appears reasonable, based on comparison with other published data.

The concentration of free radicals in the dry wool gradually decreases on standing, with a corresponding fall off in the intensity of the ESR spectrum. Studies of the changes in the spectrum show that more than one free radical type was present in the wool. Preirradiated wool immersed in styrene/dioxane/methanol solution also showed changes in the ESR spectrum, indicating a drop off in free radical concentration wtih time. When holding the styrene concentration in dioxane constant at 3.06 M, the gradual increase in methanol added from 0 to 35% caused a lowering in the maximum free radical concentration attainable and a slowdown in the rate of reaching the maximum concentration.

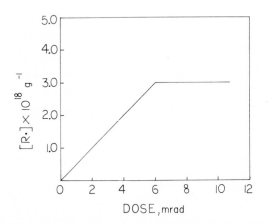

FIG. 10-1.   Generalized curve indicating how radical concentration in wool depends on dose [10].

Even in styrene/dioxane but without added methanol, the maximum free radical concentration was less than that attained in untreated dry wool.   The ESR spectrum for wool in styrene/dioxane
without methanol closely resembles that obtained for untreated dry
wool.   This implies that the styrene does not react with the wool
radicals in the absence of methanol and is confirmed by the fact
that the styrene does not graft on the wool in the absence of
methanol.

When 10, 18, or 35% methanol is present, the ESR signal appears to be a composite owing to both polystyrene radicals and wool
radicals [10].   This would correspond to the presence of growing
polystyrene chains and unreacted wool radicals.

In the postirradiation behavior of the solution containing
10 and 18% methanol, the reciprocal of free radical concentration
has been plotted versus time to produce straight-line graphs, as
would be expected for radical disappearance by a second-order
recombination reaction.

Campbell et al. [10] based their rate constant computations
on the following equation for the rate of increase in free radical
concentration:

$$d(R\cdot)_{dt} = k_1 I - k_t (R\cdot)^2 \tag{10-1}$$

in which k_1 represents the rate constant for free radical production, I is radiation intensity, and k_t is the rate constant for the second-order termination step. Integration of Eq. (10-1) yields

$$(R\cdot) = (k_1 I/k_t)^{\frac{1}{2}} \tanh (k_1 k_t I)^{\frac{1}{2}} \tag{10-2}$$

When the steady state has been reached, Eq. (10-2) becomes

$$(R\cdot) = (k_1 I/k_t)^{\frac{1}{2}} \tag{10-3}$$

For dry wool and for 0% methanol, the value of k_1 was computed directly with no correction for the very small value of the termination rate. From k_1 the value of $G(R\cdot)$ can be computed as follows, where irradiation intensity is expressed in electron volts per gram per second and concentration is in number of radicals per gram of wool.

$$G(R\cdot) = 100\ k_1 \tag{10-4}$$

Values of $G(R\cdot)$ computed in this way were 0.80, 0.35, and 0.37 for dry wool, 10% methanol, and 18% methanol, respectively. (For the methanol solution, corrections were made for rate of free radical decay.)

The decay of free radicals in dry wool gave a poor fit to a second-order decay curve, but an approximate value for k_t of 3×10^{-24} g radical^{-1} sec^{-1} was indicated. The k_t values in 10 and 18% methanol were computed to be 1.9×10^{-22} and 4.8×10^{-22} radical^{-1} sec^{-1}, respectively. Thus, termination in methanol was two orders of magnitude faster than in the absence of methanol.

Hence an increase in methanol from 10 to 18% increases the termination rate by a factor of 2.5, although the grafting yield in 18% methanol is about four times as high as in 10% methanol. These results can be explained by postulating a much higher rate of monomer diffusion in the more swollen fiber, corresponding to a much higher monomer concentration which is responsible for the higher grafting rate.

IV. GRAFTING ON CELLULOSE ACETATE FIBERS

A considerable amount of research has been done on the grafting of vinyl monomers on cellulose acetate. Some of the monomers
that have been employed include acrylic acid, allyl acrylate, vinyl
chloride, propylene, vinyl acetate, stryrene, methyl methacrylate,
divinyl sulfone, butadiene, and ethylene. The rate of grafting is
relatively unaffected by the presence of sensitizers, which is
quite different from the behavior of cotton.

Work by Hayakawa et al. [24] indicated that the mutual technique of grafting produces higher grafting yields and higher
molecular weight of grafted styrene than does the preirradiation
technique. The molecular weight distribution of grafted styrene
appears similar to that of radical-initiated polystyrene, according
to the results of Matsuzaki et al. [25]. The presence of aromatic
hydrocarbons or carbon tetrachloride has been found to decrease
the molecular weight of the grafted polystyrene chains [24]. Activation energies and transfer constants in styrene grafting were
about the same as for styrene polymerization in the homogeneous
liquid phase.

According to Rutherford and co-workers [19] no significant
modification of the physical properties of cellulose acetate was
produced by the grafting on of butadiene, acrylonitrile, ethyl
acrylate, methyl acrylate, or acrylic acid-allyl acrylate. Some
work has been done on comparing properties of acrylic ester grafts
on cellulose acetate with those of mixtures of cellulose acetate
and polyacrylates [26], It was observed that increased acrylic
content in the graft produced increased elongation, whereas the
elongation of the blend did not increase as more polyacrylate was
added.

In other studies of a similar sort, it was observed that the
dynamic elastic modulus of the graft copolymers was always lower

than that of the related physical blend. The graft copolymers of
styrene or methyl methacrylate on cellulose acetate were found less
soluble than either of the homopolymers [12]. Tensile properties
of styrene grafts on cellulose acetate fibers have been studied by
Hayakawa et al. [27].

Apparently the polymerization of styrene is faster in the
presence of cellulose acetate [28]. The grafting efficiency increased as the dose rate increased in the presence of benzene (a
nonsolvent for cellulose acetate). On the other hand, in the
presence of cellulose acetate solvents such as acetone and formamide, the grafting efficiency of styrene on cellulose acetate
decreased with increasing dose rate. It was postulated that benzene facilitated ease of monomer diffusion to the free radical
sites in the cellulose acetate. Decreased efficiency in formamide
and acetone was ascribed to the chain-transfer activity of these
solvents. Degree of acetylation of the cellulose was observed to
have some effect, with grafting efficiency showing a maximum at
about 55% acetylation.

Molecular weights and yields of the polystyrene side chains
have been found interdependent and controlled by the rates of the
growth and termination steps [13]. Using the mutual irradiation
method, a fiftyfold increase in dose rate increased the grafting
rate by only 31%. A reduction in film thickness from 0.025 to
0.01 mm was found to increase the rate of grafting at high dose
rate, indicating the usual effect of diffusion on grafting rate.
Hydrolysis experiments showed 77.6% of the grafted styrene attached
to cellulose acetate polymer backbone and the other 22.4% attached
to ester groups.

Sobue et al. [99] made a study of styrene grafting on cellulose acetate films, employing the preirradiation technique followed
by placing the films in styrene/methanol solution. Infrared was
used to identify homopolymer formed within the film, and the
percent of grafted polystyrene attached to acetyl groups was

determined by acid hydrolysis. The yield of grafting paralleled
the molecular weight of the grafted side chains.

Stannett [30] studied the effect of radiation degradation
during styrene grafting on cellulose acetate. It was found that
heterogeneously prepared grafts have avery high molecular weights
because of the slowness of the termination reaction. Studies were
made of polystyrene-cellulose acetate grafts prepared by irradiation of the cellulose acetate proor to adding styrene/pyridine
solution, as compared to graft copolymers synthesized by the mutual
irradiation of cellulose acetate immersed in styrene/pyridine
solution. The copolymeric products from the different techniques
contained about the same percentage of polystyrene. After separation of the side chains from the polymer backbone by acid hydrolysis, the resulting polystyrenes were fractionated in a column. It
was found that the polystyrene product from the preirradiation
technique had a two-peak distribution with approximately 80% of the
product having a very high molecular weight of narrow distribution.
The product from the mutual irradiation technique had a single
broad distribution peak of a rather low molecular weight.

Very broad molecular weight distributions are believed
characteristic of a diffusion-limited grafting reaction in which
the polymeric chain length (of the grafted side chain) may be
dependent on the distance of the propagation site from the film
surface. On the other hand, the formation of a large amount of
high molcular weight material of narrow distribution may result
from entrapment of free radicals in crystalline regions, followed
by growth of these until terminated by solvent addition at the
close of the run.

Wellons and Stannett [31] have postulated that cleavage of
cellulose by γ-rays takes place after a hydrogen atom has been
abstracted at the 1, 4, or 5 position. He further postulates that
such cleavage is possibly prevented by the alternative reactions
of monomer addition or transfer to solvent. An estimate was made

of the comparable amounts of side-chain formation and cleavage by
running identical grafting experiments in styrene and α-methylstyrene. It is known that styrene grafts on the polymer chain,
whereas α-methylstyrene adds to free radical sites but undergoes
no further polymerization. The amount of cleavage was determined
by comparing molecular weight of the original cellulose acetate
with that irradiated in α-methylstyrene. The number of grafted
chains was estimated by determining the molecular weight of the
grafted polystyrene which was split from the cellulose acetate
by hydrolysis.

Irradiation of films immersed in the swelling solution (80/20
styrene/pyridine) was found to produce four times as many side
chains as cleavages. When the film was irradiated prior to placing
in the swelling solution, there were more cleavages than side
chains produced. Other work indicated that approximately one-half
of the grafted side chains were attached to ester groups of the
cellulose acetate.

V. GRAFTING ON POLYOLEFIN FIBERS

Grafting on polyethylene fiber is of interest because of the
considerable amount of work done on the subject, and because the
grafting behavior illustrates the effect of crystallinity, orientation, branching, and polymer degradation.

Lamb and Weak [32] investigated the effect of pressure on the
grafting of various vinyl monomers on irradiated polyethylene. The
rate of grafting became larger as pressure increased in the styrene/
polyethylene system up to about 3000 atm, above which the grafting
became slower. Poor reproducibility was obtained using methyl
methacrylate grafting, possibly because polyethylene does not swell
appreciably in this monomer. Attempts to graft tetrachloroethylene,

cis- and *trans-*dichloroethylene, and vinylidene chloride were unsuccessful up to 5000 atm.

Wu investigated the effect of antioxidants on grafting of
vinyl monomers to polypropylene and polyethylene fibers [33]. It
was found that ferrous ion retards the rate of homopolymerization
in the presence of such fibers and also reduces the rate of
grafting.

Studies by Armstrong *et al.* [19] showed that the grafting of
vinyl monomers on to polypropylene fibers is not influenced by
sensitizers such as methanol or acetic acid. Factors which do
modify the rate of grafting include orientation, degree of crystallinity, and polymer microstructure. The slow penetration of monomers into polypropylene results in grafting near the surface for
most monomers. Rutherford investigated areas of grafting by the
use of dyed fiber cross sections [19]. Matsuo *et al.* [34] analyzed
styrene concentration across polypropylene film thickness by
microinterferometry during radiolytic grafting and observed styrene
concentration curves similar to a catenary in shape. Odor and
Geleji [35] observed that at high and increasing dose rates of
120 krads/h or more the grafting rate of acrylic acid on polypropylene film from aqueous solution approached a constant upper
limit. At lower dose rates, a greater percentage of grafting was
observed under the same experimental conditions, possibly due to
a more uniform distribution of grafting deeper into the film.

Grafting of vinyl acetate on polypropylene fiber was observed
to cause cracks in the fiber [36]. However, such cracking was not
caused by grafting styrene, methyl methacrylate, or acrylonitrile
on polypropylene. The elongation of grafted polypropylene was
less, for most monomers, than that of ungrafted polypropylene.
The thermal contraction of the grafted fibers was increased.

Several investigators have observed that radiation-grafted
polypropylene has better dyeability properties [19, 37]. However,
the wash fastness of dyed, grafted polypropylene fibers was found
unsatisfactory at times [19].

Grafting of acrylonitrile, butadiene, methyl acrylate, or ethyl acrylate on polypropylene fiber caused no significant change in the physical properties of the fiber. Grafting of methyl acrylate or acrylonitrile on polypropylene caused the melting point to increase [19].

REFERENCES

1. Arthur, J. C., Jr. and Blouin, F. A., *Amer. Dyestuff Reptr.*, 51, 1024 (1962).

2. Holahan, F. S. and Levi, D. D., Picatinny Arsenal, Dover, N. J., Tech. Rept. No. 3567, p. 61.

3. Immergut, E. H., in *Encyclopedia of Polymer Science and Technology* (H. F. Mark, N. G. Gaylord, and N. M. Bihales, eds.), Vol. 35, Interscience, New York, 1965, p. 242.

4. Arthur, J. C., Jr., Mares, T., and Hinojosa, O., *Textile Res. J.*, 36, 630 (1966).

5. Baugh, P. J., Hinojosa, O., and Arthur, J. C., Jr., *J. Appl. Pol. Sci.*, 11, 1139 (1967).

6. Arthur, J. C., Jr., *Textile Res. J.*, 28, 204 (1958).

7. Arthur, J. C., Jr. and Daigle, D. J., *Textile Res. J.*, 34, 653 (1964).

8. Arthur, J. C., Jr. and Demint, R. J., *Textile Res. J.*, 30 505 (1960).

9. Arthur, J. C., Jr., Blouin, F. A., and Demint, R. J. *Amer. Dyestuff Reptr.*, 49, 383 (1960).

10. Campbell, D., Williams, J. L., and Stannett, V., *Advance in Chemistry Series* (R. F. Gould, ed.), Vol. 66, American Chemical Society Publications, Washington, D. C., 1967.

11. Krassig, H. A. and Stannett, V., *Fortschr. Hochpolymer. Forsch.*, 4, 111 (1965).

12. Yasuda, H., Wray, J. A., and Stannett, V., *J. Pol. Sci., Pt. C.*, 2, 387 (1963).

13.   Stannett, V., Wellons, J. D., and Yasuda, H., *J. Pol. Sci.*,
      Pt. C., 4, 551 (1964).

14.   Armstrong, A. A. and Rutherford, H. A., *Text. Res. J.*, 33(3),
      264 (1963).

15.   Arthur, J. C., Jr., Markezich, A. R., and McSherry, W. F.,
      Text. Res. J., 33(11), 896 (1963).

16.   Bernard, O. R., Gagnaire, D., and Servoz-Gavin, P., *J. Chim.
      Phys.*, 60, 1348 (1963).

17.   Blouin, F. A., Morris, N. J., and Arthur, J. C., Jr., *Text.
      Res. J.*, 36(4), 309 (1966).

18.   Razikov, K.Kh., Usmanov, Kh.U., and Azizov, U., *Pol. Sci.*,
      USSR, Engl. Transl., 6(11), 1959 (1964).

19.   Armstrong, A. A., Walsh, W. K., and Rutherford, H. A., USAEC
      Report No. NCSC-2477-11, North Carolina State College,
      Oct. 31, 1963.

20.   Azizov, U., Usmanov, Kh.U., and Sadykov, M. U., *Pol. Sci.*,
      USSR, Engl. Transl, 1(1), 18 (1965). *Vysokomol. Soedin.*,
      7(1), 19 (1965).

21.   Blouin, F. A., Morris, N. J., and Arthur, J. C., Jr., *Text.
      Res. J.*, 36(4), 309 (1966).

22.   Kadonaga, M., *Denki Tsushin Kenkyusho Kenku Jitsuyoka Hokoku*,
      12(11), 1449 (1963). (*Chem. Abstr.*, 62, 10537g.)

23.   Usmanov, Kh.U., and Azizov, U., *J. Pol. Sci.*, *Pt. C*, 4, 579
      (1964).

24.   Hayakawa, K., Kawase, K., and Matsuda, T., *Kobunshi Kagaku*,
      20, 612 (1963). (*Chem. Abstr.*, 60, 1461g.)

25.   Matsuzaki, K., Komagata, H., and Sobue, H., *Kogyo Kagaku
      Zasshi*, 67(11), 1949 (1964).

26.   Ide, F., Handa, R., Uchida, Y., and Makatsuka, K., *Kobunshi
      Kagaku*, 21, 750 (1964). (*Chem. Abstr.*, 62, 1060f.)

27.   Hayakawa, K., Kawase, K., and Matsuda, T., *Nagoya Kogyo
      Gijutsu Shikenso Hokoku*, 15(1), 21 (1966).

28.   Sumitomo, H., Takamuku, S., and Hachihama, Y., *Kogyo Kagaku
      Zasshi*, 66(2), 269 (1963).

29. Wellons, J. D., Schindler, A., and Stannett, V., *Polymer*, 5, 499 (1964).

30. Stannett, V., Report AD-439593 (AROD-3630-6), Research Triangle Institute, Camille Dreyfus Laboratory, 1964.

31. Wellons, J. D. and Stannett, V., *J. Pol. Sci.*, *Pt. A*, 3, 847 (1965).

32. Lamb, J. A. and Weak, K. F., *J. Pol. Sci.*, *Pt. A*, 2, 2875 (1964).

33. Wu, J., Stasyak, Kh.A., Kocherginskaya, L. A., Rozenblyum, N. D., Konkin, A. A., and Rogovin, Z. A., *Khim. Volokna*, No. 5, 12 (1963). (*Chem. Abstr.*, 60, 707g.)

34. Matsuo, H., Iino, K., and Kondo, M., *J. Appl. Pol. Sci.*, 7(5), 1833 (1963).

35. Odor, L. and Geleji, F., *Radiation Chemistry*, Symposium Proceedings, Tihany, Hungary, 1962, pp. 255-260.

36. Torikai, S. and Kubo, M., *Kobunshi Kagaku*, 21, 132 (1969). (*Nucl. Sci. Abstr.*, 18, 31612.)

37. Odor, L. and Geleji, F., *Magy. Textiltech.*, 17(3), 121 (1965). (*Chem. Abstr.*, 62, 1643h.)

38. Kenny, P. and Nicholls, C. H., Proceedings of International Wool Conference, Paris, July, 1965.

39. Vlasov, A. V., Mikhailov, N. V., Tokareva, L. G., Razikov, D. R., Tsetlin, B. L., and Glazunov, P. Y., *Khim. Volokna*, No. 6, 24 (1963).

40. T. Muenzel to Huberlein and Co., Ger. 1,237,056, March, 1967.

41. Okamura, S., Iwasaki, T., Kobayashi, Y., and Hayashi, K., *Large Radiation Sources in Industry Conference, Warsaw*, 1, 459 (1959).

42. Robalewski, A., Zielinski, W., and Adromatowicz, T., *Polimery*, 11(9), 419 (1966).

43. Tsujitol, W., *Sen-i-Gakkaishi*, 23(7), 327 (1967).

44. Magat, E. E., Miller, I. K., Tanner, D., and Zimmerman, J., *J. Pol. Sci.*, *Pt. C.*, 4, 615 (1964).

45. Wu, J., Stasyuk, K., Kocherginskaya, L., Rozenblyum, N., Kankin, A., and Rogovin, Z., *Khim. Volokna*, No. 5, 12 (1963). (*Chem. Abstr.*, 60, 707g.)

46. Shinohara, Y. and Tomioka, K., *J. Pol. Sci.*44, 195 (1960).

47. Sakurada, I., Okada, T., and Eiko, K. K., *Large Radiation Sources in Industry Conference, Warsaw*, 1, 423 (1959).

48. Bessonov, A. I., Vatushkin, N. I., and Glazunov, P. Y., *Plasticheskie Massy*, No. 5, 3-4 (1965).

49. Arthur, J. C. and Grant, J. N., *Text. Res. J.*, 36(10), 934 (1966).

50. Goryaev, V., Ryabchikova, G., and Tarasova, Z. N., *Radiat. Khim. Polim. Mater. Simp.*, p. 171 (1966).

51. Tsetlin, B. L., *Industrial Uses of Large Radiation Sources, Conference Proceedings, Salzburg, 1963*, Vol. I, International Atomic Energy Agency, Vienna, 1963 (STI/PUB/75), p. 289.

52. Khakimov, I., Usmanov, K., and Sadykov, M., *Khim. i. Fiz.-Khim. Prirodn. Sintetich. Polimerov, Akad. Nauk Uz. SSR, Inst. Khim. Polimerov*, No. 2, 29 (1964). Chem. Abstr., 61, 13517a.)

53. Matsuzaki, K. and Miyota, T., *Kami-pa Gikyoshi*, 19(12), 575 (1965). (*Chem. Abstr.*, 64, 8471d.)

54. Arthur, J. C. and Blouin, F. A., *Text. Res. J.*, 34(8), 733 (1964).

55. Imamura, R. and Washijima, S., *Seni-i Gakkaishi*, 21(7), 349 (1965). (*Nucl. Sci. Abstr.*, 20, 29294.)

56. Armstrong, A. A. and Rutherford, H. A., *Industrial Uses of Large Radiation Sources Conference Proceedings, Salzburg, 1963*, Vol. I, p. 257 (STI/PUB/75).

57. Horio, M., Imamura, R., and Mizukami, M., *Bull. Inst. Chem. Res., Kyoto Univ.*, 41(1), 17 (1963). (*Nucl. Sci. Abstr.*, 17, 27426.)

58. Stannett, V., Araki, K., Gervasi, J. A., and McLeskey, S. W., *J. Pol. Sci., Pt. A*, 3, 3763 (1965).

59. Sakurada, I., Okada, T., Hatakeyama, S., and Kimura, F., *J. Pol. Sci., C.*, No. 4, 1233 (1964).

60. Tsuji, W., Imai, M., Kadono, Y., *Bull. Inst. Chem. Res., Kyoto Univ.*, 43(1), 94 (1965).

61. Furuhashi, A. and Kodonaga, M., *J. Appl. Pol. Sci.*, 10(1), 141 (1966).

62. Kawase, K. and Hayakawa, K., *Pachat. Res.*, 30(1), 116 (1967). (Chem. Abstr., 66, 2917k.)

63. Kadonaga, M. and Uada, K., *Kobunshi Kagaku*, 21, 657 (1964). (Nucl. Sci. Abstr., 19, 9192.)

64. Iwakura, Y., Kurosaki, T., Nagakubo, K., Takeda, K., and Miura, M., *Bull. Chem. Soc. Japan*, 38(8), 1349 (1965).

65. Razikov, K. K., Usmanov, K. U., and Azizov, U., *Pol. Sci. USSR, Engl. Transl.*, 7(10), 1980 (1965).

66. Razikov, K. K. and Usmanov, K. U., *Pol. Sci., USSR, Engl. Transl.*, 8(3), 421 (1966).

67. Matsuda, T. and Maeda, N., *Nagoya Kogyo Gijutsu Shikensho Kokoku*, 12, 455 (1963). (*Nucl. Sci. Abstr.*, 18, 8419.)

68. Razikov, K. K., Azizov, U., and Usmanov, K. U., *Uzbeksh. Khim. Zh.*, 8(2), 66 (1964). (*Chem. Abstr.*, 61, 4530b.)

69. American Cyanamid, Neth. Appl. 6,605,345 (1966).

70. Azizov, U., Usmanov, K. U. and Sadykov, M. U., *Khim. i Fiz.-Khim. Prirodn. Sintetich. Polimerov, Akad. Nauk Uz. SSR, Inst. Khim. Polimerov*, No. 2, 23 (1964).

71. Livshits, R. M., Frolova, A. A., Kozlov, P. U., and Rogovin, Z. A., *Pol. Sci., USSR, Engl. Transl.*, 6(3), 634 (1964).

72. Sakurada, I., Okada, T., and Eiko, K., *Large Radiation Sources in Industry Conference, Warsaw*, 1, 447 (1960).

73. Cline, E. T., U. S. Pat. 2, 956, 899 (1960).

74. das Gupta, S., *Amer. Dyestuff Reptr.*, 55, 211 (1966).

75. Yasuda, H., Wray, J. A., Stannett, V., *J. Pol. Sci., Pt. C*, 2, 387 (1963).

76. Geleji, F. and Odor, L., *J. Pol. Sci., Pt. C*, 4, 1223 (1964).

77. Usmanov, K. U., *Celuloza Hirtie (Bucharest)*, 14(7-8-9), 363 (1965). (*Chem. Abstr.*, 64, 5283h.)

78. das Gupta, S., *Can. Textile J.*, 81(13), 57 (1964).

79. Kurilenko, A. I., Smetanina, L. V., Aleksandrova, L. B., and
     Karpov, V. L., *Pol. Sci. USSR, Engl. Transl.*, 7(11), 2123
     (1965).
80. Kurilenko, A. I., Smetanina, L. V., Aleksandrova, L. R.,
     Shiriaeva, G. V., and Kaysov, V. C., *Akad. Nauk, SSSR, Dokl.*,
     156(2), 372 (1964).
81. Kesting, R. E. and Stannett, V., *Makromol. Chem.*, 65, 248
     (1963).
82. Bernard, O. R. and Gagnaire, D., *Electronic Magnetic Resonance
     and Solid Dielectrics, Proceedings of the 12th Coloque Ampere,
     Bordeaux*, p. 269 (1964).
83. Okada, T. and Kimura, F., *Nippon Hoshas Hobunshi Kenkyu
     Kyokai Nempo*, 5, 39 (1964). (*Nucl. Sci. Abstr.*, 19, 26420.)
84. Arthur, J. C., Jr. and Blouin, F. A., *Text. Res. J.*, 34(8),
     733 (1964).
85. Arthur, J. C., Jr. and Blouin, F. A., *J. Appl. Pol. Soc.*,
     8(6), 2813 (1964).
86. Blouin, F. A. and Arthur, J. C., Jr., *Text. Res. J.*, 33(9),
     727 (1963).
87. Huang, R. Y. M. and Rapson, W. H., *J. Pol. Sci., Pt. C*, 2,
     169 (1963).
88. Huang, R. Y. M., *J. Appl. Pol. Sci.*, 10(2), 325 (1966).
89. Pikler, A., *Textile*, 20(10), 372 (1965). (*Chem. Abstr.*, 64,
     5240c.)
90. Tsuji, W., Imai, Ml, and Kadono, Y., *Bull. Inst. Chem. Res.,
     Kyoto Univ.*, 42(1), 68 (1964).
91. Imamura, R. and Taga, T., *Seni-i Gakkaishi*, 21(6), 303 (1965).
     (*Chem. Abstr.*, 64, 3751a.)
92. Imamura, R., Taga, T., and Inagaki, H., *Seni-i Gakkaishi*,
     21(6), 311 (1965). (*Chem. Abstr.*, 64, 8365a.)
93. Kaeppner, W. M. and Huang, R. Y. M., *Text. Res. J.*, 35(6),
     504 (1965).
94. Gotoda, M., Ueno, T., Kusama, Y., and Matsuda, O., *Kobunshi
     Kogoku*, 23, 229 (1966). (*Chem. Abstr.*, 66, 3055r.)

95. Pikler, A., Lodesova, D., and Svitek, J., *Textile*, 20(12), 480 (1965).

96. Hayakawa, K., Kawase, K., and Matsuda, T., *Kobienshi Kagaku*, 20, 540 (1963).

97. Hayakawa, K. and Kawase, K., *Nagoya Kogyo Gijutsu Shikensho Hokoku*, 13(3), 118 (1964).

98. Kawase, K. and Hayakawa, K., *Nagoya Hogyo Gijutsu Shikensho Hokoku*, 13, 448 (1964).

99. Sobue, H., Matsuzaki, K., Komagata, H., and Ishida, A., *J. Pol. Sci., Pt. C*, 2, 415 (1963).

100. Dobo, J. and Czvikovszky, *J. Pol. Sci., Pt. B*, 2(10), 939 (1964).

101. Majumdar, S. K. and Rapson, W. H., *Text. Res. J.*, 34(12), 1007 (1964).

102. Biber, B. L. and Konkin, A. A., *Khim. Volokna*, No. 2, 54 (1966). (*Chem. Abstr*, 65, 845g.)

103. Biber, B. L. and Konkin, A. A., *Khim. Volokna*, No. 4, 29 (1965). (*Chem. Abstr.*, 63, 13465h.)

104. Torikai, S. and Mukoyako, E., *Kobunshi Kagaku*, 21, 120 (1964). (*Nucl. Sci. Abstr.*, 18, 31611.)

105. Hayakawa, K. and Kawase, K., *J. Pol. Sci., Pt. A1*, 5(3), 434 (1967).

106. Usmanov, K. U., Azizov, U., and Sadykov, M. U., *Radiat. Khim. Polim. Mater. Simp., Moscow*, p. 153 (1964).

107. Schwarz, E. and Roudeix, H., *Schriftenreihe Bundesmin. Wiss. Forsch., Radionukl. (Munich)*, No. 4, 85 (1964).

108. Roberts, R., *Text. Rec.*, 79(946), 59 (1962).

109. Zielinski, W. and Achmatowicz, T., Polish Report PAN-490/XVII (1963).

110. Tsuji, W. and Ikeda, T., *Seni-Gakkaishi*, 23(7), 335 (1967).

111. Lynn, J. W. and Skraba, W. J., to Union Carbide, U.S. Pat. 3,326,788 (1967).

112. Mertrichenkov, E. E. and Kachan, A. A., *Sin. Eiz.-Khim. Polim. Akad. Nauk Ukr. SSR, Inst. Khim. Vysokomol. Soedin.,* No. 3, 35 (1966).

113. Kachan, A. A., Chenvyatsova, L. L., Konnyev, K. A., Mentvichenko, E. F., and Gnyp, N. P., *J. Pol. Sci., Pt. C,* 16, 3033 (1967).

114. Tomioka, K., to Toyo Rayon Co., Japan 18,329 (1967).

115. Pesek, M. and Jarkovsky, J., *Chem. Prum.,* 17(4), 199 (1967).

116. Hayakawa, K. and Kawase, K., *J. Po. Sci., Pt. B* , 6(1), 33 (1968).

117. Hayakawa, K., Kawase K., and Matsuda, T., *Radioisotopes,* 17(8), 368 (1968).

118. Okada, T. and Suzuki, Y., *Nippon Genshirgoku Kenkyusho Chosa Hokoku,* JAERI-5018, 5-14 (1968).

119. Spinks, J. W. T. and Woods, R. J., *An Introduction to Radio-chemistry,* John Wiley, New York, 1964.

120. Platzer, N. A., in *Irradiation of Polymers; Advances in Chemistry Series, Number 66* (R. F. Gould, ed.), American Chemical Society Publication, 1967, p. 11.

121. Greene, R. E., Allen, R. S., Ransohoff, J. A., and Woordard, D. G., *Isotopes Radiat. Tech.,* 9(1), 92 (1971).

122. Harmer, D. E., Anderson, L. C., and Martin, J. J., *Nuclear Engineering, Part I, Chem. Engr. Progr., Symp. Ser.,* 50, No. 11, (1954) (AECU 2981).

123. Westendorp, W. F., in *Radiation Sources* (A. Charlesby, ed.), Macmillan, New York, 1964.

124. Miller, C. W., in *Radiation Sources* (A. Charlesby, ed.), Macmillan, New York, 1964.

125. Crowley-Milling, M. C., *Proc. Instr. Elect. Engr.,* 107A, (1960).

126. Luniewski, R. S. and Brenner, W., *Plastics Technology,* p. 35, October 1970.

127. Rotkirch, E., *Large Radiation Sources for Industrial Processes, Symposium Proceedings, Munich, 1969,* International Atomic Energy Agency, Vienna, 1969 (STI/PUB/236).

128. Hoffman, A. S., *Isotopes Radiat. Tech.,* 9, No. 1, (1971).

APPENDIX

A. <u>Radiolytic Grafting of Monomers on Fibers</u>

Table 10-A summarizes published literature on radiolytic grafting of monomers on fibers for the period 1958 to 1968, inclusive. The coverage is extensive, but not complete in the sense of listing every article published in this period. For certain monomer/polymer combinations, such as styrene/cellulose, there was an extremely large number of publications. In all such cases the coverage has been restricted to a reasonable number of arbitrarily selected articles.

The table shows that cobalt-60 was the radiation source employed in almost all cases. Where other sources were used, the nature of the source is indicated.

563

TABLE 10-A

Radiolytic Grafting of Monomers on Fibers

Monomer	Fiber	Fiber type	Radiation source	Investigator and Reference
Acenapthalene	Polypropylene	-	Cobalt-60	Hayakawa [116]
Acetylene	Polyamide	Nylon-6	Cobalt-60	Vlasov [39]
	Polyethylene	-	Cobalt-60	Vlasov [39]
	Polypropylene	-	Cobalt-60	Vlasov [39]
Acrylamide	Cellulose	Cotton	-	Muenzel [40]
	Polyamide	Nylon-6	Cobalt-60	Okamura [41]
	Cellulose	Rayon	Cobalt-60	Armstrong [19]
Acrylic acid	Polyethylene terephthalate	-	-	Robalewski [42]
	Polyvinyl chloride	-	Cobalt-60	Tsuji [110]
	Polypropylene	-	Cobalt-60	Tsujitol [43]
	Polyamide	Nylon-6	Cobalt-60	Vlasov [39]
	Polyamide	Nylon-66	Cobalt-60	Magat [44]
	Polyamide	Nylon-66	Cobalt-60	Armstrong [19]
	Polyester	-	Cobalt-60	Armstrong [19]
	Celluse	Cotton	Cobalt-60	Armstrong [19]

	Cellulose	Rayon	Cobalt-60	Armstrong [19]
	Cellulose acetate	-	Cobalt-60	Armstrong [19]
	Polyethylene	-	Cobalt-60	Vlasov [39]
	Polypropylene	-	Cobalt-60	Odor [35]
	Polypropylene	-	Cobalt-60	Wu [45]
	Wool	-	Cobalt-60	Armstrong [19]
	Acrylic	-	Cobalt-60	Armstrong [19]
Acrylonitrile	Polyethylene	-	Cobalt-60	Shinohara [46]
	Polyester	-	-	Mertrichenkov [112]
	Polyvinyl alcohol	-	Cobalt-60	Sakurada [47]
	Polyamide	-	-	Kachan [113]
	Polyamide	Nylon-6	Linear accelerator	Bessonov [48]
	Polyethylene terephthalate	Lavsan	Linear accelerator	Bessonov [48]
	Cellulose	Cotton	Cobalt-60	Arthur [49]
	Caprylamide	-	-	Goryaev [50]
	Polyamide	Nylon-6	Cobalt-60	Vlasov [39]
	Polyamide	Nylon-6	x-Ray	Tsetlin [51]
	Polyamide	Nylon-6	Cobalt-60	Krassig [11]
	Polyamide	Nylon-66	Cobalt-60	Armstrong [19]
	Polyester	-	Cobalt-60	Armstrong [19]
	Cellulose	-	Cobalt-60	Khakimov [52]

Monomer	Substrate	Fiber	Radiation source	Reference
Acrylonitrile (continued)	Cellulose	—	—	Lynn [111]
	Cellulose	—	—	Matsuzaki [53]
	Cellulose	Cotton	Cobalt-60	Armstrong [19]
	Cellulose	Cotton	Cobalt-60	Armstrong [19]
	Cellulose	Cotton	Cobalt-60	Arthur [54]
	Cellulose	Rayon	—	Pesek [115]
	Cellulose	Rayon	Cobalt-60	Imamura [55]
	Cellulose acetate	—	Cobalt-60	Armstrong [56]
	Cellulose acetate	—	Cobalt-60	Armstrong [19]
	Polyethylene	—	Cobalt-60	Tsetlin [51]
	Polyethylene	—	Cobalt-60	Vlasov [39]
	Polyethylene	—	Cobalt-60	Vlasov [39]
	Polypropylene	—	—	Tomioka [114]
	Polypropylene	—	Cobalt-60	Wu [45]
	Polypropylene	—	Cobalt-60	Armstrong [19]
	Polypropylene	—	Cobalt-60	Tsetlin [51]
	Wool	—	Cobalt-60	Armstrong [56]
	Wool	—	Cobalt-60	Horio [57]
	Wool	—	Cobalt-60	Stannett [58]
	Cotton (cyanoethylated)	—	Cobalt-60	Arthur [15]
	Acrylic	—	Cobalt-60	Armstrong [19]

Monomer	Polymer	Substrate	Radiation source	Reference
Allyl acetate	Wool	-	Cobalt-60	Stannett [58]
Allyl acrylate	Polyamide	Nylon-66	Cobalt-60	Armstrong [19]
	Cellulose	Cotton	Cobalt-60	Armstrong [19]
	Cellulose	Rayon	Cobalt-60	Armstrong [19]
	Cellulose acetate	-	Cobalt-60	Armstrong [19]
	Polypropylene	-	Cobalt-60	Armstrong [19]
Allyl benzoate	Wool	-	Cobalt-60	Stannett [58]
Allyl stearate	Wool	-	Cobalt-60	Stannett [58]
Butadiene	Cellulose	-	Cobalt-60	Sakurada [59]
	Cellulose	-	-	Lynn [111]
	Polyvinyl alcohol	-	Cobalt-60	Lynn [111]
	Polyethylene	-	Cobalt-60	Lynn [111]
	Polyamide	Nylon-6	Vlasov [39]	Armstrong [56]
	Polyamide	Nylong--	Cobalt-60	Armstrong [19]
	Polyamide	Nylon-66	Cobalt-60	Armstrong [19]
	Polyeater	-	Cobalt-60	Armstrong [56]
	Cellulose	Cotton	Cobalt-60	Armstrong [19]
	Cellulose	Cotton	Cobalt-60	Tsuji [60]
	Cellulose	Cotton	-	Tsuji [60]
	Cellulose	Rayon	-	Armstrong [19]
	Cellulose	Rayon	Cobalt-60	Armstrong [19]

Monomer	Polymer	Fiber	Source	Reference
Butadiene (continued)	Cellulose acetate	–	Cobalt-60	Armstrong [56]
	Cellulose acetate	–	Cobalt-60	Armstrong [19]
	Polyethylene	–	Cobalt-60	Furuhashi [61]
	Polyethylene	–	Cobalt-60	Vlasov [39]
	Polypropylene	–	Cobalt-60	Armstrong [56]
	Polypropylene	–	Cobalt-60	Kadonaga [22]
	Polypropylene	–	Cobalt-60	Vlasov [39]
	Wool	–	Cobalt-60	Armstrong [56]
	Wool	–	Cobalt-60	Armstrong [19]
	Acrylic	–	Cobalt-60	Armstrong [19]
1,3-Butylene dimethacrylate	Polyamide	Nylon-66	Cobalt	Armstrong [19]
	Cellulose	Rayon	Cobalt-60	Armstrong [19]
	Cellulose	–	Cobalt-60	Armstrong [19]
	Polypropylene	–	Cobalt-60	Armstrong [19]
Butyl methacrylate	Polypropylene	–	–	Kawase [62]
Iso-Butyl methacrylate	Polypropylene	–	–	Kawase [62]
Dichloroethylene	Polyethylene	–	Cobalt-60	Kadonaga [63]
Divinyl benzene	Caprylamide	–	–	Goryaev [50]
Divinyl sulfone	Polyamide	Nylon-66	Cobalt-60	Armstrong [19]
	Polyester	–	Cobalt-60	Armstrong [19]

	Substrate	Source	Reference
	Cellulose	Cotton	Cobalt-60	Armstrong [19]
	Cellulose	Rayon	Cobalt-60	Armstrong [19]
	Cellulose acetate	-	Cobalt-60	Armstrong [19]
	Polypropylene	-	Cobalt-60	Armstrong [19]
	Wool	-	Cobalt-60	Armstrong [19]
	Acrylic	-	Cobalt-60	Armstrong [19]
Ethyl acrylate	Polyamide	Nylon-66	Cobalt-60	Armstrong [56]
	Polyamide	Nylon-66	Cobalt-60	ARmstrong [19]
	Polyester	-	Cobalt-60	Armstrong [19]
	Cellulose	Cotton	Cobalt-60	Armstrong [56]
	Cellulose	Cotton	Cobalt-60	Armstrong [19]
	Cellulose	Rayon	Cobalt-60	Armstrong [19]
	Cellulose Acetate	-	Cobalt-60	Armstrong [19]
	Polypropylene	-	Cobalt-60	Armstrong [56]
	Polypropylene	-	Cobalt-60	Armstrong [19]
	Wool	-	Cobalt-60	Armstrong [19]
	Acrylic	-	Cobalt-60	Armstrong [19]
Ethylene	Polyamide	Nylon-6	Cobalt-60	Vlasov [39]
	Polyamide	Nylon-6	x-Rays	Tsetlin [51]
	Polyamide	Nylon-66	Cobalt-60	Armstrong [19]
	Polyester	-	Cobalt-60	Armstrong [19]
	Cellulose	Cotton	Cobalt-60	Armstrong [19]

Monomer	Polymer	Fiber	Source	Reference
Ethylene (continued)	Cellulose	Rayon	Cobalt-60	Armstrong [19]
	Cellulose acetate	-	Cobalt-60	Armstrong [19]
	Polyethylene	-	Cobalt-60	Tsetlin [51]
	Polyethylene	-	Cobalt-60	Vlasov [39]
	Polypropylene	-	Cobalt-60	Tsetlin [51]
	Polypropylene	-	Cobalt-60	Vlasov [39]
	Polypropylene	-	Cobalt-60	Armstrong [19]
	Wool	-	Cobalt-60	Armstrong [19]
	Acrylic	-	Cobalt-60	Armstrong [19]
Ethylene dimeth-acrylate	Polyamide	Nylon-66	Cobalt-60	Armstrong [19]
	Cellulose	Rayon	Cobalt-60	Armstrong [19]
	Cellulose acetate	-	Cobalt-60	Armstrong [19]
	Polypropylene	-	Cobalt-60	Armstrong [19]
2-Ethyl hexyl-acrylate	Cellulose	Rayon	Cobalt-60	Armstrong [19]
Glycidyl methacrylate	Polypropylene	-	Cobalt-60	Iwakura [64]
	Polyvinyl chloride	-	Cobalt-60	Iwakura [64]
Isoprene	Cellulose	-	-	Lynn (111)
	Polyamide	-	-	Kachan [113]
Maleic acid	Polyamide	Nylon-66	Cobalt-60	Magat [44]
Methacrylamide	Cellulose	-	Cobalt-60	Khakimov [52]

Monomer				Reference
Methacrylic acid	Cellulose	Cotton	Cobalt-60	Razikov [18, 66]
	Cellulose	Cotton	Cobalt-60	Usmanov [23]
	Polyamide	Nylon-66	Cobalt-60	Matsuda [67]
	Polyethylene terephthalate	-	Cobalt-60	Matsuda [67]
	Cellulose	-	Cobalt-60	Razikov [18]
	Cellulose	Cotton	Cobalt-60	Razikov [68]
	Cellulose	Cotton	Cobalt-60	Usmanov [23]
	Polyethylene	-	Cobalt-60	Matsuda [67]
	Acrylic	-	Cobalt-60	Matsuda [67]
	Polyethylene	-	-	Robalewski [42]
Methyl acrylate	Polytetrafluoroethylene	-	-	[69]
	Polyamide	Nylon-66	Cobalt-60	Armstrong [14]
	Polyamide	Nylon-66	Cobalt-60	Armstrong [19]
	Polyester	-	Cobalt-60	Armstrong [19]
	Cellulose	-	Cobalt-60	Azizov [70]
	Cellulose	-	-	Livshits [71]
	Cellulose	-	-	Matsuzaki [53]
	Cellulose	Cotton	Cobalt-60	Armstrong [56]
	Cellulose	Cotton	Cobalt-60	Armstrong [19]
	Cellulose	Cotton	Cobalt-60	Razikov [65]
	Cellulose	Rayon	Cobalt-60	Armstrong [19]

Monomer	Fiber		Radiation source	Reference
Methyl acrylate (continued)	Cellulose acetate	–	Cobalt-60	Armstrong [19]
	Polypropylene	–	Cobalt-60	Armstrong [19]
	Wool	–	Cobalt-60	Armstrong [19]
	Acrylic	–	Cobalt-60	Armstrong [19]
Methyl methacrylate	Polyvinyl alcohol	–	Cobalt-60	Sakurada [72]
	Polyamide	Nylon-66	Electron accelerator	Cline [73]
	Polyethylene terephthalate	–	Electron accelerator	Cline [73]
	Polyamide	Nylon-6	Cobalt-60	Vlasov [39]
	Polyamide	Nylon-6	x-Rays	Tsetlin [51]
	Polyamide	Nylon-66	Cobalt-60	Armstrong [19]
	Polyester	–	Cobalt-60	Armstrong [19]
	Cellulose	–	Cobalt-60	Azizov [70]
	Cellulose	–	–	Matsuzaki [53]
	Cellulose	Cotton	Cobalt-60	Armstrong [19]
	Cellulose	Cotton	–	Blouin [21]
	Cellulose	Cotton	Cobalt-60	das Gupta [74]
	Cellulose	Rayon	Cobalt-60	Armstrong [19]
	Cellulose acetate	–	Cobalt-60	Yasuda [75]
	Cellulose acetate	–	Cobalt-60	Armstrong [19]
	Polyethylene	–	Cobalt-60	Kadonaga [63]

Monomer	Substrate		Radiation	Reference
	Polyethylene	–	Cobalt-60	Tsetlin [51]
	Polyethylene	–	Cobalt-60	Vlasov [39]
	Polypropylene	–	Cobalt-60	Geleji [76]
	Polypropylene	–	Cobalt-60	Tsetlin [51]
	Polypropylene	–	Cobalt-60	Vlasov [39]
	Wool	–	Cobalt-60	Horio [57]
	Wool	–	Cobalt-60	Armstrong [19]
	Acrylic	–	Cobalt-60	Armstrong [19]
Methylolacrylamide	Cellulose	Cotton	Cobalt-60	Armstrong [19]
α-Methylstyrene	Cellulose	Cotton	Cobalt-60	Usmanov [77]
2-Methyl-5-vinyl pyridine	Polyethylene	–	Cobalt-60	Wu [33]
	Polypropylene	–	Cobalt-60	Wu [33]
Phenylacetylene	Polyamide	Nylon-6	Cobalt-60	Vlasov [39]
	Polypropylene	–	Cobalt-60	Vlasov [39]
	Polyethylene	–	Cobalt-60	Vlasov [39]
Propargyl alcohol	Polyamide	Nylon-6	Cobalt-60	Vlasov [39]
	Polyethylene	–	Cobalt-60	Vlasov [39]
	Polypropylene	–	Cobalt-60	Vlasov [39]
Propylene	Polyamide	Nylon-6	Cobalt-60	Clasov [39]
	Polyamide	Nylon-6	x-Rays	Tsetlin [51]

Monomer	Polymer	Fiber	Source	Reference
Propylene (continued)	Polyamide	Nylon-66	Cobalt-60	Armstrong [19]
	Polyester	-	Cobalt-60	Armstrong [19]
	Cellulose	Cotton	Cobalt-60	Armstrong [19]
	Cellulose	Rayon	Cobalt-60	Armstrong [19]
	Cellulose	-	Cobalt-60	Armstrong [19]
	Polyethylene	-	Cobalt-60	Tsetlin [51]
	Polyethylene	-	Cobalt-60	Vlasov [39]
	Polypropylene	-	Cobalt-60	Tsetlin [51]
	Polypropylene	-	Cobalt-60	Vlasov [39]
	Polypropylene	-	Cobalt-60	Armstrong [19]
	Wool	-	Cobalt-60	Armstrong [19]
	Acrylic	-	Cobalt-60	Armstrong [19]
Iso-Propyl methacrylate	Polyporpylene	-	-	Kawase [62]
Propyl methacrylate	Polypropylene	-	-	Kawase [62]
Styrene	Polyethylene	-	Cobalt-60	Shinohara [46]
	Cellulose	Rayon	Cobalt-60	Okamura [41]
	Linen	-	Cobalt-60	Okamura [41]
	Cellulose	Cotton	Cobalt-60	Okamura [41]
	Polyvinyl alcohol	-	Cobalt-60	Sakurada [72]
	Cellulose	-	Cobalt-60	Sakurada [59]
	Polyvinyl alcohol	-	Cobalt-60	Sakurada [59]

Polyethylene	–	Cobalt-60	Sakurada [59]
Cellulose	Cotton	–	das Gupta [78]
Polyamide	Nylon-6	Cobalt-60	Vlasov [39]
Polyamide	Nylon-6	Cobalt-60	Kurilenko [79]
Polyamide	Nylon-6	Cobalt-60	Kurilenko [80]
Polyamide	Nylon-66	Cobalt	Armstrong [19]
Polyethylene terephthalate	Lavsan	Cobalt-60	Kurilenko [79]
Polyester	–	Cobalt-60	Armstrong [19]
Cellulose	–	Cobalt-60	Kesting [81]
Cellulose	–	Cobalt-60	Bernard [82]
Cellulose	–	Cobalt-60	Okada [83]
Cellulose	Cotton	Cobalt-60	Arthur [84]
Cellulose	Cotton	Cobalt-60	Arthur [85]
Cellulose	Cotton	Cobalt-60	Blouin [86]
Cellulose	Cotton	Cobalt-60	Huang [87, 88]
Cellulose	Cotton	Cobalt-60	Pikler [89]
Cellulose	Cotton	Cobalt-60	Tsuji [90]
Cellulose	Rayon	Cobalt-60	Huang [87]
Cellulose	Rayon	Cobalt-60	Imamura [91, 92]
Cellulose	Rayon	Cobalt-60	Kaeppner [93]
Cellulose	Rayon	–	Gotoda [94]
Cellulose	Rayon	–	Pikler [95]

Monomer	Substrate		Radiation source	Reference
Styrene (continued)	Cellulose acrtate	-	Cobalt-60	Hayakawa [96, 97]
	Cellulose acetate	-	Cobalt-60	Kawase [98]
	Cellulose acetate	-	Cobalt-60	Sobue [99]
	Cellulose acetate	-	Cobalt-60	Wellons [29]
	Polyethylene	-	Xobalt-60	Dobo [100]
	Polyethylene	-	Cobalt-60	Kadonaga [63]
	Polyethylene	-	Cobalt-60	Vlasov [39]
	Polypropylene	-	Cobalt-60	Matsuo [34]
	Polypropylene	-	Cobalt-60	Vlasov [39]
	Polypropylene	-	Cobalt-60	Armstrong [19]
	Wool	-	Cobalt-60	Horio [57]
	Wool	-	Cobalt-60	Stannett [58]
	Jute	-	Cobalt-60	Majumdar [101]
Tetrachloroethylene	Polyethylene	-	Cobalt-60	Kadonaga [63]
Trichloroethylene	Polyethylene	-	Cobalt-60	Kadonaga [63]
Vinyl acetate	Polyvinyl aclcohol	-	Cobalt-60	Sakurada [72]
	Polyamide	-	-	Kachan [113]
	Polyamide	Nylon-6	Cobalt-60	Vlasov [39]
	Polyamide	Nylon-66	Cobalt-60	Armstrong [56]
	Polyamide	Nylon-66	Cobalt-60	Armstrong [19]
	Polyester	-	Cobalt-60	Armstrong [19]
	Cellulose	-	Cobalt-60	Azizov [70]

Cellulose	–	–	Biber [102]
Cellulose	–	–	Biber [103]
Cellulose	Cotton	Cobalt-60	Armstrong [19]
Cellulose	Cotton	–	Blouin [21]
Cellulose	Rayon	Cobalt-60	Armstrong [19]
Cellulose acetate	–	Cobalt-60	Armstrong [19]
Polyethylene		Cobalt-60	Vlasov [39]
Polypropylene	–	Cobalt-60	Torikai [104]
Polypropylene	–	Cobalt-60	Vlasov [39]
Polypropylene	–	Cobalt-60	Armstrong [19]
Wool	–	Cobalt-60	Armstrong [19]
Acrylic	–	Cobalt-60	Armstrong [19]
Vinyl chloride — Polyamide	Nylon-66	Cobalt-60	Armstrong [19]
Polypropylene	–	–	Hayakawa [117]
Polypropylene	–	Cobalt-60	Hayakawa [105]
Polyester	–	Cobalt-60	Armstrong [19]
Cellulose	Cotton	Cobalt-60	Armstrong [19]
Cellulose	Rayon	Cobalt-60	Armstrong [19]
Cellulose acdtate	–	Cobalt-60	Armstrong [19]
Polypropylene	–	Cobalt-60	Armstrong [19]
Wool	–	Cobalt-60	Armstrong [19]
Acrylic	–	Cobalt-60	Armstrong [19]

Monomer	Substrate	Fiber type	Radiation source	Reference
Vinylidene chloride	Polypropylene	-	Cobalt-60	Hayakawa [105]
	Polypropylene	-	-	Hayakawa [117]
	Cellulose	-	Cobalt-60	Usmanov [106]
	Cellulose	Cotton	Cobalt-60	Usmanov [77]
	Polyethylene	-	Cobalt-60	Kadonaga @63]
α-Vinyl pyridine	Polyamide	Nylon-6	Linear acelerator	Bessonov [48]
	Polyamide	Caprylamide	-	Goryaev [50]
	Polyamide	-	-	Kachan [113]
	Cellulose	-	Cobalt-60	Usmanov [106]
	Polyester	-	-	Okuda [118]
4-Vinyl pyridine	Polyvinyl chloride	-	Cobalt-60	Schwarz [107]
	Polyester	Polyethylene terephthalate	Cobalt-60	Roberts [108], Zielinski [109]
	Polyester	Polyethylene terephthalate	-	Okada [118]
n-Vinyl pyroolidone	Polyamide	Nylon-6	-	Torikai [104]
	Polyamide	Nylon-66	Cobalt-60	Armstrong [19]
	Polyester	-	Cobalt-60	Armstrong [19]
	Cellulose	Cotton	Cobalt-60	Armstrong [19]
	Cellulose	Rayon	Cobalt-60	Torikai [104]
	Cellulose	Rayon	Cobalt-60	Armstrong [19]

	Cellulose acetate	—	Cobalt-60	Armstrong [19]
	Polypropylene	—	Cobalt-60	Torikai [104]
	Polypropylene	—	Cobalt-60	Armstrong [19]
	Wool	—	Cobalt-60	Armstrong [19]
	Acrylic	—	Cobalt-60	Armstrong [19]
Vinyl toluene	Celluslose	Cotton	Cobalt-60	Usmanov [77]

B.   Irradiation Cost Estimates

While the main emphasis in this book has been placed on
fundamental principles, a short summary of the elements involved
in irradiation cost is included at this point for the sake of completeness.

The two main types of radiation sources are: (1) radioisotopes and (2) machines.   Radioisotopes, which will be discussed
first, have the advantage of supplying deeply penetrating radiation.   Such deeply penetrating radiation can be used to polymerize
suitable vinyl monomers that are impregnated in wood or concrete.
Gamma-rays from certain isotopes can also be used to penetrate
metal reaction vessels (instead of glass) in carrying out various
chemical reactions, thus improving mechanical safety as well as
heat transfer to or from the vessel.   Gamma-radiation from isotopes
also has the advantage of providing a relatively uniform radiation
intensity in the reaction vessel.   Less desirable factors in the
use of radioisotopes include the need for annual or semiannual
replenishment and the generally higher estimated cost per radiation
dose for radioisotopes than for electron beams from accelerators.

Comparative costs for different radioisotopes are presented
in Table 10-B1.   The costs range from \$1.00/Ci for cobalt-60 to
\$20,000/Ci for radium-226.   Cobalt-60 is used more commonly than
the other isotopes shown in the table.   Though the prices indicated
are of the right order of magnitude, the cost of an isotope source
will also be a function of its size because the cost of fabricating
and handling a small source may exceed the cost of the isotope
itself.   It has been predicted [120] that cobalt-60 may drop to as
low as 20 cents/Ci in the future, corresponding to \$13,000/kilowatt
output.   This would compare with \$5,000-10,000/kW output for
electron accelerators, the price being higher at higher voltage.

As an example of a radioisotope cost study, costs were estimated [121] for the use of cobalt-60 in the pilot-plant emulsion

TABLE 10-B1

Cost of Isotopes Commonly Used as Sources of Radiation[a]

Isotope	Cost per curie, dollars
Cesium-137	5
Cobalt-60	1-4
Hydrogen-3	7-30
Polonium-210	150
Radium-226	20,000
Strontium-90 + Yttrium-90	7.50
Sulfur-35	1.50

[a]Taken from Ref. [119].

polymerization of vinyl acetate copolymers. The economics for radiation catalysis were compared with those for a plant based on conventional thermal catalysts (initiators). Four different sizes of plant were studied--15, 25, 50, and 100 x 10^6 lb/year, all on a wet basis. Only a minor fraction of the capital cost was associated with polymerization equipment. Capital costs of the two systems were similar, but the radiation-plant capital costs in the 50 and 100 x 10^6 lb/year sizes were about 5% lower than those for the conventional plant. The estimated operating profit was between 5 and 10% higher for the radiation plant in all cases studied, the difference in catalyst cost being a principal contributing factor.

Using an annual cash flow discount of 10%, radiation plants were written off somewhat earlier than conventional plants. It was recognized that this was a comparison of an undeveloped plant for radiation catalysis with a rather reliable estimate of a conventional plant based on present practice. The differences in process economics were therefore within the margin of error of the

estimates. However, it was noted that even with the costs of
shielding and the purchase and replacement of the radioisotope, the
radiation-catalysis production appeared to be at least competitive
with conventional latex production.

Another example of the use of γ-radiation to catalyze a chemical reaction has been discussed by Spinks and Woods [119]. Gamma-
benzene hexachloride, which is one of the six stereoisomers formed
by the addition of chlorine to benzene, is usually manufactured by
photochlorination using glass vessels. Pilot-plant work has shown
that it can also be made using ionizing radiation to initiate the
chlorination, with a G value for the product of about 10^5. Use
of γ-radiation allows a metal vessel to be used in place of a glass
one. The arrangement of the pilot plant using radioisotopes has
been described by Harmer et al. [122], Some of the work was done
using a mixture of fission products obtained from spent fuel rods.
The cost of the benzene hexachloride was estimated to be $1.51 and
$1.60/kg, using mixed fission products and cesium-137, respectively.
This is in line with the cost of the commercial γ-benzene hexachloride produced photochemically, which is quoted at $1.90 to
$3.00/kg.

Westendorp [123] has carried out detailed analyses of the
costs of using resonant-trnasformer electron-beam generators. The
costs involved were classified and analyzed separately by class.
The capital cost of the apparatus was computed from the quantity
of material to be given a certain millirep treatment, knowing the
kilowatt output per generator and the total number of generators
required. Building and shielding costs included the floor space
necessary for the installed equipment and the materials handling
space. The cost of the equipment to convey the product through
the radiation field was charged to the process. Other costs were
maintenance cost for equipment, labor costs of process attendants,
power costs, and return on investment.

In obtaining his estimates, Westendorp considered the factors
affecting the efficiency of the process. For example, the amount

of beam overtravel beyond the edges of the product was considered,
as well as penetration of the product as determined by voltage of
the electron source, whether the irradiation was one-sided or two-
sided, etc.

Westendorp presented conclusions for a 2-MeV resonant transformer (10 kW) used at 8000 h/year. The mrep pound treatment unit
is by definition the absorption of 0.00117 kW h of radiation. This
10-kW machine would treat 8500 mreps lb/h at 100% utilization
efficiency, or for a more realistic value of 75% efficiency, 6450
mreps pounds/h. The unit treatment cost would be $15.59/6450 lb,
or 0.24 cents/mrep lb. It was reported that further machine improvements will result in even lower costs in the coming years.

Miller [124] has discussed the advantages of the various
electron accelerator types for different operations. In any irradiation process the two important quantities are the dose required and the rate at which material is to be treated. If the
electron beam is fully utilized, a 1-kW power source can treat
800 lb of material/h at a dose of 1 Mrad. In many cases it is not
possible to utilize the electron beam fully, and then some utilization factor must be taken into account. For example, although
bulk powders may allow a utilization factor approaching 100%, the
treatment of molded articles or pharmaceuticals in packages may
reduce this factor to 25%.

In choosing a suitable installation, one of the major considerations is the cost per kilowatt of appropriate equipment. For
low energy machines, where energy does not exceed 2 MeV, the linear
accelerator is less economical than the Van de Graaff or resonant
transformer equipment unless short pulse operation is required.
However, when larger energies are needed, the linear accelerator
provides a suitable solution. In small magnetron-powered units
with energy in the 3-6 MeV range, the cost per installed kilowatt
may be approximately $96,000, whereas for larger klystron driven
units providing powers from 5 kW upward and energies greater than
4 MeV, the cost is reduced to about $24,000/installed kW.

Possibly the most detailed analysis has been given by Crowley-Milling [125], in which he considers the use of a klystron-driven accelerator with a 5-kW beam output. Several typical processes with appropriate dose and utilization factors were considered, and both cost and capacity were computed and are presented in Table 10-B2. In every case shown, it was assumed that operation was for a single 8-h shift, 5 days/week. Continuous use of the equipment would reduce the equipment costs shown by a factor of 1.7.

TABLE 10-B2

Costs of Irradiation Processing[a]

Process	Typical dose, Mrad	Estimated utilization, percent	5-kW klystron-driven accelerator	
			Capacity, lb/h	Cost of treatment per pound, dollars
Treatment	25	80 (in bulk)	130	0.12
		25 (moldings)	40	0.40
Curing of rubbers	5	80 (in bulk)	640	0.025
		25 (moldings)	200	0.080
Steriliza-tion of drugs	2	80 (in bulk)	1600	0.01
		5 (in phials)	100	0.16

[a]Taken from M. C. Crowley-Milling, *Proc. Inst. Elect. Engrs.*, 107A, (1960). Cost figures shown converted from British coinage units given in original Table.

An alternative figure often used in estimating costs is the price to be paid per kilowatt hour of the electron beam. On this basis, the costs per kilowatt hour for the equipment of Table 10-B2 would be as follows.

1. For a 4-MeV 600-W machine hired at $120.00/day, cost per kilowatt hour is $24.96.

2. For a 5-kW klystron-driven machine operated at 40 h/week, cost per kilowatt hour is $3.24.

3. For a 5-kW klystron-driven machine operating continuously, cost per kilowatt hour is $1.53.

Luniewski and Brenner [126] have estimated radiation-cross-linking cost for polyethylene-insulated wire as 0.94 to 2.26 cents/lb, compared to 0.95-1.9 cents/lb for such insulation chemically cross-linked with benzoyl peroxide. The same authors quote 20 cents/100 ft^2 for a 2-Mrad coating cure using an 0.3-MeV accelerator. They also report a possible high volume application involving a high-speed cold-stampable fiber-reinforced thermosetting B-stage sheet whose cure is completed by exposure to ionizing radiation at ambient temperature. Production rates of 50-60 parts/min were considered possible, compared to 2-3 parts/min using conventional chemical curing.

Hoffman [128] has studied the economics of the radiation-induced curing of coatings. He notes that while the major operating costs in the radiation curing of coatings are the costs of the coatings themselves (and the substrates), this is also true for chemical curing or thermal drying of other systems. Rotkirch [127], in a detailed analysis of the coating of fiberboard panels with unsaturated polyester-styrene mixtures using either radiation or chemical curing, found radiation (or catalyst) costs to be minor. His results showed that electron coating becomes more economical than chemical methods as the coating thickness and/or production speed increases. Other work showed that the operating cost of a radiation-curing process running at line speeds of 200 ft/min may be less than one-half the costs involved in the conventional oven-curing process.

GLOSSARY

absorbed dose rate   The absorbed dose rate is the absorbed dose per
　　unit time and is expressed in units of rads per unit time, or
　　electron volts per gram per unit time.

active radical sites   In grafting reactions, these are the free
　　polymeric radicals formed inside polymeric films or fibers to
　　which the monomer must diffuse before grafting can take place.

air-equivalent wall   A wall surrounding an ionization chamber
　　having a mean atomic number close to that of air, such as
　　nylon.

air-wall chamber   A chamber for absolute ionization measurements
　　containing air, where the "sensitive volume" irradiated by the
　　beam is surrounded by air, that is, an "air-wall."

alpha-rays   Rays consisting of α-particles, which are doubly
　　charged helium nuclei emitted by unstable nuclei.

amorphous phase   A term used in reference to a polymer that is
　　partially crystalline and partially "amorphous." The mole-
　　cules in the amorphous phase have no particular ordered
　　arrangement or pattern and do not exhibit an x-ray diffraction
　　pattern.

anionic polymerization   There are several types of anionic poly-
　　merization, one of which involves the addition of butyl
　　lithium catalyst to the monomer, followed by propagation.
　　Such mechanisms often have no termination reaction.

annihilation radiation   The radiation emitted when a positron com-
　　bines with an electron, the two particles being replaced by
　　two 0.51-MeV γ-rays emitted in opposite directions.

587

antineutrino A fundamental particle of no charge and almost no mass, which is emitted when a neutron decays to give a proton and a β-particle.

antistatic properties In polymer technology, certain films and fabrics have the undesirable property of picking up an electric charge, which makes them cling to themselves and other objects. Materials without this undesirable property are said to have "antistatic properties."

atomic absorption coefficient The atomic absorption coefficient is the absorption coefficient per atom and can be computed from the linear absorption coefficient.

ATR infrared spectroscopy ATR means "attenuated total reflectance." Essentially, a beam of infrared light passing through glass or crystal to reach a planar interface of crystal/plastic will be reflected, but will be partially absorbed in the process of reflection, thus allowing the IR absorption spectrum of the plastic to be studied.

autoxidation A term commonly applied to the oxidation of unsaturated hydrocarbons by a chain mechanism.

beta-plus-particle A positive β-particle, sometimes called a positron.

beta-rays Sometimes called β-particles, consisting of electrons emitted by radioisotopes.

Bragg curve A graphical plot of the specific ionization produced by a charged particle versus the distance from the end of the particle track.

bremsstrahlung x-Rays produced whenever high-speed electrons are rapidly decelerated, as when they pass through the field of an atomic nucleus.

bulk polymerization Polymerization of liquid monomer without the addition of solvent or water.

cage effect   When a molecule separates into two fragments, the
    fragments may be held close together by a sort of "cage" of
    surrounding molecules, with the result that the fragments re-
    combine before they can drift further apart; also referred to
    as the Franck-Rabinowitz effect.

carbonium ion   In cationic polymerization, the propagating species
    is an ion pair consisting of a positive carbonium ion and a
    negative gegen ion. The carbonium ion is usually symbolized
    by HM^+, where M represents monomer.

catalyst-cocatalyst complex   In cationic polymerization, the
    catalyst reacts with the cocatalyst to form a complex. For
    example, BF_3 reacts with H_2O to form $H^+(BF_3OH)^-$.

cationic mechanism   A polymerization mechanism in which the propa-
    gating species in an ion pair consisting of a positive carb-
    onium ion and a negative gegen ion. A typical catalyst is
    BF_3, and a cocatalyst such as water must also be present.

"c" axis   For descriptive purposes, crystal unit cells are con-
    sidered to be oriented about a, b, and c axes. In the case
    of a hexagonal crystal, the c axis is the one that is perpen-
    dicular to the plane of the hexagon.

chain reaction   In polymer chemistry, the successive additions of
    monomeric units to a free radical is a chain reaction, the
    addition of each unit constituting one step in the chain.

chain transfer   The transfer of the reactivity of a polymer radical
    to another species, which then adds successive monomer units
    to form another polymer chain.

characteristic x-rays   These are produced by electron bombardment
    of target nuclei and have energies corresponding to the pos-
    sible electron transitions from shell to shell of the target
    nuclei.

cocatalyst   In cationic polymerization, both a catalyst and co-
    catalyst must be used. Water is an example of a cocatalyst.

coherent (Rayleigh) scattering A process in which a photon is
    scattered with little loss of energy, the scattered radiation
    being in-phase (coherent) with the incident radiation.

collision complex In radiation chemistry, an ion formed by the
    collision of an ion and a molecule which may be the final
    product, or which may dissociate into fragments.

collision of the second kind A collision in which energy is trans-
    ferred from an electronically excited molecule to a molecule
    in its ground state.

columnar ionization When densely ionizing radiation, such as α-
    rays, passes through matter, a *column* of ions and excited
    species is formed about the particle track.

Compton scattering In the Compton effect, a photon intereacts with
    an electron, which may be loosely bound or free, so that the
    electron is accelerated and the photon deflected with reduced
    energy.

condensation In polymer chemistry, a reaction in which the mono-
    meric units link together with the simultaneous elimination of
    a small molecule such as water.

conversion In a polymerization reaction, "conversion" refers to
    percent conversion, meaning the percent of original monomer
    that has been converted to polymer.

copolymer A product formed by the addition of two or more differ-
    ent monomers to a growing polymer chain.

coulombic attractive force According to Coulomb's law, the
    attractive force between unlike-charged particles is
    proportional to the product of the charges and inversely
    proportional to the dielectric constant and the square of the
    distance between the particles. Hence, such an attractive
    force is large in organic solvents of low dielectric constant.

critical thickness In the radiation-induced grafting of monomer on
    a polymeric film, the critical thickness is the thickness
    above which the grafting rate per unit film weight falls off
    rapidly.

crystallite Polyethylene resin of a given density contains a
certain percentage of amorphous material, the remainder being
crystalline in nature. The crystalline portion is generally
considered to be made up of very small crystals known as
crystallites.

crystal replication In solid-state polymerization, a process
whereby the crystalline form and structure of the polymer
produced tends to "replicate" or duplicate the crystalline
monomer's form and structure.

degradative chain transfer In free radical polymerization, unimolecular termination of the propagating species is referred
to as "degradative chain transfer," and results in the production of a stable radical that will not propagate.

depolymerization This term may be used in a general sense to
indicate any sort of decrease in the molecular weight of a
polymer. At times it is used specifically to indicate decomposition of the polymer to yield monomer molecules.

differential thermal analysis A technique by which phase transitions or chemical reactions can be followed by observing a
strip-chart recording of the heat absorbed or liberated.

diffusion coefficient The flux of a component in a certain
direction is proportional to its concentration gradient in
that direction, the proportionality constant being equal to
the diffusion coefficient D.

dilatometer A device to measure change in volume by movement of
a liquid along a capillary; in polymer chemistry, employed to
measure decrease in volume as polymerization occurs, the rate
of volume decrease usually being proportional to rate of
polymerization.

direct grafting method On radiation chemistry, the direct grafting
method consists of irradiating the polymer in the presence of
the monomer.

domains In order to explain the high rate of exciton transfer, it has been postulated that small groups of ordered molecules (domains) exist in liquid hydrocarbons.

DPPH   Diphenylpicryl hydrazyl, a substance sometimes used in radiation chemistry to react with and titrate (scavenge) free radicals.

du Nuoy ring   A small ring generally of platinum used to indicate surface tension by the force required to raise it from the surface of a liquid.

durable-press materials   These are fabrics which, after pressing once with a hot iron or steam press, retain the creases formed and pressed shape for a very long period of time.

dyeability   In polymer technology, most fabrics made from natural fibers can be dyed readily, and the dye adheres strongly to the fabric. Such is not the case with fabrics of synthetic polymer fibers, and such fibers may have to be specially treated to improve the dye adherence or "dyeability."

electronic absorption coefficient   The electronic absorption coefficient is the absorption coefficient per electron and can be computed from the linear absorption coefficient.

electronic equilibrium   In a free-air chamber for absolute ionization measurement, the sensitive volume (in which ionization is measured) is surrounded by an air wall. Some ions produced in the sensitive volume arise from secondary electrons produced outside the sensitive volume, but these are compensated for by electrons that are lost from the sensitive volume.

electron volt   An amount of energy equal to that acquired by an electron accelerated by a potential difference of one volt.

energy (Compton) absorption coefficient   This quantity equals the total electronic Compton absorption coefficient minus the Compton scattering coefficient.

ESR spectroscopy   Spectroscopy involving energy levels produced by application of a magnetic field to a substance which contains unpaired electrons.

e value In copolymerization theory, the reactivity ratios can be
expressed in terms of Q values and e values, where the Q
value for each monomer is related to its reactivity, and the
e value for each monomer is related to the polarization of its
double bond.

exciton transfer A method of energy transfer whereby a quantity of
energy called an "exciton" moves rapidly from one molecule to
another, remaining for a very brief time (less than the period
of one vibration) in each.

exposure dose rate The exposure dose rate is the exposure dose per
unit time and is expressed in roentgens per unit time.

first generation species When an ionizing particle passes through
matter, the first excited and ionized species formed along its
path are known as the "first generation species."

first-order transition In thermodynamics, a transition involving a
discontinuity in a primary thermodynamic quantity such as
volume or heat content.

flux The flux of particles or photons at a point is the number of
particles or photons entering a sphere of unit cross-sectional
area at that point in unit time. The units of flux are
particles per square centimeter second.

flux depression The depression of the flux density of neutrons in
a section or volume of a nuclear reactor caused by the introduction of foreign objects or substances.

fragmentation pattern A pattern of types and abundances of products from the fragmentation of a molecule. Fragmentation of
a molecule can result from electron bombardment in a mass
spectrometer or from irradiation with ionizing radiation.

Franck-Condon transition A transition between two molecular electronic energy levels, during which it is a good approximation
to assume that the nuclei do not change their relative
positions; sometimes referred to as a "vertical transition."

Franck-Rabinowitz effect When a molecule separates into two
fragments, the two fragments may be held close together by a

sort of "cage" of surrounding molecules, so that the fragments
recombine before they can drift further apart; also referred
to as the "cage effect."

Fricke dosimeter The reaction involved in the Fricke dosimeter is
the oxidation of an acid solution of ferrous sulfate to the
ferric salt, in the presence of oxygen and under the influence
of radiation.

galvinoxyl A radical scavenger consisting of a hindered phenoxyl
radical prepared by oxidizing 3,3',5,5'-tetra-t-butyl-4,4'-
dihydroxydiphenylmethane.

gamma-rays Electromagnetic radiation of very short wavelength,
emitted by a nucleus in transition from an upper to a lower
energy state.

gas chromatography More properly called gas-liquid chromatography:
a process in which an inert carrier gas containing one or more
volatile components is passed through a column containing a
nonvolatile liquid held as a thin layer on solid particles.
The components emerge one at a time and are detected by a
suitable detector.

gegen ion In cationic polymerization, the propagating species is
an ion pair consisting of a positive carbonium ion and a
negative gegen ion.

gel effect An acceleration or autoacceleration of the polymerization rate owing to increased viscosity causing a decreased
termination rate resulting from the decreased rate of diffusion of polymer radicals through the medium (often called the
Trommsdorff effect).

geminate pairing When a molecule separates into two fragments, the
fragments may be kept close together by a sort of "cage" of
surrounding molecules, with the result that the fragments
recombine before they can drift further apart; also referred
to as the "cage effect."

glass fibers In polymer technology, glass fiber is used as filler

to strengthen and toughen plastic laminates, especially polyester laminates.

glass-transition temperature Equals the temperature corresponding
to a sudden change in slope of the volume versus temperature
curve. Below this temperature a rubber or plastic becomes
glassy and stiff. The transition that takes place is sometimes called a second-order transition.

glassy state A state in which polymerization may be carried out,
usually obtained by dissolving the monomer in a mixture of
solvents, which turns glassy but does not freeze solid when
subjected to very low temperatures.

glassy-state polymerization A type of polymerization carried out
after dissolving the monomer in a mixture of solvents, which
turns glassy but does not freeze solid when subjected to very
low temperatures.

graft copolymer A graft copolymer is made up of a backbone chain
consisting entirely of monomer X attached to one or more side
chains composed entirely of monomer Y.

grafting efficiency The grafting efficiency is defined as the
ratio of the weight of grafted monomer to the total weight of
monomer consumed in both grafting and homopolymerization.

grafting ratio The grafting ratio is the weight ratio of the
grafted polymer to the original ungrafted polymer.

gram rad The gram rad is the unit of integral absorbed dose and is
equal to 100 ergs or 6.24×10^{13} eV.

G_R *value* In radiation chemistry, the G_R value equals the number of
radicals formed per 100 eV of radiation absorbed.

G value In radiation chemistry, the number of molecules changed
for each 100 eV of energy absorbed.

heat of polymerization Amount of heat evolved per mole of monomer
converted to polymer.

homogeneous polymerization A polymerization in which only a single
phase is involved, as in bulk polymerization or solution polymerization.

homopolymerization In the radiation-induced grafting of a monomer
on a polymer, a certain fraction of the monomer may undergo
polymerization with itself (homopolymerization), rather than
undergo grafting on the polymer.

host molecule When a crystal contains an impurity or minor component, the molecules making up the bulk of the crystal are
the "host" molecules.

hot radicals Highly excited radicals formed by the dissociation
of excited molecules, and in other ways.

Huggins constant The constant k in the equation,

$$\eta_{sp}/c = [\eta] + k[\eta]^2 c$$

where η_{sp} is specific viscosity, $[\eta]$ is intrinsic viscosity,
and c is concentration.

hydrophilicity The property of attracting and holding water.
Polymeric surfaces, which attract and absorb water, are said
to be hydrophilic.

incoherent (Compton) scattering Scattering where there is a random
phase relationship between the incident and scattered radiation.

Indirect action When a solute/solvent pair is irradiated, the
solvent may absorb radiant energy, pass it on to the solute,
and cause the solute to be decomposed in this way by "indirect
action."

indirect grafting method In radiation chemistry, this involves
irradiation of the polymer, which may or may not be stored for
a time after irradiation, followed by bringing the polymer
into contact with the monomer.

inductive resonance A process by which energy is transferred by
quantum-mechanical resonance from a solvent molecule to a
distant solute molecule, molecules intermediate between the
two playing no part in the transfer.

inelastic collisions Inelastic collisions between charged particles and stopping material take place when the charged

particles produce excitation and ionization in the stopping
material.

inelastic scattering of neutrons Inelastic scattering occurs if
the neutron is absorbed by the nucleus and then a neutron with
lower energy reemitted, leaving the original nucleus in an
excited state.

infusible Impossible to melt. An infusible material will char or
decompose instead of melting.

initiation step In free radical polymerization, the initiation
step is the breakdown of a substance such as a peroxide to
form free radicals.

initiator In polymer chemistry, a substance capable of initiating
or starting the polymerization reaction, or of providing
active species which initiate the polymerization; most commonly used to refer to peroxides or hydroperoxides, which break
down thermally to yield active free radicals which initiate
the polymerization.

ionic mobility The velocity of an ion in a unit electric field is
called the ionic mobility.

ionic yield In radiation chemistry, the ratio of the number of
molecules undergoing change to the number of ion pairs formed.

ionizing radiation Particles or photons having sufficiently high
energy to ionize the substrate being traversed.

ion-pair yield In radiation chemistry, the ratio of the number of
molecules undergoing change to the number of ion pairs formed.

isotactic polymer A stereoregular polymer, generally a carbon atom
chain, in which all the substituent groups (other than hydrogen) lie above or below the plane of the main chain.

K_α *radiation* x-Rays due to the transition of an electron from the
L shell to the K shell of an atom.

kinetic chain length The average number of molecules polymerized
per initiating radical (in free radical polymerization).

kinetic chain lifetime In polymerization, the total length of time
that the polymer chain grows before becoming terminated.

laminates   In polymer technology, the term "laminate" refers to a
    flat sheet or section usually cured under conditions of heat
    and pressure. It may contain several layers of paper or
    fabric to add strength.

latex   In polymer chemistry, a dispersion of small polymer parti-
    cles in water, commonly called a polymer "emulsion"; generally
    results from an emulsion polymerization reaction.

light scattering photometer   An instrument to measure the scatter-
    ing of a beam of light by a polymer solution, from which the
    weight-average molecular weight of the polymer can be computed.

linear electron accelerator   A device whereby electrons are in-
    jected in pulses into a straight, segmented wave guide and
    accelerated by the electric field of an electromagnetic wave
    that travels down the tube.

linear energy transfer (LET)   The linear rate of loss of energy
    (locally absorbed) by an ionizing particle traversing a
    material medium.

living ends   In anionic polymerization there may be no effective
    mechanism of terminating the growing polymer chains, in which
    case the growing ends of the chains are referred to as
    "living" ends.

living polymer   In anionic polymerization there may be no effective
    mechanism of terminating the growing polymer chains, in which
    case the polymer is often referred to as a "living" polymer.

macroscopic cross section   The total cross section for neutron
    capture of all the nuclei in 1 cm^3 of stopping material.

mass absorption coefficient   The mass absorption coefficient equals
    the linear absorption coefficient divided by the density of
    the absorbing material.

mass stopping power   The mass stopping power of a material for
    electrons or other charged particles equals the energy loss
    per unit path length divided by the density of the material.

mers   Repeating units in a polymer chain having the same or similar
    composition to that of the monomer.

microscopic cross section The microscopic cross section for
   neutron capture is the cross section for a single nucleus of
   the stopping material.

molecular products Those products which are formed as molecules
   rather than free radicals and do not undergo a reduction in
   yield when radical scavengers are added.

molecular weight distribution A graphical plot of the weight of
   polymer of a given size versus the molecular weight.

most probable distribution In polymer chemistry, the molecular
   weight distribution for which $\overline{M}_w/\overline{M}_n$ equals 2.

mutual method In the mutual method of radiolytic grafting of
   monomers on polymeric fibers, the monomer is padded onto the
   fibers, which are then subjected to irradiation.

negative polaron The solvated electron, $(H_2O)^-$.

neutrino A fundamental particle having no charge and a mass less
   than 1% of the mass of the electron.

neutralization A reaction between a positive and negative reactant,
   for example, a combination of a positive ion with an electron
   or negative ion.

nonradiative resonance A process by which energy is transferred by
   quantum-mechanical resonance from a solvent molecule to a
   distant solute molecule, molecules intermediate between the
   two playing no part in the transfer.

normal kinetics In polymer chemistry, normal kinetics is said to
   apply when the classic rate equation holds true, indicating
   rate dependence on the first power of monomer concentration,
   the half-power of initiation rate, etc.

nuclide Any species of atom characterized by the number of neutrons and protons in its nucleus.

number-average degree of polymerization Equals the number-average
   molecular weight divided by the molecular weight of the
   monomer.

number-average molecular weight Equals the total weight of polymer

divided by the sum over all molecular species of the number of
moles of each species present.

orientation A stretching operation sometimes performed on polymers
such as polypropylene, which tends to align a major proportion
of the polymeric chains in the direction of stress.

osmotic pressure From the measured osmotic pressure of a polymer
solution it is possible to compute the number-average molecular weight of the polymer.

outgassed monomer The outgassing of a monomer is the removal from
the monomer of dissolved gases, mainly oxygen. It is accomplished by placing the monomer in a trap connected to a vacuum
pump and subjecting it to several successive cycles of freezing down, pumping off, thawing again, etc.

pair production The absorption of a photon in the vicinity of an
atomic nucleus with the production of two particles, an electron and a positron.

parent ion In mass spectrometry, the ion produced by a molecule
through simple loss of an electron; no fragmentation is
involved.

partial acetylation The hydroxyl groups of cellulose fibers may be
esterfied with acetic acid (acetylated). When only a fraction
of the total number of hydroxyl groups has been acetylated,
the cellulose is said to be "partially acetylated."

penetration depth In the radiation-induced grafting of monomer on
a thick polymeric film, the monomer penetrates into the surface to a certain depth, the penetration depth, before becoming consumed by the grafting reaction.

percentage of grafting The percentage of grafting equals 100 times
the weight gain during grafting divided by the weight of the
original ungrafted polymer sample.

peroxide grafting technique In the peroxide technique of grafting
monomer on fibers, the fiber is first irradiated in air to
form polymeric peroxides, which are later employed to decompose and initiate grafting when the fiber is immersed in
liquid or gaseous monomer.

photochemistry The study of reactions brought about by the absorption of light which produces excited molecules or atoms but does not ionize the absorber.

photoelectric effect In this effect, the entire energy of the incident photon is transferred to a single atomic electron, which is ejected from the atom with an energy equal to the difference between the photon energy and the binding energy of the electron within the atom.

plasticizer A low molecular weight substance added to a plastic; usually employed to increase flexibility or softness.

polarizing microscope A device for viewing polymers with polarized light, which aids in revealing textural features of solid polymers such as spherulites.

positive holes When electrons are raised to the "conduction band" and wander through a solid, they leave behind positive holes which may also be mobile, although less mobile than the electrons.

positive polaron The solvated positive ion, $(H_2O)^+$.

positron A fundamental particle having a single positive charge and a mass equal to that of the electron.

posteffect In the radiation chemistry of polymers, the posteffect is any polymerization that continues to take place after the source of radiation has been removed.

postirradiation technique In solid-state polymerization, a technique whereby the monomer is irradiated for a limited time, removed from the irradiation field, and allowed to continue polymerizing in the absence of radiation.

preirradiation method In the preirradiation method of grafting monomer on polymeric fibers, the fiber is first irradiated and then brought into contact with liquid or gaseous monomer.

pressed disk technique A technique for measuring the infrared absorption of solid substances, whereby the solid is ground into a fine powder, mixed with potassium bromide, and pressed in a die in the form of a thin small disk. The disk is the sample through which the IR radiation is passed.

propagation In polymer chemistry, the propagation step in the
    mechanism of polymerization consists of the addition of a
    monomer molecule to a free radical.

protective colloid A soluble, lyophilic colloid added to a dis-
    persion of polymer particles ("polymer emulsion") to help
    prevent coagulation of the particles.

pulse radiolysis A technique whereby pulses or bursts of short
    duration and high intensity from an electron accelerator are
    allowed to impinge upon the substrate being studied. The
    short-lived species that are formed may be studied by optical
    absorption or other suitable means.

quantum yield In photochemistry, the number of molecules that
    react chemically per quantum absorbed.

quenching The conversion of an excited molecule to the ground
    state, or to a more stable excited state, by the influence of
    an additive (quencher).

Q value In copolymerization theory, the reactivity ratios can be
    expressed in terms of Q values and e values, where the Q value
    for each monomer is related to its reactivity, and the e value
    for each monomer is related to the polarization of its double
    bond.

radical entrapment In polymer chemistry the entrapment of a free
    radical in a highly crystalline or cross-linked portion of a
    polymer, thus prolonging the life of the radical and prevent-
    ing it from undergoing reaction.

radical products Those products which are formed as free radicals
    rather than as molecules and therefore undergo a reduction in
    yield when radical scavengers are added.

radical titration A method of free radical detection and measure-
    ment involving the addition of a selected substance which
    scavenges (titrates) the free radicals.

randomization An example would be the random replacement of H by

D, or of D by H, in the reaction complex formed in a
radiation-induced reaction.

reaction cross section An expression of reaction probability in
units of area. The greater the cross section, the higher the
probability of reaction (or capture).

reactor radiation A substance placed in a hole leading into a reactor is subjected to reactor radiation, consisting of a mixture of α-, β-, and γ-rays, and other types of radiation.

relativistic neutrons Neutrons with energies greater than 10 MeV.

reactivity ratios In copolymerization, these are defined as

$$r_1 = k_{11}/k_{12} \quad \text{and} \quad r_2 = k_{22}/k_{21}$$

where the rate constants pertain to the reactions,

$$M_1\cdot + M_1 \xrightarrow{k_{11}} M_1\cdot$$
$$M_1\cdot + M_2 \xrightarrow{k_{12}} M_2\cdot$$
$$M_2\cdot + M_1 \xrightarrow{k_{21}} M_1\cdot$$
$$M_2\cdot + M_2 \xrightarrow{k_{22}} M_2\cdot$$

where M_1 and M_2 are the monomers undergoing copolymerization.

resonance capture An example is the resonance "capture" of electrons by water molecules, so-called because the process takes
place at certain electron energies but not at others; sometimes referred to as a "resonance reaction."

resonance peak For neutrons in the intermediate energy range,
the cross section for capture by a nucleus frequently passes
through a series of peaks known as resonance peaks at certain
definite energies.

resonance process A resonance process or reaction involves the
reaction of one reactant having a certain amount of energy
with another reactant able to absorb exactly that amount of
energy, or very close to it. An example is the capture of
an electron by a molecule having a strong electron affinity,

such as oxygen or a halogen.

roentgen One roentgen is an exposure dose of x- or γ-radiation such that the associated corpuscular emission per 0.001293 g of air produces, in air, ions carrying 1 eu of quantity of electricity of either sign.

rotating sector technique Method of generating light flashes at a controlled flashing rate for irradiation of a photochemical reaction involving radical combination. The mean lifetime of the radicals can be calculated from a theoretical relation between the flashing rate and the reaction rate.

sampling technique A technique for determining the lifetime of a free radical by observing the ESR signal of the radical at a given delay time after each pulse of irradiation, allowing the enitre radical decay curve to be obtained by employing several delay times.

scavenger A substance added to remove or "scavenge" some other component of a reaction, such as a free radical scavenger, an ion scavenger, etc.

scintillator A substance which emits a small flash, or scintillation when struck by one particle or quantum of ionizing radiation.

second-order transition A discontinuous change in the volume-temperature curve of a polymer, more properly called the "glass transition."

sedimentation From the rate of sedimentation or settling of a polymer in solution during ultracentrifugation, the Z-average molecular weight can be computed.

sensitizer In the original, photochemical sense, this term indicated a molecule that absorbs radiant energy then passes the energy to a second molecule which reacts. Used in a broader sense in radiolytic grafting on fibers, it means any added substance that promotes grafting, such as a solvent capable of swelling the fibers.

short-term effects In radiation chemistry, these are the immediate effects of radiation on matter such as the production of ionized and excited molecules and atoms.

specific energy loss The specific energy loss for electrons and other charged particles equals the energy loss per unit path length.

specific ionization In radiation chemistry, the total number of ion pairs produced in a gas per unit length of track (of an ionizing particle).

spontaneous termination In cationic polymerization, this is a type of termination characterized by separation of the polymer chain from the catalyst-cocatalyst complex. The complex then initiates another polymer chain.

spur As an ionizing ray or particle passes through a gas (or liquid in some cases), small groups of ions known as "spurs" are formed along the particle's path.

steady state In free radical polymerization kinetics, a steady state exists when the rate of radical formation equals the rate of radical disappearance, so that radical concentration is steady or constant.

stereospecific reaction An example would be the preferential loss of hydrogens from the 2,3-positions of butane ions during H_2 transfer.

sticky collision A collision between an ion and a molecule which leads to the formation of a long-lived collision complex.

stereoregular polymers Polymeric chains containing asymmetric carbon atoms having a definite repeated spatial arrangement of substituents on the carbon atoms of the chain.

stopping power The stopping power for electrons and other charged particles equals the energy loss per unit path length.

subexcitation electrons Low-energy long-life electrons having energies of 0.5 to 4 eV.

superexcited molecule A molecule that has received energy in excess of its lowest ionization potential without immediate loss of an electron.

tandem spectrometer A double-focusing mass spectrometer, involving
   passing the ion beam first through an electrostatic field then
   through a magnetic field. Much higher resolution is obtained
   than with a single-focusing instrument.

termination In polymer chemistry, the step in the polymerization
   mechanism that terminates or ends the polymerization chain
   reaction.

thermal neutrons Neutrons in thermal equilibrium with their
   surroundings, and having energies of the order of 0.025 eV
   at room temperature.

thermosetting In polymer chemistry, a thermosetting resin is one
   which when heated reacts to form a three-dimensional, insoluble, infusible network.

topochemical reaction A topochemical reaction is one which may or
   may not take place (or take place only slowly), depending on
   the relative geometric arrangements of the reactants in space.

topotactic reaction A solid-state reaction in which the crystal
   orientations of the product are correlated with the crystal
   orientations of the reactants.

total linear absorption coefficient This equals the sum of the
   linear absorption coefficients for the photoelectric, Compton,
   and pair-production processes.

Torr A unit of pressure equal to 1 mm of mercury.

triplet excited state An excited state having a net spin equal to
   one, so that the spin angular momentum vector may have three
   orientations in a magnetic field. Triplet excited states
   have longer lifetimes than singlet excited states, because
   transition from a triplet state to a singlet ground state
   involves spin inversion, which is forbidden by selection
   rules.

true Compton absorption coefficient This quantity equals the total
   electronic Compton absorption coefficient minus the Compton
   scattering coefficient.

urea-clathrate complex   This is a complex of urea having a honeycomb pattern, with a long hexagonal hole in the center. Butadiene fits in the hole in such a way that a stereoregular
polymer of butadiene is formed when the urea-clathrate complex
of butadiene is irradiated.

Van de Graaff accelerator   A device in which an electrostatic
charge is carried to a high voltage electrode by a rapidly
moving belt, and the potential difference between the electrode and the ground used to accelerate electrons or positive
ions to high velocities.

vapor grafting technique   In the radiation-induced grafting of
monomer on polymer, this technique consists in surrounding the
polymer sample with monomer vapor, either during or after
irradiation of the polymer.

viscosity molecular weight   More accurately, viscosity-average
molecular weight, which equals

$$M_v = \left(\sum_{i=1}^{\infty} w_i M_i^{\,a} \right)^{1/a}$$

where w_i is the weight in grams of the polymeric species of
molecular weight, M_i. The value of a is defined by

$$[\eta] = K' M^a$$

where $[\eta]$ is intrinsic viscosity and K' is a constant.

vulcanization   In rubber, a process causing cross-linking and
resulting in increased tensile strength and other property
changes.

wash fastness   The ability of a dyed fabric to retain its dye and
not fade, usually after a prescribed wash cycle under a
certain set of conditions.

weight-average degree of polymerization   Equals the weight-average
   molecular weight divided by the molecular weight of the monomer.

Z-average degree of polymerization   This is equal to

$$\overline{DP}_Z = \frac{\int_0^\infty n^2 W(n)\ dn}{\int_0^\infty nW(n)\ dn}$$

   where $W(n)$ is the fractional weight of polymer with degree of
   polymerization equal to n.

Ziegler catalyst   A catalyst mixture commonly consisting of a
   metal alkyl and a titanium compound, employed in the polymerization of ethylene at low pressure.

AUTHOR INDEX

Numbers in parentheses are reference numbers and indicate that an author's work is referred to although his name is not cited in the text. Underlined numbers give the page on which the complete reference is listed.

A

Abkin, A. D., 303, 306, 310, 314, 315, 316, 318

Abramson, F. P., 133(9,11), 134, 180, 202(55,58), 247, 248

Achmatowicz, T., 561

Acker, I., 514, 523(49), 524(49, 50), 525(50), 530(49), 532(49), 533(49)

Adachi, T., 397(90), 249

Adams, G. E., 226(92), 249

Adicoff, A., 349(33), 353

Adler, G., 322(4,44), 327(4,5,6), 328(28), 349(32), 351, 352, 353, 358, 488(9), 502(9), 512, 528(9)

Adromatowicz, T., 557, 564(42), 571(42)

Alberino, L. M., 415, 416, 428

Aleksandrova, L. B., 560, 575 (79,80)

Aleksandrov, I. V., 408(72), 515

Alexander, P., 380, 382(31,32), 426, 427

Alfrey, T., Jr., 436(8), 457(27), 474

Allen, A. O., 111(27), 115, 186 (12), 189(15), 214(74), 237 (117), 245, 248, 250, 279 (29), 302

Allen, G., 264(15), 301

Allen, R. S., 443(18), 473, 562, 580(21)

Amagi, Y., 322(41), 353

Amjad, M., 303, 310(58)

Ander, P., 327(2), 328(2), 351, 359(2)

Anderson, A. R., 147(44,45), 155, 159(68), 160, 161(60), 162(68,76), 163, 165(70), 182, 183, 184

Anderson, L. C., 562, 582(122)

Angier, D. J., 514, 526

Aquilanti, V., 140, 182

Araki, K., 440(16), 473, 558, 566(58), 567(58), 576(58)

Armstrong, A. A., 542(14), 543 (19), 544, 550, 554, 555(19), 556, 558, 564-579

Arthur, J. C., Jr., 538(1,4,5), 539(6), 540(4), 541(7,8,9), 542(15,17), 543, 555, 556, 558, 560, 565, 566, 572, 575, 577

Ashkin, J., 59(2), 60, 81, 114

Aso, C., 307, 316

Atsushi, A., 303, 310

Auer, E. E., 286(62), 304, 440 (15), 473

Auerbach, I., 403, 427

Ausloos, P., 136, 137, 138(26, 27), 139(23,28,33,35), 140, 141(48), 143(39,40,41), 181, 182, 189(24,25), 229(96), 230, 233(11), 246, 249, 250

Autio, T., 463(31), 474

Azizov, U., 542(18), 543(20), 544, 556, 559, 561, 571(18, 23,65,68,70), 572(70), 576 (70), 578(106)

B

Bach, N., 214(76), 248

Back, R. A., 147(46), 182, 194 (34), 246

609

Beta-rays, definition, 19
Bimolecular termination, 266
Bimolecular termination, in
    polyesters, 454
Boron trifluoride, 75
Bragg curve, 65
Bremsstrahlung, 21
Burton irradiator, 12
Butadiene polymerization, 312

C

Caging, 188
Calcium acrylate polymerization,
    343
Calcium acrylate, X-ray study,
    344
Capture cross section, 35
Carbanion formation, 401
Carbonium ion, 258
Carbonyl compound radiolysis,
    229
Carboxylic acid radiolysis, 227
Carboxymethylation, 543
Castable elastomers, 464
Catalyst-cocatalyst complex,
    257
Cationic polymerization, 257
Cellulose acetate fiber,
    grafting on, 550
Cellulose acetate powder, 523
Cellulose depolymerization, 540
Cellulose fiber grafting, 535
Cesium-137, 10
Chain lifetime, in solution
    polymerization, 271
Charge transfer, 130
Chain transfer, in grafting, 483
Chain transfer to counter ion,
    258
Chain transfer to monomer, 259
Charlesby-Pinner function, 385
Chlorine-containing additives,
    463
Cis-trans-isomerization, 218
Clay, grafting on, 505
Cluster theory, 219
Coating techniques, 459
Cobalt-59, 9
Cocatalyst, 257
Co-crosslinking, 392
Co-crosslinking, of water-

soluble polyers, 397
Co-crosslinking theory, 394
Coherent (Rayleigh) scattering,
    88
Columnar ionization, 95
Compton effect, 82
Condensation ion-molecule
    reaction, 132
Conduction band, 129
Co-network, 394
Copolymerization, solid state,
    339
Cost of irradiation, general,
    580
Cotton fiber grafting, 538
Critical concentration, 395
Cross-linking density, 396
Cross-link density, from stress
    relaxation, 465
Cross-linking mechanism, for
    polyesters, 452
Crosslinking of hydrocarbons,
    207
Crosslinking polymer list, 371
Crosslinking sequence
    (Charlesby), 375
Crossover effect, 452
Cross section for capture, 78
Crystal lattice geometry, 326
Crystallinity loss, 376
Curie, 99
Curing rate, polyester resin,
    446
Curtain coating, 462
Cyanoethylated cotton fiber, 566
Cyanoethylation, 543
Cyclohexane radiolysis, 213
Cyclooctatetraene, 324

D

Degradation by scission, 372
Degree of polymerization, for
    anionic polymer, 263
Dehydrogenation, of cellulose,
    539
Delta-rays, 16
Derivative curve, ESR, 544
Deuteride transfer, 140
Differential thermal analysis,
    346
Diffusion constant, 487